Molecular biology techniques have considerably improved the understanding of mechanisms underlying the development of flowers. These processes can be critically important in managing the yield of agricultural crops. The book covers molecular aspects in the control of flower morphogenesis in particular species, including the role and regulation of gene expression in gamete development and the variation in the longevity of flowers and the pattern of senescence.

The book will appeal to students and researchers in floral physiology.

SOCIETY FOR EXPERIMENTAL BIOLOGY
SEMINAR SERIES: 55

MOLECULAR AND CELLULAR ASPECTS OF PLANT REPRODUCTION

SOCIETY FOR EXPERIMENTAL BIOLOGY SEMINAR SERIES

A series of multi-author volumes developed from seminars held by the Society for Experimental Biology. Each volume serves not only as an introductory review of a specific topic, but also introduces the reader to experimental evidence to support the theories and principles discussed, and points the way to new research.

MOLECULAR AND CELLULAR ASPECTS OF PLANT REPRODUCTION

Edited by

R.J. Scott
Department of Botany, University of Leicester

A.D. Stead
Department of Biology, Royal Holloway, University of London

CAMBRIDGE
UNIVERSITY PRESS

CAMBRIDGE UNIVERSITY PRESS
Cambridge, New York, Melbourne, Madrid, Cape Town, Singapore, São Paulo

Cambridge University Press
The Edinburgh Building, Cambridge CB2 8RU, UK

Published in the United States of America by Cambridge University Press, New York

www.cambridge.org
Information on this title: www.cambridge.org/9780521455251

© Cambridge University Press 1994

First published 1994
This digitally printed version 2008

A catalogue record for this publication is available from the British Library

Library of Congress Cataloguing in Publication data
Molecular and cellular aspects of plant reproduction / edited by R. J.
  Scott, A. D. Stead. Society for Experimental Biology
    p.   cm.   – (Seminar series; 55)
  Includes index.
  ISBN 0–521–45525–1 (hc)
      1. Flowers—Morphogenesis—Molecular aspects—Congresses.
2. Plants—Reproduction—Molecular aspects—Congresses.   3. Plant
cellular control mechanisms—Congresses.   I. Scott, R. J. (Rod J.)
II. Stead, A. D.   III. Series: Seminar series (Society for
Experimental Biology (Great Britain)); 55.
QK653.M68   1995
581.1'6–dc20   94–17256   CIP

ISBN 978-0-521-45525-1 hardback
ISBN 978-0-521-05048-7 paperback

Contents

viii   *Contents*

Contributors

ANGENENT, G.C.
Department of Developmental Biology, DLO – Centre for Plant
Breeding and Reproduction Research (CPRO-DLO), PO Box 16, 6700
AA Wageningen, The Netherlands
ANTHONY, R.G.
Department of Molecular Biology, Horticulture Research
International, Worthing Road, Littlehampton, West Sussex, BN17
6LP, UK
ATWAL, K.K.
Wolfson Laboratory for Plant Molecular Biology, School of Biological
Sciences, University of Birmingham, Edgbaston, Birmingham, B15
2TT, UK
BARBACAR, N.
Institute of Genetics, Academy of Science of the Moldavian Republic,
Chisinau, Republic of Moldavia
BUSSCHER, M
Department of Developmental Biology, DLO – Centre for Plant
Breeding and Reproduction Research (CPRO-DLO), PO Box 16, 6700
AA Wageningen, The Netherlands
CALLOW, J.A.
School of Biological Sciences, The University of Birmingham,
Birmingham B15 2TT, UK
COLUMBO, L.
Department of Developmental Biology, DLO – Centre for Plant
Breeding and Reproduction Research (CPRO-DLO), PO Box 16, 6700
AA Wageningen, The Netherlands
FERRANT, V.
Université Paris-XI, Centre d'Orsay, Centre de Researches sur les
Plantes, URA 1128, Biologie du Développment des Plantes, F-91405,
Orsay Cédex, France

FRANKEN, J.
Department of Developmental Biology, DLO – Centre for Plant
Breeding and Reproduction Research (CPRO-DLO), PO Box 16, 6700
AA Wageningen, The Netherlands
FRANKLIN, F.C.H.
Wolfson Laboratory for Plant Molecular Biology, School of Biological
Sciences, University of Birmingham, Edgbaston, Birmingham B15
2TT, UK
FRANKLIN-TONG, V.E.
Wolfson Laboratory for Plant Molecular Biology, School of Biological
Sciences, University of Birmingham, Edgbaston, Birmingham B15
2TT, UK
GREEN, J.R.
School of Biological Sciences, The University of Birmingham,
Birmingham B15 2TT, UK
HESLOP-HARRISON, J.
137 Bargates, Leominster, Herefordshire HR6 8QS, UK
HESLOP-HARRISON, Y.
137 Bargates, Leominster, Herefordshire HR6 8QS, UK
HIRD, D.L.
Department of Botany, University of Leicester, University Road,
Leicester LE1 7RH, UK
HODGE, R
Department of Botany, University of Leicester, University Road,
Leicester LE1 7RH, UK
JAMES, P.E.
Department of Molecular Biology, Horticulture Research
International, Worthing Road, Littlehampton, West Sussex BN17 6LP,
UK
JORDAN, B.R.
Department of Molecular Biology, Horticulture Research
International, Worthing Road, Littlehampton, West Sussex BN17 6LP,
UK
Present address: New Zealand Institute for Crop and Food Research
Limited, Levin Research Centre, Kimberley Road, Private Bag 4005,
Levin, New Zealand
KREIS, M.
Université Paris-XI, Centre d'Orsay, Centre de Researches sur les
Plantes, URA 1128, Biologie du Développment des Plantes F-91405,
Orsay Cédex, France
LARSEN, P.B.
Department of Horticulture, Purdue University, West Lafayette, IN
47907-1165, USA

MOZLEY, D.
Molecular Biology Department, Horticulture Research International,
Worthing Road, Littlehampton, West Sussex BN17 6LP, UK
NEGRUTIU, I.
Institute of Molecular Biology, St Genesius Rode, Belgium
or Phytotec SA, Chaussee Romaine 77, B-5030 Gembloux Belgium
PAUL, W.
Department of Botany, University of Leicester, University Road,
Leicester LE1 7RH, UK
RIDE, J.P.
Wolfson Laboratory for Plant Molecular Biology, School of Biological
Sciences, University of Birmingham, Edgbaston, Birmingham B15
2TT, UK
SCOTT, R.J.
Department of Botany, University of Leicester, University Road,
Leicester LE1 7RH, UK
SMARTT, S.
Department of Plant Sciences, Oxford University, South Parks Road,
Oxford OX1 8RB, UK
STAFFORD, C.J.
School of Biological Sciences, The University of Birmingham,
Birmingham B15 2TT, UK
STEAD, A.D.
Department of Biology, Royal Holloway, University of London,
Egham Hill, Egham, Surrey TW20 OEX, UK
TEN HAVE, A.
Department of Horticulture, Purdue University, West Lafayette, IN
47907-1165, USA
THOMAS, B.
Molecular Biology Department, Horticulture Research International
Worthing Road, Littlehampton, West Sussex BN17 6LP, UK
TWELL, D.
Department of Botany, University of Leicester, University Road,
Leicester LE1 7RH, UK
VAN DOORN, W.G.
Agrotechnological Research Institute (ATO-DLO), PO Box 17, 6700
AA Wageningen, The Netherlands
VAN TUNEN, A.J.
Department of Developmental Biology, DLO – Centre for Plant
Breeding and Reproduction Research (CPRO-DLO), PO Box 16, 6700
AA Wageningen, The Netherlands
VAN WENT, J.
Wageningen Agricultural University, Plant Cytology and Morphology

Department, Arboretumlaan 4, NL-6703 BD Wageningen, The Netherlands
WHITEHEAD, C.S.
Department of Botany, Randse Afrikaanse Universiteit, 2000 Johannesburg, South Africa
WOLTERING, E.J.
Agricultural Research Institute, Haagsteeg 6, PO Box 17, NL-6700 AA Wageningen, The Netherlands
WOODSON, W.R.
Department of Horticulture, Purdue University, West Lafayette, IN 47907-1165, USA
WORRALL, D.
Department of Botany, University of Leicester, University Road, Leicester LE1 7RH; UK
WRIGHT, P.J.
School of Biological Sciences, The University of Birmingham, Birmingham B15 2TT, UK

J. HESLOP-HARRISON

Introduction: 'Where do we go from here?'

The articles in this volume provide a stimulating overview of several aspects of current research in plant reproductive biology. Some convincingly illustrate how age-old problems of plant reproductive systems are proving to be accessible to the awesomely powerful methods of molecular biology, while others show how this approach can complement – and be complemented by – other new techniques for investigating intracellular processes and cellular interactions at levels beyond that of the gene. In the light of these many current developments, it is surely not overly optimistic to suggest that we are entering a new era in the exploration of plant reproductive phenomena.

So, in what directions might research be expected to progress now? While defining priorities is always fraught, it is at least permissible to attempt to identify a few parts of this very wide subject where there is a real promise that some of the outstanding problems could yield to the application of the new methodologies. In this connection it is important for the plant scientist to keep an eye on what is happening in animal and microbial cell biology; after all it is from these much more fully populated and much better resourced fields that we borrow most of our technology. We need also to bear in mind that sexual reproduction in plants involves essentially all the basic processes of development: cell nutrition, division, growth and communication, tissue differentiation and organ morphogenesis. So it is assured that most of the basic mechanisms involved are likely to be common to other tissues, and cannot therefore be considered altogether in isolation: in this sense the processes of reproduction can be looked upon as a sub-set of the wider spectrum of plant developmental processes in general. Yet when they provide relatively uncluttered situations where specific phenomena are conveniently open for experimental investigation and analysis, the results could help to illuminate other areas

Society for Experimental Biology Seminar Series 55: *Molecular and Cellular Aspects of Plant Reproduction*, ed. R.J. Scott & A.D. Stead.
© Cambridge University Press, 1994, pp. 1–7.

of plant biology. Examples are offered by the cellular interactions basic to sexual reproduction in plant groups from the fungi and algae to the angiosperms. The molecular mechanisms concerned may indeed exemplify types of interaction operating more generally in *somatic* tissues: but they are rather more accessible for experimentation when the interactions are between separate, independent cells, perhaps originating from physiologically different individuals. The sexual recognition systems already investigated in the brown and green algae provide perhaps the simplest models, although their heterogeneity warns against premature generalisation. The common feature is the involvement of cell surface receptor systems, postulated for both species-specific and mating-type specific recognitions. In each situation protein–carbohydrate interactions are suspected, a concept owing something to animal-cell models. There has been some enthusiasm for extending the idea to higher plant groups, with the suggestion that sperm–egg recognition in the angiosperms may also depend upon the interaction of specific membrane- or wall-held factors. In actuality there is as yet little hard evidence for this interpretation: but opportunities may nevertheless become available for rigorous testing of the proposition, with the development of methods for extracting male gametes from pollen tubes and for isolating embryo sacs and even the egg cells. This is certainly an exciting prospect, but there are still formidable difficulties to be overcome, and a good deal more basic groundwork needs to be done before the interactions associated with fertilisation in the *intact* system can be said to be understood. Structural studies by electron microscopy of embryo sacs and fertilisation itself have progressed with some speed in recent years, but still no wholly satisfactory general account of the disposition of male-originating and female-originating membranes at the time of fertilisation is available. It seems certain that the male gamete as seen in the pollen tube is not directly involved in its intact state in the interaction with the egg – perhaps only the male nucleus. In this event, the question remains open as to between what surfaces and in what sites 'recognition' – if indeed there be such – takes place. Yet resolving these problems is surely not beyond the reach of available methods once the questions are properly defined.

Undoubtedly embryo sacs, because of their relative inaccessibility, have hitherto been the target of rather less physiological research than their role in angiosperm reproduction would merit. The situation is likely to change rapidly now with the advent of new techniques, including notably X-ray microanalysis, immunofluorescence and new classes of molecular probes, some potentially capable of use with the living system. Attempts to elucidate the organisation of the actin and microtubule cytoskeletons of the embryo sac have not been altogether satisfactory to date, yet these must without doubt have significant functions in governing nuclear dispositions during the development of the sac, and probably also in controlling the specific directional movements of the male nuclei during the characteristic double fertilisation. Could the disposition of the actin cytoskeleton rather than postulated recognition events determine the actual nuclear fusions?

Then, another age-old question – following upon the nuclear fusions, the zygote divides and embryogenesis is initiated: what is it that activates the previously dormant egg? Fertilisation also stimulates the primary endosperm nucleus into division, but none other in the sac or in the surrounding tissues is affected, indicating that the influences must be strictly localised. This question has exercised generations of cytologists, but the physiological problems involved have been largely neglected. Comparable events in animal fertilisation have been attributed to the release of second message after the contact of the sperm with the egg surface, and fertilisation in algae and fungi, more fully investigated than in the higher plants, provides other models. But the analogies may not be especially informative when it is not cells but nuclei that are involved in the first interactions. It is pertinent here that there are records from orchids with a simplified type of embryo sac of the egg nucleus dividing after the advent of the male nucleus but before physical contact, as though some form of pheromone were involved. Cytokinins? Probably not: they would seem to lack the specificity required. Evidently further experimentation aimed at establishing the actual nature of the stimulus is overdue, especially considering the wider importance of attaining some general understanding of the factors that control nuclear division in intact plant tissues.

Events in the embryo sac lie at the end of a long cascade of interactions starting with the capture of pollen on the stigma surface. Pollen–pistil relationships have attracted much research in recent years, but many aspects remain enigmatic. On the face of it, the phenomena fall into two categories: (a) generalised interactions concerned with pollen capture, hydration, germination, pollen tube penetration and growth through the style; and (b) specific interactions related to the governance of the breeding system, including the control of self- and cross-compatibility. A hindrance in the conduct of research is that the two categories overlap in so many ways that it is often difficult to determine with which set of phenomena one is dealing. Nevertheless, with the background of more than 60 years of classical genetical research, major advances have been made in analysing the molecular basis of phenomena in the second category, those concerned with self-incompatibility regulated by S-gene systems. The commoner of the two principal types of mechanism involves control on the pollen side through the haploid genome, and the other, govern-ance through the diploid genome of the pollen parent. In each, recogni-tion of genetic identity in pollen and pistil at the loci concerned leads to rejection, resulting in effective promotion of outbreeding. The dramatic achievement has been the sequencing of the S-genes from certain species and the identification and characterisation of S-specific glycoproteins. While there is as yet no fully satisfactory explanation of the biochemical and physiological basis of the inhibition in any species, the development of *in vitro* bioassays for some species is already simplifying the attack by eliminating effects related to other, perhaps irrelevant, interactions on the stigma surface. However, evidence already available indicates that there is variation in the operation of ostensibly similar systems, and also shows that rejection in some species probably involves more than one single inhibitory event but may depend on disruption at several points in the normal pattern of pollen–stigma interactions. Dissecting the control pathways is likely to prove a challenging, but not impossible, task.

Stigma cells and tissues of transmitting tracts respond to the advent of pollen and pollen tubes in various ways, in some species ways akin to the hypersensitive reactions associated with pathogen invasion. Several striking effects are well documented, but so far lack convincing physiolo-gical explanations. Pollination in some instances induces secretion, pos-sibly thereby yielding products important for pollen-tube growth. The secretion-stimulating factor remains to be identified; but it may be the same as that which re-directs intracellular movement in the receptor cells and leads, in incompatible combinations, to localised deposition of cal-lose in the wall. In other species, pollination disables receptive cells by increasing the permeability of their membranes. Again, the causal factor remains unidentified, but it is interesting that pollen extracts can increase the permeability of the membranes of HeLa cells, a possible clue to one of the interactions with the mucous membrane. These all present approachable problems, now that the phenomenology is reasonably well documented.

One intriguing aspect of the pollen–pistil interaction concerns pollen-tube guidance. How does a single, tip-growing cell, originating on the stigma surface, find its way to the tiny aperture of the micropyle of an ovule that may be as much as 15 cm away? The concept of chemotropic control, borrowed largely from fungal lore, has held sway for many years, but is now seen as inadequate to explain all features of tube growth. Control seems to depend upon three factors: mechanical constraint imposed by the morphology of the pollen-tube pathway; tactile stimuli re-directing growth in response to contact with physical obstacles, and reaction to specific chemical signals. But, in no instance, has the whole

pathway been plotted and analysed to the point where it can be claimed
that all components of the guidance system are fully understood. Perhaps
one of the most significant questions concerns the role of calcium, at one
time identified as a possible chemotropic factor. By the use of
Ca^{2+}-specific fluorescent probes and X-ray microanalysis the ion has
already been shown to be specifically associated with the pollen-tube
pathway, and the possibility that calcium provides directional control by
affecting the disposition of the actin cytoskeleton and in consequence the
growth vector remains open for testing.

The angiosperms have been the most resourceful of all groups, plant
and animal, in exploiting the potential diversity of their reproductive
systems to achieve ecological adaptation and so evolutionary success. In
many instances this has involved perversion of the sexual system, and
even complete abandonment of its central function of gene recombination, as in the various apomictic groups. The task of disentangling the
genetics of aposporous apomixis has made a very promising start, and
currently molecular techniques are being applied to achieve a more
detailed understanding of the gene activity lying at the base of the main
phenomena. The grasses, with more than 25% of aposporous species in
some tribes, provide a special challenge, not least because of their economic significance. Apospory involves the transfer of the potential for
embryo sac formation from the haploid spore to a diploid somatic cell of
the nucellus, and can, in some instances, be regulated by environmental
manipulation. The switch is in effect a sporophyte–gametophyte transition without a prior meiosis. The nominated nucellar cell or cells can
first be identified by a rapid accumulation of cytoplasmic RNA, but it
remains to be established whether the cells undergo earlier differentiation
comparable with that of the meiocyte itself, another opportunity for the
application of molecular methods. Diplospory, more diverse in its cytological manifestations in the various groups where it is habitual, involves
the coordination of several developmental aberrations, and is probably
less accessible to techniques currently available. Yet there are aspects
which do invite early attention, notably the control of parthenogenesis,
an essential part of an apomictic cycle. In both diplosporous and aposporous systems, a previously dormant diploid egg is activated autonomously in the absence of fertilisation, so initiating embryogenesis. The
contrast with the situation in sexual embryo sacs, discussed above, is
noteworthy.

Male sterility, long exploited by plant breeders as a tool for the production of hybrids, has latterly been subjected to intensive research. Genetical studies dating back half-a-century established that the control in
onions, maize and other species was extranuclear, and the principal out-

come of recent work has been the proof that the defect blocking normal pollen development is in the mitochondrial genome. The story is by no means complete, however, and much remains to be learned about the functions of restorer genes, and not least about the physiology of the response itself. There is an interesting link here with sex determination in polyoecious species. Here the sterilisation of male flowers may involve a whole range of blockages, from the inhibition of inception of primordia to late pollen abortion, matched on the female side by a similar range of aberrations.

As we have noted in the introductory paragraphs, fundamental processes common to all plant tissues necessarily underlie much of reproductive development. Yet there are some ontogenetic events not matched at all elsewhere in the plant body. One such is the formation of patterned pollen grain walls, involving morphogenetic processes and associated specialised syntheses peculiar to the reproductive phase of angiosperms and without close parallels in other groups. Here we see a single haploid cell moulding a refractory polymer, sporopollenin, to generate a structured wall which bears a genetic imprint so specific that it can often be used to identify family, genus and even, in some instances, species. Earlier work revealed much about the mechanics of wall formation and about the parts played by the cytoskeleton and membranes of the haploid cell itself, as well as indicating some of the likely functions of the contiguous diploid tissues of the parent sporophyte. Now genetic manipulation is throwing new light on causal links. Coupled with the application of the newer methods for exploring cell structure and chemistry, this approach holds the promise of finally revealing how the determinants conveyed in the genome are ultimately translated into unique, taxonomically specific, structures. Surely to achieve this would represent an advance significant for the whole of biology.

The present conspectus reflects the contents of the rest of the volume in its heavy bias towards the angiosperms. In doing so it typifies the overall orientation of research on plant reproduction: much less attention is given to the reproductive systems of other vascular groups and to the lower plants than to the flowering plants. The reasons for this are obvious enough; the angiosperms are not only the most important group economically, but contribute so much more than others to the totality of the human environment. Yet much of the lesser body of research on other groups has been of the very highest quality, and has already shown how much they have to offer in opportunities for the application of the new armoury of methods in investigating processes often less accessible in the angiosperms. In another sense they can help to reveal the pathways whereby the often reduced and specialised functions of flowering plant reproduction are likely to have evolved, a point not lost upon the great morphologists and cytologists of the nineteenth century and the early years of the present.

A.J. VAN TUNEN, M. BUSSCHER,
L. COLUMBO, J. FRANKEN
and G.C. ANGENENT

Molecular control of floral organogenesis and plant reproduction in *Petunia hybrida*

Introduction

Flower development has intrigued man since the dawn of time. Firstly flowers are very attractive for their ornamental value and fragrance. Secondly, the flower forms the place in which the plants' reproductive organs and cells are formed. Therefore, development of efficient reproductive organs and cells is of vital interest for the maintenance of life on earth. For these reasons, flowers have also a considerable commercial value both in the area of the production of cut flowers and pot plants and in the area of plant breeding.

A typical dicot flower consists of four concentric whorls of floral organs. The outermost whorl contains the sepals. The main function of these floral organs is the protection of the vulnerable young flowerbud. The second whorl consists of the petals which are often very showy and serve an important function in the attraction of pollinators. Whorl three consists of the stamens in which the male gametes are formed. Whorl four consists of the carpels which are often fused to form a pistil. In the ovary of the pistil the egg cells are present within the ovules. Upon fertilisation, the pollen tube grows towards the ovules and the egg cells, and in a double fertilisation event the egg cell and the central cell are fertilised by the two male reproductive cells.

Floral induction, floral development and plant reproduction have been studied already for a long time especially at morphological and physiological levels. For a number of years these reproductive processes have also been investigated at the molecular and molecular-genetic level. These studies were especially performed using *Arabidopsis thaliana* and *Antirrhinum majus*, and to a lesser extent *Petunia hybrida* and *Nicotiana tabacum*, as model systems.

Society for Experimental Biology Seminar Series 55: *Molecular and Cellular Aspects of Plant Reproduction*, ed. R.J. Scott & A.D. Stead.
© Cambridge University Press, 1994, pp. 9–16.

Flowering and the ABC model

Flowering is induced by different external or internal factors. For instance light, temperature, plant hormone levels and/or age of the plant are important for the developmental switch of the juvenile phase into the generative phase. In this process a set of class I regulatory genes are induced and the vegetative meristem starts to form the inflorescence meristems (Koornneef, Hanhart & Van der Veen, 1991; Figure 1). Subsequently, class II genes are induced which trigger the formation of a floral meristem from the inflorescence meristem. After that, the floral meristem forms the various floral organ primordia which is controlled by the class III floral regulatory genes. Finally, late developmental processes are induced which lead to the formation of the different sub-organs, tissues and specialised cell types within the flower organs. Genes belonging to class IV might regulate these developmental programmes.

Genes belonging to class II and III have been isolated and characterised recently (for review see Coen & Meyerowitz, 1991). Most of these genes encode potential transcriptional activators which contain a common DNA binding sequence known as the MADS box (Sommer *et al.*, 1990). Furthermore, the class III genes have been hypothesised to regulate the determination of the four types of floral organ primordia according to a so-called ABC model. This model has been developed for

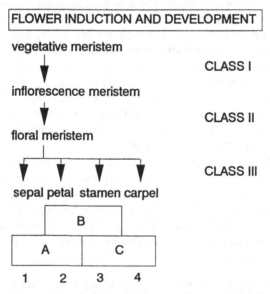

Fig. 1. Induction of flowering and floral organogenesis.

Arabidopsis and *Antirrhinum* (Coen & Meyerowitz, 1991). According to this model, type A gene expression determines the formation of sepals in whorl one. If A and B genes are active, petals are formed. Expression of gene B and C results in the formation of stamens and expression of gene C alone results in the formation of carpels. Finally the model postulates that gene A and C are mutually exclusively expressed: if gene A is active gene C is suppressed and vice versa.

Isolation of MADS box genes from *Petunia hybrida*

To investigate the molecular control of floral morphogenesis and plant reproduction in *Petunia hybrida*, and to test the ABC model in another plant species than *Arabidopsis* or *Antirrhinum* we decided to characterise and genetically modify MADS box gene expression in *Petunia hybrida*. These genes were also studied in normal and mutated plants. We therefore used a PCR approach to isolate the first MADS box gene designated *fbp1* (floral binding protein gene 1) using mRNA isolated from petal tissue of young flowerbuds. A full-size *fbp1* cDNA was isolated from a petal-specific cDNA library and subsequently this library was rescreened with this full-size probe containing the conserved MADS box region under low stringency conditions (Angenent *et al.*, 1992; Angenent and van Tunen, unpublished observations). This yielded a set of cDNA clones: *fbp2–fbp5* (see also Fig. 2). Finally a mixed probe corresponding to the MADS box region of *fbp1* and *fbp2* was used to screen a carpel-specific library under low stringency conditions. This yielded an additional set of MADS box cDNAs: *fbp6–fbp15* (Angenent & van Tunen, unpublished observations). Using 3′ gene-specific probes the expression of the various MADS box genes was determined by RNA blotting. The results are summarised in Fig. 2. Subsequently the expression and function of *fbp1* and *fbp2* were studied in more detail.

Function of *fbp1*

To study the cell-specific expression of *fbp1* we have isolated an *fbp1* genomic clone. The organisation and structure of the *fbp1* gene was determined by sequence analysis. A high sequence homology in the *fbp1* coding region with that of the *globosa* (*glo*; Tröbner *et al.*, 1992) gene from *Anthirrhinum* was observed. RNA blotting revealed that the *fbp1* gene is specifically active in petals and stamens and not in sepals, carpels or any vegetative parts. *Fbp1* expression was also investigated in a Green Petals mutant (van Tunen, Gerats & Mol, 1990; Gerats, 1991; Fig. 3). This is a petunia mutant with the petals homeotically transformed into sepaloid organs. RNA blot analysis revealed that *fbp1* is expressed in

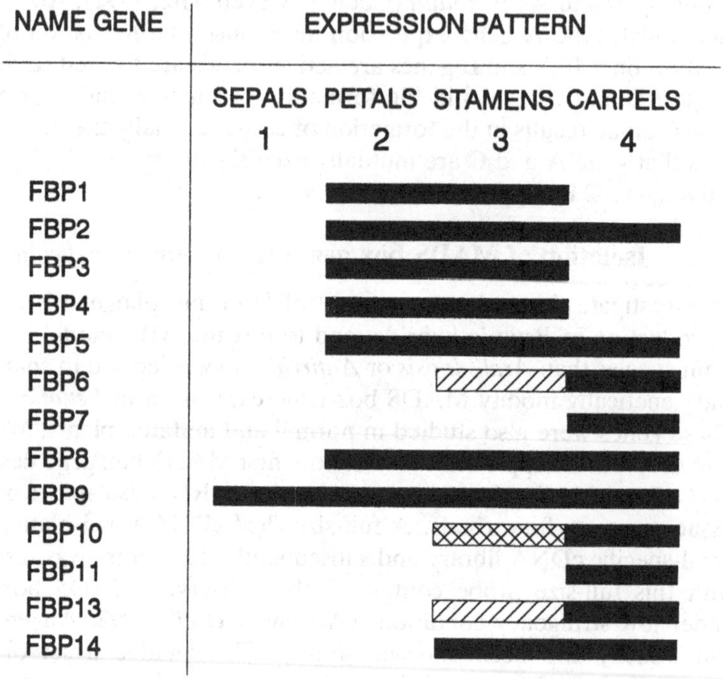

NAME GENE	EXPRESSION PATTERN

	SEPALS	PETALS	STAMENS	CARPELS
	1	2	3	4
FBP1		▓▓▓▓	▓▓▓▓	
FBP2		▓▓▓▓	▓▓▓▓	▓▓▓▓
FBP3		▓▓▓▓	▓▓▓▓	
FBP4		▓▓▓▓	▓▓▓▓	
FBP5		▓▓▓▓	▓▓▓▓	▓▓▓▓
FBP6		▨▨▨	▓▓	▓▓▓▓
FBP7			▓▓	▓▓▓
FBP8			▓▓▓▓	▓▓▓▓
FBP9	▓▓▓▓	▓▓▓▓	▓▓▓▓	▓▓▓▓
FBP10		▧▧▧	▓▓	▓▓▓
FBP11			▓▓	▓▓▓
FBP13		▨▨▨	▓▓	▓▓▓
FBP14		▓▓▓▓	▓▓▓▓	

Fig. 2. Expression patterns of *Petunia hybrida* MADS box genes designated *fbp1* to *fbp14*. The expression patterns were determined using a 3' probe specific for each gene.

the stamens of the Green Petals mutant but not in the sepaloid organs. In contrast to this the *fbp2* gene is active in whorl three and four but also at normal levels in the transformed whorl two. This indicates that the gene *gp* directly or indirectly regulates *fbp1* but not *fbp2* and suggests that the *fbp2* is a gene acting higher in the hierarchy than *fbp1* and/or *gp* (Angenent *et al.*, 1992).

To determine the spatial and temporal aspects of *fbp1* gene expression, the *fbp1* promoter was fused to the GUS reporter gene and transgenic petunia plants containing this gene construct were generated. Analysis of these transgenic plants demonstrated a petal and stamen-specific *fbp1*-driven GUS activity. Furthermore, the timing of *fbp1* promoter activity correlated with the onset of the determination of the petal and stamen primordia. These expression data combined with the observed homology with *glo* suggested that the *fbp1 is* a B type of homeotic gene.

Fig. 3. Phenotype of a flower from a control (A), the mutant line Green Petals (B) and a transgenic *Petunia* plant with inhibited *fbp1* gene expression (C). Note that, in C, the petals are transformed towards sepals and the stamens are replaced by carpels which are fused together.

We have investigated the function of *fbp1* in more detail; transgenic plants were made in which *fbp1* gene expression was inhibited (Angenent *et al.*, 1993). For this, the so-called sense co-suppression approach was followed. Therefore a gene construct was made with the full-size *fbp1* cDNA in a sense orientation under the control of the CaMV 35S promoter. Twenty-one transgenic plants containing this gene construct were raised. Two of the plants showed phenotypic alterations of which one was selected for further analysis. This primary transgenic plant possessed flowers with small green sectors on the edge of the petals and had small stigma/style structures on top of a relatively normal anther. This phenotype co-segregated with the presence of the transgenes (eight copies in the primary transformant) in offspring of this plant. RNA blotting demonstrated that *fbp1* mRNA was not present any more. In contrast to this normal levels of *fbp2* RNA were detected indicating that the co-suppression effect is specific. Progeny from a self-pollination of the primary transformant could be divided into three classes. Group one consists of plants with a normal control phenotype. Group two plants have a phenotype which resembles that of the primary transformant. Further testcrossing revealed that these plants are heterozygote for the transgenes. Testcrossing of group three plants revealed that these plants are homozygote for the introduced CaMV *fbp1* chimeric gene construct. The phenotype of these group three plants is more drastically affected: the transformation of petals into sepals is almost complete whereas the stamens are replaced by carpels. In a number of cases these five whorl-three carpels are fused to form a whorl-three gynoecium which is also connected with the normal whorl-four gynoecium (Fig. 3). Like the whorl-four gynoecium, the whorl-three gynoecium also contains ovules which develop normally and are fertile. In these ovules the expression of flavonoid-synthesis genes was investigated.

It is known that flavonoids not only play an important role as floral pigments but also have an important role during plant reproduction and development of the gametophytes (Ylstra *et al.*, 1992; Mo, Nagel & Taylor, 1992). Depletion of flavonoids from anthers and pollen in anti-sense chs transgenic petunia plants results renders these plants self-sterile (van der Meer *et al.*, 1992; Taylor & Jorgensen, 1992). Flavonoids and especially flavonols are also present in the female reproductive tissues. A GUS reporter gene fused to a *chalcone synthase* (*chs*; a flavonoid biosynthesis gene) promoter is specifically activated during ovule development. This gene construct can be used as a marker for ovule development (van Tunen, Ylstra and Angenent, unpublished observations). After crossing transgenic *chs*-GUS plants with the *fbp1* co-suppression plants, the resulting double transformant contained both transgenes. In this plant it was shown that the *chs*-GUS reporter gene is specifically active in both the whorl four and three ovules.

Taken together, it can be concluded from our experiments that a suppression of *fbp1* results in homeotic transformation of both whorl two and three in such a way that petals are transformed into sepals and stamens into carpels. This strongly indicates that *fbp1* is a B type of homeotic gene and represents the cognate homologue of *globosa*. It also demonstrates that with respect to *fbp1* the ABC model also holds for *Petunia hybrida*.

Future

It was shown that the *fbp1* gene is a B type of homeotic gene which, according to the ABC model regulates the determination of whorl two and three primordia into respectively petal and stamen primordia. Furthermore, it was demonstrated that, in the *Petunia* variety mutated for the gene *green petals*, the *fbp1* gene is downregulated. The *gp* mutation has been isolated several times in independent ways. The mutant is only affected in whorl two. Since the ABC model predicts that a mutation in an ABC gene results in phenotypic alterations in two adjacent whorls this is different. Using RFLP linkage analysis we have shown that *fbp1* is different from *gp*. Therefore, it will be interesting to clone the *gp* gene and to determine if, and how, *fbp1* and *gp* interact with each other. In the *Antirrhinum globosa/deficiens* situation it was found that the products of these two B type genes indeed interact *in vitro* and bind as a heterodimer to the *deficiens* promoter (Tröbner *et al.*, 1992).

Furthermore, several other aspects of the MADS box gene family are interesting. It will be interesting to characterise the function of other MADS box genes. Some of these might be potential ABC genes. In this context the action of *fbp2-6, 8, 10, 13* and *14* will be investigated. The MADS box genes which are expressed in the female reproductive organs will also be studied. Especially *fbp7* and *fbp11* are of interest since these two genes are only expressed in whorl four. We hypothesise that these genes might represent class IV MADS box genes. Maybe these genes are involved in the regulation of the development of sub-organs or tissues within the pistil. Experiments have been designed to downregulate *fbp7* and *fbp11* gene expression following the sense co-suppression technique developed for *fbp1*.

References

Angenent, G.C., Busscher, M., Franken, J., Mol, J.N.M. & van Tunen, A.J. (1992). Differential expression of two MADS box genes in wild-type and mutant petunia flowers. *The Plant Cell,* **4**, 983–93.

Angenent, G.C., Franken, J., Busscher, M., Colombo, L. & van Tunen, A.J. (1993). Petal and stamen formation in petunia is regulated by the homeotic gene fbp1. *The Plant Journal,* **4**, 101–12.

Coen, E.S. & Meyerowitz, E.M. (1991). The war of the whorls: genetic interactions controlling flower development. *Nature,* **353**, 31–7.

Gerats, A.G.M. (1991). Mutants involved in floral and plant development in *Petunia*. *Plant Science,* **80**, 19–25.

Koornneef, M., Hanhart, C.J. & van der Veen, J.H. (1991). A genetic and physiological analysis of late flowering mutants in *Arabidopsis thaliana. Molecular and General Genetics,* **229**, 57–66.

Mo, Y., Nagel, C. & Taylor, L.P (1992). Biochemical complementation of chalcone synthase mutants defines a role for flavonols in functional pollen. *Proceedings of the National Academy of Sciences, USA,* **89**, 7213–17.

Sommer, H., Beltràn, J.-P., Huijser, P., Pape, H., Lönnig, W.-E., Saedler, H. & Schwarz-Sommer, Z. (1990). Deficiens, a homeotic gene involved in the control of flower morphogenesis in *Antirrhinum majus:* the protein shows homology to transcription factors. *EMBO Journal,* **9**, 605–13.

Taylor, L.P. & Jorgensen, R. (1992). Conditional male fertility in chalcone synthase-deficient petunia. *Journal of Heredity,* **83**, 11–17.

Tröbner, W., Ramirez, L., Motte, P., Hue, I., Huijser, P., Lönnig, W.-E., Saedler, H., Sommer, H. & Schwarz-Sommer, Z. (1992). *GLOBOSA:* a homeotic gene which interacts with *DEFICIENS* in the control of *Antirrhinum* floral organogenesis. *EMBO Journal,* **11**, 4693–704.

van der Meer, I.M., Stam, M.E., van Tunen, A.J., Mol, J.N.M. & Stuitje, A.R. (1992). Inhibition of flavonoid biosynthesis in petunia anthers by antisense approach results in male sterility. *The Plant Cell,* **4**, 253–62.

van Tunen, A.J., Gerats, A.G.M. & Mol, J.N.M. (1990). Flavonoid gene expression follows the changes in tissue development of two *Petunia* homeotic flower mutants. *Plant Molecular Biology Reporter,* **1**, 50–60.

Ylstra, B., Touraev, A., Benito Moreno, R.M., Stöger, E., van Tunen, A.J., Vicente, O., Mol, J.N.M. & Heberle-Bors, E. (1992). Flavonols stimulate development, germination and tube growth of tobacco pollen. *Plant Physiology,* **100**, 902–7.

B.R. JORDAN, R.G. ANTHONY
and P.E. JAMES

Control of floral morphogenesis in cauliflower (*Brassica oleracea* L. var. *botrytis*): the role of homeotic genes

Introduction

Cauliflower (*Brassica oleracea* L. var. *botrytis*) is a member of the genus *Brassica* in the family *Brassicaceae*. It is a major economic crop which is marketed as a large, white, pre-floral, compact curd. The cauliflower curd is, however, susceptible to a number of morphological defects which can seriously reduce its commercial value (King, 1990). The genetic determinants that control floral meristem development and flower morphogenesis in cauliflower have not been approached at the molecular level until recently. Medford, Elmer and Klee (1991) have now isolated and partially characterised a number of meristematic genes expressed in the immature cauliflower curd. In addition, Anthony, James and Jordan (1993) have isolated a homologue of the *Antirrhinum flo* gene (Coen *et al.*, 1990) that is thought to be involved in regulating floral initiation.

In this review, we will briefly describe the morphology and physiology of the cauliflower, with particular emphasis on floral development. In addition, we will describe the isolation of genes involved in floral initiation and discuss how these genes may interact to control cauliflower morphology.

Morphology and physiology of cauliflower floral development

Five stages of development have been recognised between vegetative growth and flowering in cauliflower (Margara & David, 1978): 1. The vegetative stage in which the small pointed shoot apex is surrounded by leaf primordia. These arise acropetally in a spiral succession and axillary branches do not develop; 2. Initiation of inflorescence results in precocious formation of axillary buds at the apex to form clusters of meristems

Society for Experimental Biology Seminar Series 55: *Molecular and Cellular Aspects of Plant Reproduction*, ed. R.J. Scott & A.D. Stead.
© Cambridge University Press, 1994, pp. 17–29.

and bracts; 3. The curd develops by the multiplication of meristems and each new meristem gives rise to a higher order; 4. Curd maturity with no flower initials; 5. Floral differentiation and elongation of some of the inflorescence branches. Figure 1 shows scanning electron micrographs of a mature curd (Fig. 1A) and an area of curd initiating floral morphology (Fig. 1B).

The morphology of the cauliflower curd has been extensively studied and a wide range of suggestions made as to the exact nature of its anatomy and developmental status (Sadik, 1962). Because of its morphology it is now generally considered to be an early stage of inflorescence development. However, some physiological responses would suggest that it is a vegetative structure, for instance the curd of var. 'Snowball' requires no vernalisation, although vernalisation is required for floral initiation (Sadik, 1962, 1967). In addition, many of the morphological and cellular changes during curd maturation would normally suggest that it was composed of a large number of floral meristems (apical broadening and increased cell division), these floral meristems constantly aborting and failing to form flowers. This rather ambiguous morphology therefore provides a particularly interesting tissue to investigate the difference between inflorescence and floral meristems.

Shoot apices of 5-week-old cauliflower can be divided into five zones: a mantle region, central mother cells, cambial-like cells, peripheral cells and rib meristem (Sadik & Ozbun, 1968). At this time, curd initiation involves the widening and flattening of the apex. The mantle layer has increased in number to 6–7 cell layers in depth, containing highly stratified flattened cells with large nuclei and small conspicuous nucleoli. The central-mother cell zone has large, vacuolated and irregularly shaped cells, while the cells of the rib meristem zone are arranged longitudinally. Numerous periclinal divisions take place to form the bracts and first order branch initials. The formation of the bracts, instead of leaves, subtending the branch primordia marks the commencement of curd formation. In addition, loss of the cambial-like zone is associated with curd initiation. The curd then develops from an extensive proliferation of apical meristems and associated branching.

Instead of a loss of zonation leading to the development of floral organs, axillary meristems are initiated (Medford, Elmer & Klee, 1991). These axillary meristems proliferate (with associated branches). This amplification leads to the formation of the curd, consisting of 10^5 to 10^6 apical meristems. The apical meristems in the cauliflower curd are usually between 100 μm and 190 μm, depending upon the stage of development and remain terminal, located in the uppermost 0.1 mm to 0.2 mm, so that only the surface cell layers are meristematic. The formation of the

(a)

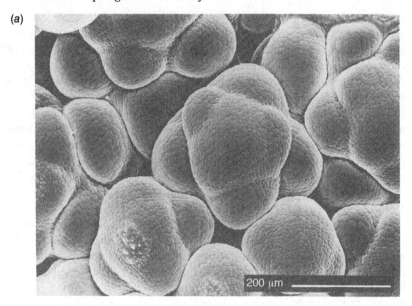

(b)

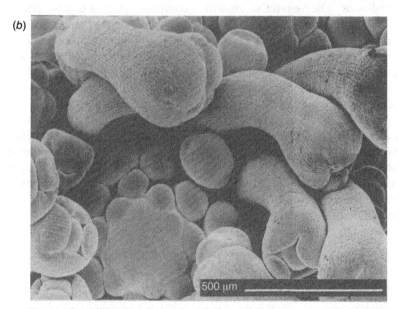

Fig. 1. Scanning electron micrograph of (*a*) mature cauliflower curd showing the multiplication of the apical meristems, (*b*) cauliflower curd showing peduncle elongation during initiation of floral morphology.

curd precedes floral initiation, and further development of the apical meristems to floral morphology is arrested. When floral initiation does commence, it takes place in distinct patches over the surface of the curd, the areas of initiation comprising about 10–20% of the total area. Other *Brassica* varieties (Broccoli: var. *italica*) arrest floral development at different stages and this leads to the diverse range of morphology.

The size and shape of the cauliflower curd is genetically determined, and curding can be either dominant or recessive (King, 1990). Curding may, however, not be polygenetically inherited. For instance, wide crosses within cauliflower show poor curding ability in the F_1 and F_2 generations. This wide diversity of curd morphology has also been accounted for by the influence of modifier genes (King, 1990; Bowman, 1992 and see below for further discussion).

Brassica show a wide variation in their response to the environment (see Friend, 1985 for detailed discussion). Most *Brassica* species show either a quantitative or qualitative response to low temperature. The photoperiodic response is usually of a quantitative long-day plant. Some biennials, however, may be day neutral after vernalisation. In cauliflower, the curd development is initiated after a period of vegetative growth that is dependent in duration upon genotype and environmental conditions. The vegetative growth is composed of two distinct phases: a juvenile stage in which curd initiation is not possible and a mature or 'inductive phase' when it can be induced. In some cauliflower cultivars seed vernalisation is effective in allowing floral initiation, presumably by shortening the juvenile phase. Chilling of mature vegetative plants is usually effective in initiating curd formation. During the juvenile phase, however, chilling delays curd formation as it retards growth. Provision of carbohydrate solutions can also partially replace the effect on curd formation of cold treatment in post-juvenile tissue (Atherton, Hand & Williams, 1987). Although the precise role of carbohydrates remains unknown, it is clear that redistribution of carbohydrates to the apical region is involved in curd initiation. This accumulation of carbohydrate at the apices could act either as an increased metabolic substrate or as a regulator of gene expression by some form of feed-back control.

Isolation and characterisation of a *Brassica oleracea flo* homologue

Recently, dramatic progress has been made in the isolation and characterisation of genes that control the transition and organogenesis of floral development (Bowman & Meyerowitz, 1991; Coen, 1991; Coen & Meyerowitz, 1991; Jordan & Anthony, 1993). These genes have been isolated by the combined use of molecular genetics and floral mutants, principally from *Antirrhinum* and *Arabidopsis*. They have been termed homeotic genes as they are involved in the spatial arrangement of cells and tissues within the organism. The *Antirrhinum majus flo* gene, and its *Arabidopsis* homologue *leafy*, represent homeotic genes which are expressed at an early stage in floral initiation, and interact in a sequential manner with other homeotic genes affecting floral organ identity (Coen *et al.*, 1990; Weigel *et al.*, 1992 and see below). The *floricaula* mutant from which *flo* was isolated initiates vegetative growth and development of inflorescence shoots in the normal manner. The mutant, however, cannot progress from inflorescence to floral meristem and produces a proliferation of inflorescence shoots in place of normally developing flowers. The FLO protein product is thus essential for floral development. The morphology of the developing cauliflower curd may be considered analogous to the *floricaula* mutant if it is truly an early stage of inflorescence development. It is therefore likely that a *flo* homologue will play an important role at some stage in cauliflower floral development.

To isolate the *flo* homologue from cauliflower, a cDNA library was constructed from florally initiating tissue of the double haploid cauliflower line DJ7032 derived from 'Nedcha' (supplied by Dr David Ockendon). The cDNA library was then screened with the *Antirrhinum flo* sequence. Approximately 55 000 plaques were screened, and three plaques showed positive hybridisation with the *flo* probe at high stringency (clones designated *bofh: brassica oleracea flo homologue*).

Analysis of *bofh* cDNA clones

Complete DNA sequencing of two *bofh* recombinants revealed them to be identical at the nucleotide level (Anthony, James & Jordan, 1993). The *bofh* cDNA contains a 1245 nucleotide open reading frame (ORF) which is capable of encoding a polypeptide of 415 amino acids. The predicted molecular weight of the cDNA-deduced BOFH protein is 46657 daltons with a predicted isoelectric point of pH 6.93. A search comparing the *bofh* DNA sequence with the EMBL and GENBANK databases detected significant homology only with the *flo* and *lfy* sequences. Alignment of the 415 amino acid BOFH protein with the 396 amino acid FLO protein (Coen *et al.*, 1990) reveals 82% identical plus conservative substitutions. Alignment of BOFH with the 424 amino acid isoform of LFY (Weigel *et al.*, 1992) reveals 92% identical residues plus conservative substitutions. Three distinctive domains can be identified. Two of these regions, a proline-rich domain and an acidic region located downstream of the proline-rich domain, indicate that FLO is a transcriptional activator (Coen *et al.*, 1990). Both of these domains are common features of transcription factors, although they are not usually present in the same protein, they are thought to be characteristic of transcriptional activating domains (Latchman, 1991; Mermod *et al.*, 1989). These domains are conserved at the amino acid level, although not at the level of primary sequence.

A third domain is found N-terminal to the acidic region and is 31 amino acids long and contains leucine residues repeating every 7/8 amino acid positions. Secondary structure predictions indicate an alpha helix. A slightly basic region is located N-terminal and adjacent to the leucine repeats. This domain thus has some of the characteristics of a leucine zipper motif (Heeckeren, Sellers & Struhl, 1992; O'Shea *et al.*, 1991). Figure 2 shows the helical wheel structure of the leucine repeating region of BOFH. The leucines and most of the hydrophobic residues are located on one hydrophobic face.

Temporal and spatial expression of *bofh*

Northern blots of poly(A)$^+$ RNA purified from different stages in the developing curd and during floral initiation were probed with pBOFH3 cDNA. The *bofh* gene was found to encode a message of approximately 1.6 kb, corresponding to the size of the deduced cDNA. This corresponds to the 1.6 kb transcripts identified in both *lfy* and *flo* (Coen *et al.*, 1990; Weigel *et al.*, 1992). The 1.6 kb RNA transcript was detected from the first formation of curd. The initial level of expression is low and continues to increase as the curd develops. Northern analysis of RNA isolated from the branches of the curd showed some *bofh* expression. In addition, no mRNA transcripts were detectable in either cauliflower leaf, stem or root.

The spatial expression of *bofh* was investigated by *in situ* hybridisation. The expression of *bofh* RNA transcripts was found to be strong within the first few cell layers of the apical meristem (Fig. 3). Expression was detected in the bracts which are located on the sides of the branches and below the curd. Similar expression in the bracts has been found for both *flo* and *leafy*. These studies also confirm that the expression of *bofh* is detectable from the very earliest stages of curd formation.

It is particularly significant that *bofh* expression is present in the curd tissue long before there is any sign of floral morphology. Thus, by analogy to *flo*, *bofh* expression should not be present in an inflorescence meristem, and this suggests that the curd tissue contains truly floral meristems. The presence of *bofh* does not, however, correlate to the initiation of truly floral morphology in the cauliflower. It is therefore

Helical wheel representation of repeating leucines in BOFH

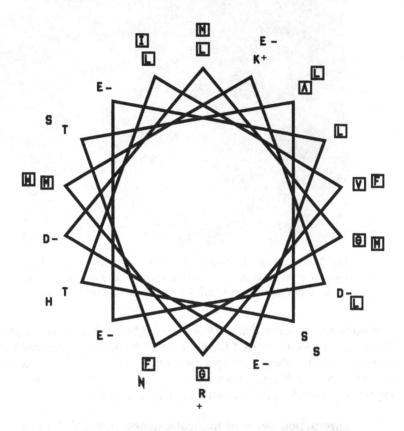

Fig. 2. Helical wheel representation of residues 69 to 99 of the BOFH protein. The view is from the NH$_2$-terminus. The inner circle represents the first five turns of the helix. Hydrophobic residues are boxed.

clear that, although *bofh* is likely to be necessary for flowering, some other factors are required before progress to full floral development takes place. It is possible that, in cauliflower, the expression of other genes may be needed (see below). It is also possible that, although transcription of *bofh* takes place, the RNA is not translated or requires some post-translational modification before it is active. It is therefore necessary to investigate the presence of the BOFH protein in relation to the time of flowering.

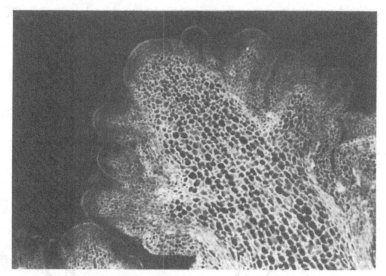

Fig. 3. *In situ* hybridisation of cauliflower curd meristem tissue showing *bofh* gene expression. The *bofh* expression can be seen as dark areas in the outer meristematic cell layers (magnification x75).

Results also indicate that *bofh* is encoded by a small family of genes. Evidence has been reported for multigene families of *bofh* homologues being present in other plant species (Southerton *et al.*, 1993). It is not known if these different members of the gene family show differential expression, and their role remains to be ascertained.

Interaction between homeotic genes to establish curd formation

It is clear from the sequence comparison of *flo*, *leafy* and *bofh* that these genes are extremely well-conserved homologous genes. It is also likely that they are all important in the early stages of floral initiation and specifically at the stage of transition from the inflorescence to the floral meristem. There are, however, clear differences in the pattern of expression of these genes, and this implies functional differences. For instance, even strong *Arabidopsis leafy* alleles still allow floral morphology to develop. In *Antirrhinum*, however, there is no floral development in the absence of *flo* expression. In addition, *leafy* expression is present in the primordia of all the developing floral organs (Weigel *et al.*, 1992). In contrast, *flo* is not expressed in the stamen primordia during organogenesis in *Antirrhinum*. Furthermore, studies with *bofh* suggest that it is expressed at all times from curd initiation until floral development. Thus, if the curd is an inflorescence structure, as has been proposed, *bofh* should not be expressed until floral initiation is actually taking place.

Another group of 'early' genes has also recently been isolated. They are represented in *Arabidopsis* by *apetala1* (*ap1*) and in *Antirrhinum* by *squamosa* (*squa*). Both of these genes belong to the MADS family of genes (Jordan & Anthony, 1993). In *ap1* mutants there are alterations in both the development of the meristem and in organ identity. Thus, in *ap1* mutants several individual flowers arise where there would only be one in a wild-type plant. These secondary flowers arise in the axils of the first whorl of the primary flower or occasionally may be found on the pedicel with no subtending organs. The development of the secondary flowers may reflect a reversion to an inflorescence meristem. This reversion to an inflorescence morphology is temperature dependent with lower temperature promoting the inflorescence. Homeotic changes to organ identity also take place, with the sepals becoming leaf-like organs and the petals failing to develop (Huijser *et al.*, 1992; Mandel *et al.*, 1992). In the case of *ap1*, the functional differences have been clearly revealed at the phenotypic level by the recessive allele *cauliflower* (*cal*) in *Arabidopsis* (Bowman, 1992). This modifier enhances the *ap1* phenotype, such that floral meristems are quite clearly transformed into inflorescence meristems. In *ap1-1 cal-1* double mutants, each meristem, that in wild-type would give rise to a single flower, behaves like an inflorescence meristem. Each meristem produces an indeterminate number of primordia which in turn act as inflorescence meristems. This process is repeated, giving rise to an inflorescence morphologically very similar to that of *Brassica oleracea* L. var. *botrytis* (Bowman, 1992). Some of the primordia eventually differentiate into flowers. The *ap1-1* interaction with *cal-1* is strongly influenced by temperature. At a low temperature (15 °C) the inflorescence morphology can be maintained without flower development for up to four months (Bowman, 1992). This clearly demonstrates the temperature regulation of the *ap1* gene. An analogous situation occurs with the cauliflower curd, in that differentiation of abortive floral primordia to functional floral primordia is primarily temperature dependent.

In controlling the transition to floral morphology, there is strong evidence that a number of homeotic genes function in combination. Thus, the *ap1* gene has been shown to interact with *lfy* in *Arabidopsis* (Bowman, 1992), giving enhancement of the *leafy* phenotype. Consequently, in the *leafy-ap1* double mutants there is a complete conversion to inflorescence morphology. In contrast, if either of these 'primary' genes is missing, only partial conversion to floral morphology is achieved.

It has been proposed that the first step in the establishment of floral meristem identity is the independent activation of *lfy* and *ap1*. Following this establishment of activity, *ap1* and *lfy* positively interact with each other. A strong *ap1-1* mutation will greatly enhance both a weak and a strong *lfy* mutant (Bowman, 1992). Other genes, such as *ap2* and *cal*, act to modify and reinforce the activity of the 'primary genes'. This work is of particular interest because it suggests that cauliflower curd morphology is controlled by a small number of interacting genes, which is likely to include *bofh*, an *ap1* homologue and the *cal* gene. In addition, the effect of temperature on curd formation may be explained in terms of regulation of gene expression. At present, however, it is not clear how the actual switch to floral morphology is controlled in the curd tissue. Thus, in cauliflower, what is the cellular process that maintains an apparently floral meristem and prevents the initiation of flowers?

Functional conservation of *flo* homologues

It is clear from the morphology and physiology of the cauliflower that the process of flowering is complex. However, *Brassica* species do provide a very useful system for investigating the molecular biology of flowering and the isolation of homeotic genes will greatly facilitate our understanding of this process.

If the homeotic genes have specific functions in the process of flowering it is likely that they should be highly conserved between species. Functional conservation of floral control genes has been demonstrated for the homeotic genes *ap3* (*apetala-3*) of *Arabidopsis* and *def* (*deficiens*) of *Antirrhinum* (58% identical at the amino acid level (Jack, Brockman & Meyerowitz, 1992). Conservation of the gene sequence is also a major feature of *flo*, *lfy* and *bofh*. This is reflected in the significant homology found between the two genes, *bofh* and *lfy* (89% identical at the amino acid level). This would be expected as *B. oleracea* and *A. thaliana* are both members of the same family, Brassicaceae. However, there is also very extensive homology between *flo* and *bofh*. *Antirrhinum majus* is a member of the most recently evolved subclass of flowering plants the Asteridae and is very distantly related to the Brassicaceae in evolutionary terms (Cronquist, 1981). From our current interpretation of evolutionary descent, these genes must therefore have evolved within the very earliest class of flowering dicotyledoneae, the Magnoliopsida. The degree of conservation that has been maintained in *flo* homologues suggests that they have been subject to very high evolutionary constraints and must play a vital role in flowering. In the case of *ap3* and *def*, mutant phenotypes and expression patterns are very similar, suggesting that they perform very similar functions in the specification of floral organ identity in the two species. In contrast, however, studies with *lfy* and *flo* have revealed functional differences (Coen *et al.*, 1990; Weigel *et al.*, 1992) and it is anticipated that expression studies with *bofh* will reveal further differences between species. Thus, if these homeotic genes act primarily as transcription factors, the variation in their function is likely to be dependent upon the specific target genes that they regulate. The identification of these genes and their functions will consequently accelerate our understanding of the molecular mechanisms involved in flower development.

Acknowledgements

The authors would like to acknowledge our colleagues at (HRI-Wellesbourne) especially Dr David C.E. Wurr for his advice on the cauliflower physiology, Colin Clay for the electron micrographs and Dr Graham King for his consistent help and advice on our research. We are grateful to Dr Enrico Coen for the provision of the *flo* gene and his assistance (with his colleagues at the John Innes Institute) in the development of our *in situ* hybridisation techniques. We are also grateful to the Agricultural and Food Research Council for financial support.

References

Anthony, R.G., James, P.E. & Jordan, B.R. (1993). Cloning and sequence analysis of a *flo/lfy* homolog isolated from cauliflower (*Brassica oleracea* L. var. *botrytis*). *Plant Molecular Biology*, **22**, 1163–6.

Atherton, J.G., Hand, H.J. & Williams, C.A. (1987). Curd initiation in the cauliflower (*Brassica oleracea* L. var. *botrytis*). In *Manipulation of Flowering*, ed. J.G. Atherton, pp. 133–45. London: Butterworths.

Bowman, J.L. (1992). Making cauliflower out of *Arabidopsis*: the specification of floral meristem identity. *Flowering News Letter*, **14**, 7–19.

Bowman, J.L. & Meyerowitz, E.M. (1991). Genetic control of pattern formation during flower development in *Arabidopsis*. In *Molecular Biology of Plant Development*, ed. G.I. Jenkins & W. Schuch. Symposium of the Society of Experimental Biology, XLV. Cambridge: The Company of Biologists Ltd.

Coen, E.S. (1991). The role of homeotic genes in flower development and evolution. *Annual Review of Plant Physiology*, **42**, 241–79.

Coen, E.S. & Meyerowitz, E. M. (1991). The war of the whorls: genetic interactions controlling flower development. *Nature*, **353**, 31–7.

Coen, E.S., Romero, J.M., Doyle, S., Elliott, R., Murphy, G. & Carpenter, R. (1990). *floricaula*: A homeotic gene required for flower development in *Antirrhinum majus*. *Cell*, **63**, 1311–22.

Cronquist, A. (1981). *An Integrated System of Classification of Flowering Plants*. New York: Columbia University Press.

Friend, D.J.C. (1985). *Brassica*. In *CRC Handbook of Flowering*, ed. A. H. Halevy, pp. 48–77. Florida: CRC Press.

Heeckeren, W.J.V., Sellers, J.W. & Struhl, K. (1992). Role of the conserved leucines in the leucine zipper dimerization motif of yeast GCN4. *Nucleic Acids Research*, **20**, 3721–4.

Huijser, P., Klein, J., Lonnig, W.-E., Meijer, H., Saedler, H. & Sommer, H. (1992). Bracteomania, an inflorescence anomaly, is caused by the loss of function of the MADS-box gene *squamosa* in *Antirrhinum majus*. *EMBO Journal*, **11**, 1239–49.

Jack, T., Brockman, L.L. & Meyerowitz, E.M. (1992). The homeotic gene *APETALA3* of *Arabidopsis thaliana* encodes a MADS box and is expressed in petals and stamens. *Cell*, **68**, 683–97.

Jordan, B.R. & Anthony, R.G. (1993). Floral homeotic genes: Isolation, characterisation and expression during floral development. In *The Molecular Biology of Flowering*, ed. B. R. Jordan, pp. 93–116. Wallingford, UK: CAB International.

King, G.J. (1990). Molecular genetics and breeding of vegetable brassicas. *Euphytica*, **50**, 97–112.

Latchman, D.S. (1991). *Eukaryotic Transcription Factors*. 270pp, London: Academic Press.

Mandel, M.A., Gustafson-Brown, C., Savidge, B. & Yanofsky, M.F. (1992). Molecular characterisation of the *Arabidopsis* floral homeotic gene *APETALA1*. *Nature*, **360**, 273–7.

Margara, J. & David, C. (1978). Les étapes morphologiques du développement du meristème de choufleur. *Brassica oleracea* L. var. *botrytis*. *Comptes Riendusde l'Academie des Sciences, Paris. Serie D*, **287**, 1369–72.

Medford, J.I., Elmer, J. S. & Klee, H.J. (1991). Molecular cloning and characterisation of genes expressed in shoot apical meristems. *Plant Cell*, **3**, 359–70.

Mermod, N., O'Neill, E. A., Kelly, T.J. & Tjian, R. (1989). The proline-rich transcriptional activator of CTF/NF-1 is distinct from the replication and DNA binding domain. *Cell*, **58**, 741–53.

O'Shea, E.K., Klemm, J.D., Kim, P.S. & Alber, T. (1991). X-ray structure of the GCN4 leucine zipper, a two-stranded, parallel coiled coil. *Science*, **254**, 539–44.

Sadik, S. (1962). Morphology of the curd of cauliflower. *American Journal of Botany*, **49**, 290–7.

Sadik, S. (1967). Factors involved in curd and flower formation in cauliflower. *Proceedings of the American Society for Horticultural Science*, **90**, 252–9.

Sadik, S. & Ozbun, J.L. (1968). Development of the vegetative and reproductive apices of cauliflower, *Brassica oleracea* var. *botrytis*. *Botanical Gazette*, **129**, 365–70.

Southerton, S.G., Strauss, S.H., Olive, M.R. & Dennis, E.S. (1993). Cloning of the floral homeotic gene homolog of *leafy/floricaula* in *Eucalyptus globulus*. *Journal of Cellular Biochemistry*, supplement 17B, D227.

Weigel, D., Alvarez, J., Smyth, D.R., Yanofsky, M.F. & Meyerowitz, E.M. (1992). LEAFY controls floral meristem identity in *Arabidopsis*. *Cell*, **69**, 843–59.

B. THOMAS and D. MOZLEY

Isolation and properties of mutants of *Arabidopsis thaliana* with reduced sensitivity to short days

Introduction

Daylength is the most powerful environment factor in the regulation of the onset of flowering (Vince-Prue, 1975). Changes in photoperiod through the year provide an unambiguous index of seasonal progression. Through the ability to detect and respond to daylength, plants can tailor flowering to seasonal changes in climate, and can also ensure synchrony of flowering to facilitate outbreeding. Plants are classified as short-day plants (SDP), long-day plants (LDP), day neutral plants (DNP) or combinations of these. SDP flower in response to days shorter than a *critical daylength,* LDP flower in response to days longer than a critical daylength and DNP flower irrespective of daylength. Some plants show a requirement for particular sequences of daylengths or other combinations of the above basic types (Thomas & Vince-Prue, 1984). Photoperiodic requirements may themselves be the only determining factors, but frequently they are overlaid on other developmental or environmental factors, the most important being juvenility and vernalisation (Thomas, 1993).

Photoperiodic mechanisms

Daylength is generally accepted as being perceived by the leaves in both SDP and LDP, rather than in the apex where the transition to flowering occurs. When a permissive daylength is perceived, a semi-stable change in the properties of the leaves occurs. This can be demonstrated by the ability of such leaves to cause flowering when grafted to plants maintained in non-permissive daylengths (for example, Zeevaart, 1969). The change in the leaf is called *induction* and the molecular basis of the change is unknown. As a consequence of induction a transmissible signal, sometimes called florigen, passes to the apex leading to *evocation,* the changes leading to floral initiation and development.

Daylength perception requires a photoreceptor to distinguish between

Society for Experimental Biology Seminar Series 55: *Molecular and Cellular Aspects of Plant Reproduction*, ed. R.J. Scott & A.D. Stead.
© Cambridge University Press, 1994, pp. 31–37.

light and darkness. For most plants phytochrome fulfils this role, but in addition, a blue light photoreceptor is important in the *Cruciferae*, which includes *Arabidopsis* (Sage, 1992). The timekeeper is believed to be a circadian rhythm in light sensitivity which is both phased by and interacts with light–dark cycles (Lumsden, 1991).

While the physiology of photoperiodism is now well established, little is known at the molecular or biochemical level about induction or signal transmission. Resolution of these complex mechanisms requires a combination of genetic, physiological and molecular analyses within a common system. The advantages of *Arabidopsis thaliana*, a facultative LDP, for molecular and genetic studies, are well known and a number of late-flowering mutants are already isolated. In order to take advantage of this system, however, it was necessary to establish the parameters for photoperiodic responsivity in *Arabidopsis*.

Interaction of development with daylength response

Environmental control of flowering is overlaid on endogenous factors which establish the competence of plants to flower. In many species there is an imposed vegetative phase in which flowering cannot be initiated, irrespective of environmental treatments given to the plant. This is called a *juvenile* phase. In woody plants the juvenile phase may last for several years but in herbaceous plants it is rarely longer than a few days or weeks. It has been reported that *Arabidopsis* becomes sensitive to daylength immediately after germinating (Napp-Zinn, 1985).

A protocol was devised to test for sensitivity to daylength in young *Arabidopsis* seedlings. It was reasoned that, if plants could respond to daylength, it should be possible to delay flowering by introducing SD at early stages of development to plants otherwise grown in inductive LD conditions. The results of the experiments are summarised in Fig. 1. Treatments began two days after germination and for the first two days after that daylength had no effect. Over the next six days or so there was a day-for-day delay in flowering. These experiments confirmed that *Arabidopsis* can perceive daylength and respond to it from about four days after germination. This is about the length of time taken for the cotyledons to green and open. The response measured was to the formation of flower buds but, even though there was a significant effect on time to flowering, the leaf number to flowering remained constant in all of the treatments. Thus *Arabidopsis* has an obligatory vegetative phase but no measurable juvenile phase. By varying the length of the days given from 4–8 days after germination, a value for the critical daylength of between 8 and 10 hours was derived which is consistent with critical daylength as determined by other methods.

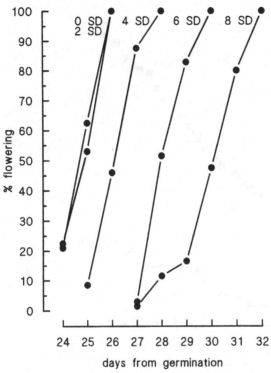

Fig. 1. Effect of exposing 2-day-old seedlings of *Arabidopsis, Landsberg erecta* to between 2 and 8 SD on subsequent time of flowering in LD.

Arabidopsis is a facultative LDP, that is it will eventually flower in SD. The photoperiod system in *Arabidopsis* can therefore be further defined with respect of the period over which it remains inducible by LD treatments. Experiments were based on essentially the same protocol as used to determine the juvenile phase but seedlings remained in SD for much longer periods before transfer to LD. After about 10–12 days in SD, further leaf production occurs and leaf number at flowering increases with the number of SD. During this phase the time of flowering remains constant from the time of transfer to LD. Eventually, however, because flowering occurs in SD, transfer to LD ceases to accelerate flowering (Fig. 2).

One interpretation of this sort of data is that flowering is the normal condition for plants and photoperiodic regulation is a mechanism by which flowering can be delayed (Gregory, 1936). Thus, eventually SD are ineffective in preventing the transition to flowering and sensitivity to daylength is lost. Alternatively, it could be that there is an increased

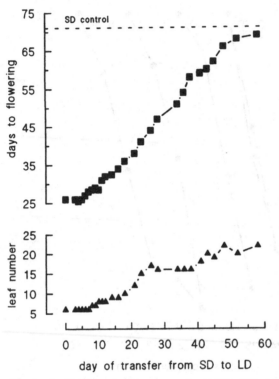

Fig. 2. *Arabidopsis* seedlings were grown in SD and transferred to LD at different times in their development. The Figure shows effect on flowering time and leaf number.

sensitivity to inductive treatments until induction can occur autonomously even in SD. An experiment to test this idea was devised in which *Arabidopsis* growing in SD were given 1, 3 or 5 LD at different times, during development and the findings are summarised in Fig. 3. They show that firstly, *Arabidopsis* does not need to be maintained in LD to flower; 5 LD given to 7-day-old plants were fully inductive. Secondly, the number of LD needed to induce flowering decreases with plant age so that by about day 20 a single LD was fully inductive. The period over which *Arabidopsis* can be induced by a single LD lasts for several weeks. In this sense, at least, flowering is more likely to be a positive consequence of a LD treatment rather than a response to withdrawal of SD.

Photoperiod responsivity in *Arabidopsis* can therefore be defined in terms of the onset of sensitivity at about four days from germination, lack of a juvenile phase, requirement for four or less LD for induction,

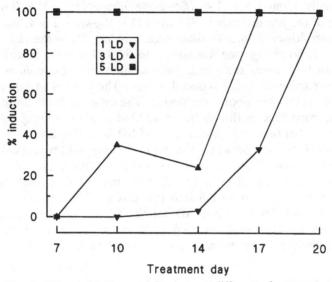

Fig. 3. Effect of 1, 3 or 5 LD given at different plant ages on photoperiodic induction in *Arabidopsis*. Plants were otherwise grown in SD and the % of plants whose flowering was advanced relative to the SD controls was determined.

increasing sensitivity to LD induction with age and eventual autonomous induction in SD. This information provides a framework for generating new mutants altered in specific components of the photoperiodic mechanism.

Mutant isolation and characterisation

Two separate strategies were employed to obtain mutants with potentially altered photoperiodic responses. In the first approach, EMS-mutated seeds were germinated in darkness and then maintained in SD from day 3 to day 8 before being transferred to LD until flowering. Under these conditions, flowering was delayed by 3–4 days compared to plants continuously in LD. Thus plants which flowered at the same time as the controls were potentially those which failed to distinguish between SD and LD. In a second screen plants were grown in SD except for receiving a single LD on day 7. Plants which flowered early in response to this treatment were potentially those with enhanced sensitivity to LD. From initial screens 45 potential mutants were identified and subjected to secondary screening. Of these, six were selected for further study and are provisionally named *fun* 1–6 (*flowering un*coupled). The flowering response of the *fun* mutants in SD and LD is shown in Fig 4. All six of the mutants flowered early to some extent under SD. In the case of *fun* 1 and *fun* 2, flowering time was similar to the LD controls in either SD or LD. In both cases the plants were smaller than the controls with leaves, in particular, being reduced in size. The flowers in *fun* 1 were small and petals were poorly developed. The other mutants flowered at about the same time as the control in LD but in SD flowering time was intermediate between the control LD and SD flowering time.

Reciprocal crosses between the mutants and wild-type (wt) lines revealed that none of the mutants was allelic and that they were recessive, with the exception of 4 which was semi-dominant. From their behaviour the mutants appear to fall into two classes. The *fun* 1 and *fun* 2 mutants not only flower early in SD but also flowered early when grown in darkness on agar supplemented with sucrose. For these mutants, flowering appears to be more or less independent from the light treat-

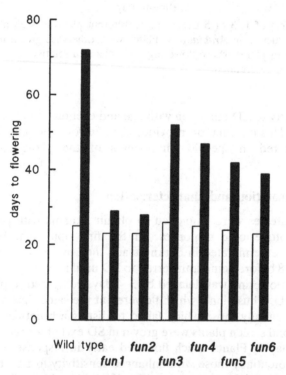

Fig. 4. Comparison of the flowering times in LD and SD of *Arabidopsis* wt (*Landsberg erecta*) and photoperiodic mutants *fun1–fun6*.

ments which they receive. They can, therefore, be regarded as constitutively flowering or autonomously induced. An interesting feature of both *fun* 1 and *fun* 2 mutants is that leaf development is impaired implying either that the mutation affects both the flowering and leaf development, or possibly that normal leaf development is required for the suppression of flowering in SD.

In the case of the remaining mutants, the effect is manifested as a quantitative response to the delaying effects of SD on flowering. In the case of *fun* 4 and *fun* 6, plants are visually similar to the controls and early flowering in SD is associated, not surprisingly, with earlier loss of sensitivity to the promoting effect of LD, that is autonomous induction occurs at an earlier stage of development. In addition, for *fun* 6 a single LD is partially inductive from day 7 and fully inductive from day 10. Photoperiodic sensitivity is retained in *fun* 4 and *fun* 6 and critical daylength is unchanged.

These mutants provide new tools for understanding LDP photoperiodic mechanisms. For *fun* 1 and *fun* 2 the flowering response has been completely separated from light or photoperiod perception, while for *fun* 4 and *fun* 6 the genes involved appear to be related to sensitivity to the photoperiodic stimulus and they may, therefore, be transduction mutants.

References

Gregory, G.G. (1936). The effect of the length of the day on the flowering of plants. *Scientific Horticulture*, **4**, 143–54.

Lumsden, P.J. (1991). Circadian rhythms and phytochrome. *Annual Review of Plant Physiology and Plant Molecular Biology*, **42**, 351–71.

Napp-Zinn, K. (1985). *Arabidopsis thaliana*. In *CRC Handbook of Flowering* Vol. 1, ed. A.H. Halevy, pp. 492–503. CRC Press.

Sage, L.C. (1992). In *Pigment of the Imagination. A History of Phytochrome Research*. San Diego: Academic Press.

Thomas, B. (1993). Internal and external controls on flowering. In *The Molecular Biology of Flowering*, ed. B.R. Jordan, pp. 1–19. C.A.B. International.

Thomas, B. & Vince-Prue, D. (1984). Juvenility, photoperiodism and vernalisation. In *Advanced Plant Physiology*, ed. M.B. Wilkins, pp. 408–39. London: Pitman Publishing Ltd.

Vince-Prue, D. (1975). *Photoperiodism in Plants*. Maidenhead: McGraw-Hill.

Zeevaart, J.A.D. (1969). *Perilla*. In *The Induction of Flowering*, ed. L.T. Evans, pp. 116–55. Melbourne: Macmillan.

N. BARBACAR and I. NEGRUTIU

Asexual mutants in *Melandrium album* (*Silene alba*): tools in cDNA cloning and analysis of an X/Y chromosome system in plants

Introduction

Mutants affecting sexual development in plants are relatively rare. Most of them concern cytoplasmic or nuclear male sterility, while a series of meiotic mutants in maize are well characterised (Kaul & Murthy, 1985). Other mutants, of the sex conversion type, have been described in maize and cucumber, two monoecious plants (Irish & Nelson, 1989; Malepszy & Niemerowicz-Szczytt, 1991).

Among flower pattern mutants, those affecting floral organs in whorls 2 and 3 exhibit various degrees of homeotic transformation of stamens into carpels e.g. *apetala3* in *Arabidopsis thaliana* and *deficiens* in snapdragon (Jack, Brockman & Meyerowitz, 1992; Schwarz-Sommer *et al.*, 1990). The corresponding genes have been cloned and represent transcription factors belonging to the MADS class.

We have recently described two asexual mutants in the dioecious *Melandrium album*, following irradiation of pollen with low doses of gamma-rays (Veuskens *et al.*, 1992). Briefly, the mutants 5K63 and 8K40 have normal perianth organs (sepals and petals), while lacking both male and female reproductive organs: vestigial stamens and a 'finger-like projection' instead of carpels are present in whorls 3 and 4 respectively. Both mutants have a deletion covering respectively 12 and 21% of the Y-chromosome.

Here we report on the use of these asexual/Y deletion mutants in cDNA subtraction cloning and exploit them in refining the current understanding of sex determination in a plant X/Y system

The experimental system

Melandrium has certain intrinsic advantages with respect to the cloning of genes specifically involved in sexual development:

Society for Experimental Biology Seminar Series 55: *Molecular and Cellular Aspects of Plant Reproduction*, ed. R.J. Scott & A.D. Stead.
© Cambridge University Press, 1994, pp. 39–48.

(i) male and female individuals have male and female flowers in which the opposite sex is arrested in its development at very early stages of respectively carpel and stamen initiation. Male or female cDNA libraries can be constructed at various stages of floral development, the early ones included, without the need to dissect out the reproductive organs prior to RNA extraction. The consequence should be a faster route to the clones of interest;

(ii) the species has a pair of heteromorphic sex chromosomes, X and Y. The Y contains functions that control defined steps in male/female sexual development, deletions on the Y chromosome result in sex conversion phenotypes (for review see Ye *et al.*, 1991) The asexual mutants represent a particular class and can play a main role in cloning experiments. They actually represent an ideal partner in the driver/pilot hybridisation system (instead of leaves or seedlings) and the appropriate controls in identifying those functions that correspond to the deleted Y domain.

Y deletion mutants in cDNA subtractive cloning: the photobiotin/streptavidin – versus the chemical cross-linking subtraction (CCLS) – approach

The first male cDNA libraries were constructed in λ GEM2 and cover two main stages of stamen formation: S_I,0–2 mm buds, corresponding to pre- and meiotic differentiation and S_{II}, 2–5 mm, the post-meiotic stage. These are relatively 'broad-stage' libraries, containing complex populations of mRNA.

Two methods of subtraction were used, according to the following procedures:

(i) Photobiotin/streptavidin, in a protocol of subtractive hybridisation and screening (Herfort & Garber, 1991) which allowed the isolation of cDNA clones corresponding to very low abundance poly(A)+ in a human cell line.

(ii) Chemical cross-linking subtraction (CCLS, cf. Hampton *et al.* (1992)) 2,5 diazividinyl-1,4-benzoquinone (DZQ) ensures the specific chemical cross-linking of the cDNA–RNA hybrids with no requirements for physical separation of these hybrids from the unic single stranded cDNA. The protocol was developed on cultured mammalian cells, with enrichment rates of up to 300-fold.

Preliminary experiments in the photobiotin/streptavidin system demonstrated that, on the basis of 1000 pfu tested, the enrichment factor was about 50 (Figs. 1 and 2). The next step consisted of performing comparative cloning experiments as outlined in Table 1. The results indicated that CCLS was more reliable than the photobiotin/streptavidin protocol. After three rounds of screening, the number of positive clones was 10- to 20-fold higher for the CCLS (Fig. 3), with half of the clones tested being confirmed in Northern analysis (Table 1). Actually, none of the 7 clones identified in the photobiotin/streptavidin system exhibited specificity in Northern blots.

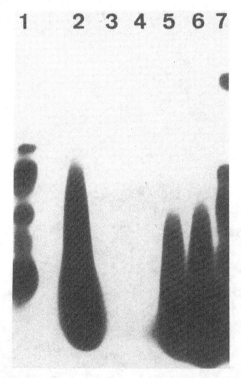

Fig. 1. Control experiment determining the label distribution during phenol extraction of a subtracted cDNA sample following cDNA/poly(A)$^+$ hybridisation in the photobiotin/streptavidin protocol Lane 1 and 7: pBR328 *Hin*fI marker; Lane 2: sample from the phenol interphase, Lane 3 and 4: samples from the phenol phase; Lanes 5 and 6: samples from the aqueous phase(subtracted pool).

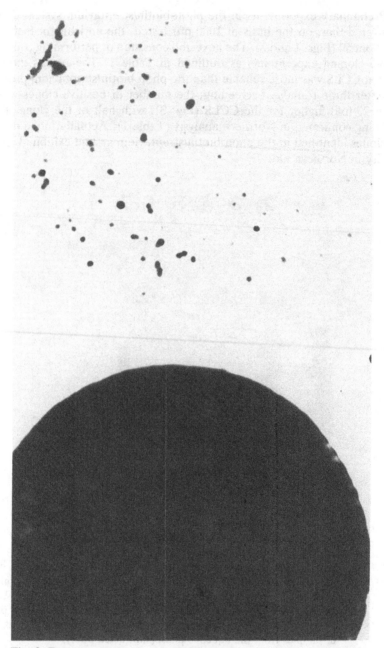

Fig. 2. Reconstruction of a photobiotin/streptavidin cDNA subtraction experiment on 1000 clones from a S_{II} male flower cDNA library in λGem 2. The two replicas were probed with respectively initial cDNA(down) and subtracted cDNA(up) probes.

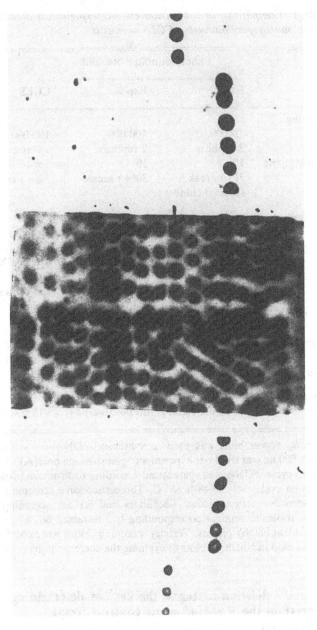

Fig. 3. Tertiary screening in the CCLS protocol: 35 clones (each as 4–6 repetitions) were tested in triplicate replicas probed with respectively labelled S_{II} 5K63(up), subtracted(centre) and leaf(down) cDNAs.

Table 1. *Comparison of subtraction cloning experiments according to photobiotin/streptavidin versus CCLS protocols*

Specification	Photobiotin/Stre ptavidin		CCLS
	Exp.1	Exp.2	
Primary Screening			
Total pfu	100 000	100 000	100 000
	2 replicas	2 replicas	3 replicas
Amount of label(cpm)	10^7	10^7	5×10^7
Yield	3(+) areas	30(+) areas	30(+) areas
	(1–2000 clones)		
Secondary Screening			
	2 replicas	3 replicas	3 replicas
Amount of label (cpm)	10^7	10^7	2×10^6 (subtr),
			2×10^7 5K63,
			leaves
Yield	8 clones	25 clones	120 clones
Tertiary Screening			
	3 replicas	4 replicas	4 replicas
Amount of label (cpm)	10^7	10^7	2×10^6; 2×10^7
Yield	3 clones	4 clones	60 clones
Size range (kb)	1.8–2.8	0.4–1.2	0.45–6
Northern Blotting			
Specific clones	no positives(0/3)	no positives(0/4)	5/11 tested

Poly(A)$^+$ from S_{II} flower buds was used to synthesise cDNA for subtraction cloning of which 500 ng was used per experiment against 10 mg poly(A)$^+$ 5K63S$_{II}$ per hybridisation cycle. cDNA was synthesised according to Promega protocols. Each hybridisation cycle was 38–48 h 65 °C. Three successive screenings were performed to identify positive clones. Secondary and tertiary screening were usually done on triplicate replicas corresponding to subtracted S_{II}, S_{II} deletion mutant 5K63 and leaf cDNA probes. Tertiary screening clones were checked for insert, which was used in Northern blots to evaluate the corresponding expression specificities.

Asexual/Y deletion mutants: the key sex determining factor(s) on the Y chromosome controls female development

Y deletions, together with normal and abnormal X–Y interchanges, have been fully exploited in man, mouse, etc., firstly, in constructing deletion maps of the Y chromosome and, secondly, in cloning ZFY and SRY genes (for review see Koopman, 1993), the latter being the long postulated testis determining factor (TDF).

The X/Y system in *Melandrium* bears some striking similarities with the one in mammals. In both cases, the only genetic difference between the two sexes is represented by the Y chromosome, which contains determinant(s) essential for the development of the male sex. The dominant effect of Y determinant(s) is maintained even at X/Y ratios >1 (Westergaard, 1958) The disparity in the number of X chromosomes between the two sexes is compensated by X inactivation (Nagai, 1992; Vyskot *et al.*, 1993).

Beyond these similarities, which suggested to Nöthiger and Steinmann-Zwicky (1987) a related strategy for sex determination, there are important differences between the animal and plant sexual developmental pathways, a few are discussed below.

In mammals, the primary (gonadal) sex is determined by the genetic sex, which in turn dictates the phenotypic (somatic) sex (Nagai, 1992). In *Melandrium*, certain somatic attributes associated with the sex appear before the 'primary' sex is determined (Ye *et al.*, 1991), which is a late developmental decision. This is a general characteristic of all higher plants; the flower module is terminal, the reproductive organs being the simultaneous product of homeotic and heterochronic processes (Lord, 1991). Such processes exhibit a high degree of genetic plasticity (e.g. chimeric structures are frequently produced). In this context, somatic sexual differentiation and germ line differentiation are late and sequential events.

In mammals, the gonadal primordium is programmed to develop into an ovary. In the presence of the Y, the primordium organises as testis (Koopman, 1993), which makes it a bipotential structure. In general, in plants the floral meristem consists of distinct primordial domains for male (whorl 3) and female (whorl 4) differentiation. Homeotic mutations, such as those mentioned above, produce sex conversions by transformation of stamen primordia into ovaries, demonstrating that stamen primordia are not necessarily predetermined for another formation.

The opposite transformation, i.e. carpels into anthers, has also been reported (Robinson-Beers, Pruitt & Gasser, 1992). Thus, both anther and carpel primordia appear to be sexually bipotential. However, in addition to sexual conversions within whorl 3 and 4, homeotic transformations into reproductive organs occur in whorl 1 and 2 as well. In addition, these organ identity transformations are frequently incomplete (i.e. chimeric), indicative of late and regional developmental fate assignment.

In mammals, the regression of the female gonadal structures is an early event limited to the Müllerian ducts (Nagai, 1992). In *Melandrium*, the suppression of the female development appears as the key process in sexual determination.

As a matter of fact, in XX plants male organ development is initiated in whorl 3 and then interrupted, producing 'vestigial' anthers In XY plants, the female whorl (4) is assigned and a 'finger-like-projection' is formed in that whorl, having neither structural nor morphological similarity with the carpel (Ye *et al.*, 1991) One may consider the finger-like-projection as a substitute for the female organ in male flowers.

Deletion of two different domains on the Y chromosome can result in extreme phenotypes within the dioecious background:

- hermaphrodite mutants, exhibiting a deletion on the (arbitrarily defined) domain I of the differential arm of the Y chromosome (Westergaard, 1958);
- asexual mutants, harbouring a deletion on the (arbitrarily defined) domain II of the differential arm of the Y chromosome (Veuskens *et al.*, 1992).

From the comparative picture of reproductive organ formation in wild-type male (XY) and female (XX) plants versus hermaphrodite and asexual mutants with deletions on defined Y domains, one can assign the following control functions to genes located within these domains:

- gene(s) ensuring anther differentiation beyond the incipient stage characteristic to XX flowers(Y_{II} domain);
- the (key) gene(s) that trigger the arrest of female development (Y_I domain).

The latter correspond(s) most likely to primary factor(s) specifically responsible for sexual determination in the *Melandrium* X/Y system. The function(s) is not supposed to exist in hermaphrodite species and, as such, not to be present in other dioecious or monoecious plants. This female suppressing function(s) may interact with the homeotic genes that set up the floral pattern in whorl 4 just after whorl assignment and primordia formation. It represents one of the possible routes in the appearance of unisexual plants among hermaphrodite species across the phylogenetic scale.

The remarkable thing about the *Melandrium* X/Y system is that is part of a whole range of sexual transformations operating within closely related species in genera such as *Silene* and *Lychnis* (Degraeve, 1980), with obvious evolutionary implications.

Acknowledgements

We are indebted to Dr John Butler (Paterson Institute for Cancer Research, Manchester) for having kindly provided us the DZQ.

References

Degraeve, N. (1980). Etudes de diverses particularités caryotypiques des genres *Silene, Lychnis et Melandrium. Boletim da Sociedade Broterian* **53**, 595–643.

Irish, E.E. & Nelson, T. (1989) Sex determination in monoecious and dioecious plants. *The Plant Cell*, **1**, 737–44.

Jack, T., Brockman, L.L. & Meyerowitz, E.M. (1992). The homeotic gene APETALA 3 of *Arabidopsis thaliana* encodes a MADS box and is expressed in petals and stamens. *Cell*, **68**, 683–97.

Herfort, M.R. & Garber, A.T. (1991). Simple and efficient subtractive hybridisation screening. *Bio/Techniques*, **11**, 600–3.

Hampton, I.N., Pope, L., Cowling, G.J. & Dexter, T.M.(1992). Chemical cross linking subtraction (CCLS): a new method for the generation of subtractive hybridisation probes. *Nucleic Acid Research*, **20**, 2899.

Kaul, M.L.H. & Murthy, T.G.V. (1985). Mutant genes affecting higher plant meiosis. *Theoretical and Applied Genetics*, **70**, 449–66.

Koopman, P. (1993). Mammalian sex determination: what we have learnt from a chromosomal design fault. In *Chromosomes Today*, vol. 11. ed. A.T. Summer & A.C. Chandley, pp. 265–75. London: Chapman and Hall.

Lord, E.M. (1991). The concepts of heterochrony and homeosis in the study of floral morphogenesis. *Flowering Newsletter*, **11**, 4–13.

Malepszy, S. & Niemerowicz-Szczytt, K. (1991). Sex determination in cucumber as a model system for molecular biology. *Plant Science*, **80**, 39–47.

Nagai Y. (1992) Primary sex determination in mammals. *Zoological Science* **9**, 475–98.

Negrutiu, I, Installé, P. & Jacobs, M. (1991). On flower design: a compilation. *Plant Science*, **80**, 7–18.

Nöthiger, R. & Steinmann-Zwicky, M. (1987). Genetics of sex determination in eukaryotes. In *Results and Problems in Cell Differentiation*, ed. W. Hennig, pp. 271–300. Berlin: Springer Verlag.

Robinson-Beers, K., Pruitt, R.E. & Gasser, C.S. (1992). Ovule development in wild-type *Arabidopsis* and two female-sterile mutants. *The Plant Cell*, **4**, 1237–49.

Schwarz-Sommer, S., Huijser. P., Nacken, W., Saedler, H. & Sommer, H. (1990). Genetic control of flower development by homeotic genes in *Antirrhinum majus. Science*, **250**, 931–6.

Veuskens, J., Lacroix, C., Truong, A.T., Hinnisdaels, S., Mouras, A. & Negrutiu, I. (1992). Genetic and molecular enrichment steps as a cloning strategy in the dioecious *Melandrium album*. In *Reproductive Biology and Plant Breeding*, ed. Y Datée, C. Dumas & A. Gallais, pp. 39–48. Berlin: Springer Verlag.

Vyskot, B., Araya, A., Veuskens, J., Negrutiu, I. & Mouras, A. (1993). DNA methylation of sex chromosomes in a dioecious plant, *Melandrium album*. *Molecular and General Genetics*, **239**, 219–24.

Westergaard, M. (1958). The mechanism of sex determination in dioecious flowering plants. *Advances in Genetics*, **9**, 217–81.

Ye, D., Veuskens, J., Wu, Y., Installé, P., Hinnisdaels, S., Truong, A.T., Brown, S., Mouras, A. & Negrutiu, I. (1991). Sex determination in the dioecious *Melandrium*. The X/Y chromosome system allows complementary cloning strategies. *Plant Science*, **80**, 93–106.

R.J. SCOTT

Pollen exine – the sporopollenin enigma and the physics of pattern

Angiosperm pollen represents the pinnacle of an evolutionary progression towards gametophyte miniaturisation that probably began with the development of a toughened wall surrounding the zygote of charphycean algae, the progenitor of the land plants, some 400 million years ago (Delwiche, Graham & Thomson, 1989). The combination of acute vulnerability and central importance in the life-cycle of sexually reproducing plants inherent in the usually brief life-cycle of the male spore has driven the assembly of various adaptive features that endow its protective capsule with remarkable properties. Foremost among these is the unparalleled combination of physical strength, chemical inertness and resistance to biological attack of the outer wall, or exine, due principally to its major structural component, sporopollenin. The evolutionary history of sporopollenin, and that of the spores of land plants, is indivisible (see Chaloner, 1976). Fossil green algae dating back to the Devonian period have been shown to contain sporopollenin (Wall, 1962) and there are reports that sporopollenin also occurs in fungi (see Shaw, 1971) indicating an origin predating the appearance of plants. In an article that fuelled the debate about the existence of extra-terrestrial life, Brooks and Shaw (1969) also reported the presence of sporopollenin in meteorites and suggested that the origin of sporopollenin could even predate life on earth.

However, the exine is often much more than a protective shield. In angiosperms, for example, the pollen wall of many species has become modified to carry specific self-incompatibility proteins that promote outbreeding. There is also recent evidence that, in common with the seed coat, pollen walls may contain compounds that protect the gametophyte against microbial attack (Paul et al., 1992; Terras et al., 1993). Furthermore, despite the unyielding nature of mature sporopollenin, in many

Society for Experimental Biology Seminar Series 55: *Molecular and Cellular Aspects of Plant Reproduction*, ed. R.J. Scott & M.A. Stead.
© Cambridge University Press, 1994, pp. 49–81.

species the exine is elaborately sculptured and patterned in a genetically determined species-specific manner; consequently exine morphology has enjoyed great utility as a taxonomic character and forms the cornerstone of a branch of botany known as palynology. Perhaps the most remarkable feature of the exine is that the often highly sophisticated sculptures are created by a single rather than a multicellular organism.

Pollen grains have attracted the attention of science ever since first observed in the early microscopes of Grew and Malpighi towards the end of the seventeenth century (Wodehouse, 1935), and the accumulated bibliography since then is truly colossal. However, the vast majority of past and present work is entirely descriptive and taxonomic in nature, and comparatively little effort has been expended to determine the mechanism of spore wall genesis. Consequently, little progress has been made toward understanding how the core programme that presumably directs the construction of the basic elements of wall architecture is modified to yield distinctive species-specific variants. The purpose of this review is to highlight the main deficiencies in our current understanding and to incorporate what new data exists into the principal model of angiosperm pollen wall ontogeny and pattern specification developed by Heslop-Harrison and Dickinson in the 1970s.

The structure of the angiosperm pollen wall

Although the diversity of size, shape and external morphology appears bewildering, the pollen grain wall in most species is constructed to the same basic plan. Pollen walls consist of a concentric series of layers, so that in cross-section the wall appears stratified. Figure 1 shows a schematic representation of a transverse section through the wall of lily pollen which engenders all the principal features found in other angiosperms. The outer layer is the exine and the inner layer the intine. Adopting the terminology of Erdtman (1969), the exine consists of an outer sculptured portion, the sexine, and a simpler inner layer, the nexine. The sculpturing of the sexine is in the form of a reticulate pattern and is made up of radially directed rods, the bacula, that have enlarged heads that fuse to form walls or muri. The nexine is subdivided into an outer nexine I, or foot-layer, and an inner nexine II. In many species, the nexine II is made up of multiple tripartite lamellae (TPL), or white lines in the mature wall, that give this layer a characteristic laminated appearance.

A major architectural feature of the pollen wall is the germinal aperture or colpus. This is the site of emergence of the pollen tube on germination and is covered only by the nexine and intine. The pollen of monocotyledons, such as lily, usually have one colpus, whilst dicotyle-

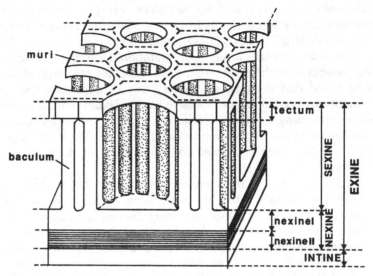

Fig. 1. Structure of the angiosperm pollen wall. Schematic representation of the main architectural features of a generalised mature pollen wall. The terminology is according to Erdtman, 1960.

donous pollen usually have three colpi, although there are numerous species with many more. Reducing or exaggerating features of the lily exine allows the derivation of most other pollen wall structural types. Pilate, or intectate exines are formed where the amount of contact between the bacula heads is reduced, giving the pollen a spiny appearance. Alternatively, increasing the connections between heads to varying degrees produces a partially (semitectate, tectate-perforate) or completely tectate (tectate-imperforate) exine (see Walker, 1976, for definitions). The tectum forms a roof over a series of interconnected voids, or vaults, traversed by roof supporting bacula, and in the most extreme form completely seals the vaulted zone from the outside world. This type of wall, and not the open reticulate exine of lily, probably represents the archaetypal form from which all others are derived (Walker & Skvarla, 1975; Walker, 1976).

Sporopollenin composition and structure, and the carotenoid red herring?

One of the principal barriers to evolving a satisfactory model for pollen wall ontogeny has been the failure to determine the biochemical composition, structure and mode of biosynthesis of sporopollenin. In 1937, Zetzsche and co-workers (for review see Shaw, 1971) determined that sporopollenin is an oxygenated hydrocarbon having a chemical composition between $C_{90}H_{130}O_{20}$ and $C_{90}H_{120}O_{30}$ and containing hydroxyl and C-methyl groups and substantial levels of unsaturation. However, further progress was hampered by the extreme chemical resistance of sporopollenin. Zetzsche found that the only satisfactory way of degrading sporopollenin without running the risk of extensive and misleading chemical rearrangements was using oxidation techniques such as ozonolysis. Under these conditions, sporopollenin yielded a mixture of simple (C_3–C_6) dicarboxylic acids, which accounted for about 40% of the total molecular structure.

Little was added to this picture until the early 1960s when Shaw and co-workers commenced a re-investigation of sporopollenin structure (Shaw, 1971). In addition to confirming the earlier findings of Zetzsche, they also found that ozonolysis gave traces of longer chain dicarboxylic acids and substantial amounts of simple straight and branched chain monocarboxylic acids, which are characteristic breakdown products of fatty acids. They also found that the comparatively aggressive technique of potash fusion yielded a mixture of phenolic acids, such as p-hydroxybenzoic, m-hydroxybenzoic and protocatechuic acid. Based on these analyses, Shaw and Yeadon (1966) proposed that sporopollenin is composed of a lipid fraction of 55–65%, consisting of molecules with a chain length of up to C_{16}, and a 'lignin' fraction representing 10–15% of the total mass.

Despite the fact that no new direct analytical data had become available, the proposition that sporopollenin was a mixed polymer of lipids and phenylpropanoid units was rapidly superseded by a chemical structure based on the polymerisation of carotenoids and carotenoid esters (Brooks & Shaw, 1968; Shaw, 1971). There were two main reasons for this revision. Firstly, Brooks and Shaw believed that in common with other polymers, sporopollenin was probably composed of a simple monomer, rather than a combination of chemically distinct monomers. Secondly, probably born out of frustration at the lack of progress, further attempts to directly determine sporopollenin structure were abandoned in favour of searching for suitable candidate monomers in developing anthers (Brooks & Shaw, 1968). In the event, carotenoids and carotenoids esterified to fatty acids emerged as strong candidates for the elusive sporopollenin precursors since these were the only compounds that possessed both the appropriate chemical characteristics (long unsaturated carbon skeleton, content of hydroxyl and C–Me groups) and a temporal pattern of accumulation that parallelled the synthesis of sporopollenin.

The only apparent inconsistency with the contemporary analytical data was the origin of the phenolic degradation product of sporopollenin.

To test the validity of their findings, Brooks and Shaw polymerised various carotenoid mixtures extracted from lily anthers and model carotenoids, such as β-carotene, under oxidative conditions in the presence of a metallic catalyst (Fig. 2). This succeeded in producing a series of 'synthetic sporopollenins' which shared many of the physicochemical properties of natural sporopollenin. For example, the molecular formulae

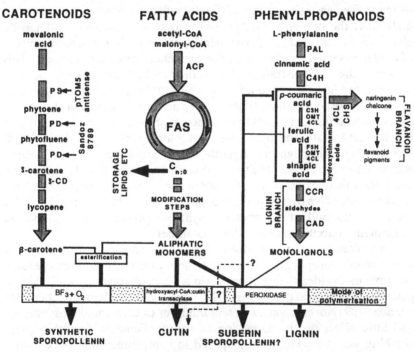

Fig. 2. The biosynthetic pathways of cutin, suberin, lignin and the potential sporopollenin monomers. The thickness of the solid lines is intended to reflect the relative contribution of each monomer to the respective polymer. The position of sporopollenin in this scheme is speculative. *Abbreviations*: PS, phytoene synthase; PD, phytoene desaturase; ACP, acyl carrier protein; FAS, fatty acid synthetase; PAL, phenylalanine ammonia-lyase; C4H, cinnamate 4-hydroxylase; 4CL, 4-coumarate: coenzyme A ligase; CHS, chalcone synthase; C3H, *p*-coumarate 3-hydroxylase; OMT, *o*-methyl transferase; F5H, ferulate 5-hydroxylase; CCR, cinnamyl coenzyme A reductase; CAD, cinnamyl alcohol dehydrogenase.

of *Lilium* pollen exine, *Lilium* carotenoid–carotenoid ester co-polymer and the β-carotene polymer were $C_{90}H_{142}O_{36}$, $C_{90}H_{148}O_{38}$ and $C_{90}H_{130}O_{30}$, respectively. Ozonolysis of the natural and synthetic compounds also yielded strikingly similar amounts of the familiar mixture of degradation products. Perhaps most significantly however, was that potash fusion of either the synthetic polymer or lily pollen exine yielded identical mixtures of the three characteristic phenolic acids (Brooks & Shaw, 1968). Consequently, Brooks and Shaw argued that these molecules resulted from molecular rearrangements (aromatisation) of the carotenoid skeleton owing to the aggressive nature of the degradation technique and, accordingly, that the concept of a lignin-like component in sporopollenin be abandoned as an artifact. Subsequently, several other persuasive pieces of evidence served to strengthen the carotenoid monomer concept. This included the demonstration that radio-labelled carotenoids became incorporated into the sporopollenin of several fungi and in the sporopollenin of angiosperm pollen (Shaw, 1971) and the observation that in the alga *Chlorella fusca,* mutants that cannot synthesise keto-carotenoids also fail to accumulate sporopollenin (Atkinson *et al.,* 1972; Burczyk & Hesse,1981). Although the hypothesis of Brooks and Shaw became widely accepted, Heslop-Harrison (1968a), expressed doubts at the time about the role of coloured carotenoids in sporopollenin biosynthesis. This was because, in his own experiments in *Lilium,* Heslop-Harrison (1968b) found that the bulk of sporopollenin synthesis preceded the main period of carotenoid accumulation within the anther.

The last ten years has witnessed a further re-evaluation of the chemical structure of sporopollenin which casts doubt on the role of carotenoids and favours the idea that sporopollenin is a mixed polymer of phenylpropanoid and fatty-acid derivatives. This began with the demonstration that Sandoz 9789 (Norflurazon), a potent inhibitor of carotenoid biosynthesis, had little effect on the formation of sporopollenin in *Cucurbita pepo* (Prahl *et al.,* 1985; Fig. 2). Compared to control material, pollen from treated plants contained only trace amounts of carotenoids beyond phytofluene and accumulated both phytoene and phytofluene, indicating the inhibition of phytoene desaturase. Although this does not rule out the involvement of saturated carotenoids, i.e. above ζ-carotene, none of these compounds was tested in artificial polymerisation experiments. Wiermann and coworkers also showed that labelled phenylalanine is incorporated into sporopollenin suggesting that phenylpropanoids constitute a genuine component of sporopollenin (Prahl *et al.,* 1985; Rittscher & Wiermann, 1988). Subsequently, nitrobenzene oxidation suggested that *p*-coumaric acid may constitute the main aromatic component of sporopollenin (Schulze Osthoff & Wiermann, 1987; Fig. 2).

This view has been strengthened by the results of more sophisticated tests including gentle degradation in AlI_3, which is capable of splitting ether bonds, and pyrolysis mass spectrometry, which together indicated that the phenolic monomers are coupled by ester bonds characteristic of polyphenolics such as lignin and suberin (Wehling *et al.*, 1989). Significantly, when 'synthetic sporopollenin' and natural sporopollenin were subjected to these and other tests (Herminghaus *et al.*, 1988), only the natural material yielded aromatic derivatives. Finally, ^{13}C NMR spectroscopy indicated that sporopollenin is not a unique substance, but a series of related biopolymers derived from largely saturated precursors such as long chain fatty acids and various, but usually modest, levels of oxygenated aromatic rings (Guilford *et al.*, 1988; Fig. 2). These authors pointed out that lignin contains such moieties, and that these serve as cross-linking agents in other plant wall components such as cutin and suberin.

Cutin, suberin, lignin and sporopollenin – a family of biopolymers?

The reassertion that sporopollenin is potentially a mixed polymer of phenolics and fatty-acid derivatives prompted Schulze Osthoff and Wiermann (1987) to propose a structural homology with cutin and suberin. Cutin and suberin are lipid polyesters found mainly in the protective outer coverings of plants. Cutin is the structural component of the plant cuticle, a layer that coats the outer surface of epidermal cells of virtually all aerial organs in liverworts, mosses, gymnosperms and angiosperms (Kolattukudy, 1980). Suberin protects the underground parts of the plant such as roots and tubers, and is also found in the Casparian band of the endodermis, the hypodermis, and the seed coat, trichomes and idioblasts of certain plant species (see Holloway & Wattendorf, 1987).

The main physiological role of cutin and suberin layers is to restrict water movement. However, waxes are actually responsible for the waterproofing, with the polymers providing a structural framework within which the soluble waxes are embedded. Lignin is also water-repellent, and may have first been exploited as a water-proofing agent in water conducting vessels, before realising its additional modern role of providing mechanical strength to the cell wall (Monties, 1989). Thus collectively, cutin, suberin and lignin function to restrict water loss, or water movement and to mechanically reinforce specialised cell walls within the sporophyte. These are precisely the functions of sporopollenous exine.

In terms of biochemical composition, cutin and lignin occupy opposite ends of a spectrum, with suberin forming a link between the two. Cutin is synthesised from oxygenated long chain fatty acids (C_{16} and C_{18}), derived mostly from oleic acid, together with a small amount of phenolic acids such as p-coumaric acid and ferulic acid (Kolattukudy, 1980). Lignin, on the other hand, is composed entirely of phenolic compounds, known as monolignols (p-coumaryl, coniferyl, and sinapyl alcohols; for recent reviews see Higuchi, 1985; Lewis & Yamamoto, 1990). Suberin has a composition intermediate between cutin and lignin, being a mixed polymer of mainly long-chain fatty acids (C_{18-30}) and a moderate content of the phenylpropanoids p-coumaric acid, ferulic acid, and tyramine, together with a small amount of monoligols (Kolattukudy, 1980; Kolattukudy & Köller, 1983; Borg-Olivier & Monties, 1993). The biosynthetic pathways for the various monomers are summarised in Fig. 2. According to a tentative model by Kolattukudy (1980, 1987), the structure of suberin consists of a lignin-like phenolic matrix on to which cutin-like fatty acids are covalently linked. The whole complex is anchored to the cell wall via covalent bonds between cell wall carbohydrates and phenolic residues within the suberin matrix, in much the same way as occurs in lignified tissues.

White lines – a hidden truth?

What evidence, in addition to direct analysis, exists to support the supposition that sporopollenin belongs to the cutin–suberin–lignin series of polymers? Such evidence as is available falls into two categories: ultrastructural and biochemical, both of which are compelling, but nevertheless circumstantial in nature.

Cutin and suberin share a common fine structure that reinforces the functional, chemical and physiological relatedness of the two polymers (Kolattukudy & Espelie, 1985). When certain cuticular membranes are prepared by a diverse range of techniques, including freeze–fracture, and visualised under the electron microscope, the cuticle proper is seen to be composed of a regular arrangement of alternate electron-opaque and electron-translucent lamellae (Wattendorf & Holloway, 1980; Holloway, 1982). Similar polylamellate layers are a common feature of suberised cell walls (Kolattukudy, 1980) and are particularly prominent in the potato tuber periderm (Schmidt & Schönherr, 1982). For comparison, Fig. 3 shows electron micrographs of the adaxial cuticle of *Iris germanica* leaf (Fig. 3A) and the suberised periderm of *Solanum tuberosum* tuber (Fig. 3B). Both are stained with osmium tetroxide which is particularly effective in revealing the presence of the lamellae. In general, the opaque layers are of variable thickness, ranging from 10–50 nm, whereas the lucent layers are of a more uniform thickness of 3 to 10 nm.

Despite the obvious homologies, suberin and cutin of the cuticle proper differ in at least one important respect: the cuticle proper forms on the outside of the cell wall some distance from the plasma membrane, whereas suberinisation occurs in close association with the plasma membrane. Thus, whatever the mechanism of lamellae formation, the participation of the plasma membrane is not a prerequisite, suggesting that some inherent property of the cutin, and possibly suberin, monomers promotes self-assembly into lamellae. It is also noteworthy that lamellae are not present in the bulk of the cuticle, which is deposited within the carbohydrate matrix of the primary wall. Thus a duality of fine structure exists in this polymer.

Lamellae do not occur in lignin, which may reflect compositional differences with cutin and suberin. Since cutin and suberin contain high levels of fatty acids, which are absent from lignin, the formation of lamellae is apparently associated with the polymerisation of aliphatic monomers. Perhaps significantly, during germination of the shoot apex of jack pine (*Pinus banksiana*) the outer surfaces of stored oil bodies gradually accumulate lamellae of the same physical appearance and dimensions as those of cutin and suberin (Cecich, 1979). Although there is no direct evidence concerning the composition of these intracellular lamellae, this observation further strengthens the association between lamellae and fatty acids.

Sporopollenin deposition is also associated with lamellae that bear a striking homology to the lamellae of cutin and suberin in several respects: cellular localisation, molecular dimensions and staining properties. However, this is not always appreciated since these lamellae are more or less obvious dependent on the species and, much more importantly, the developmental stage at which observations are made during exine deposition. Sporopollenin-associated lamellae are found in spore walls of almost all living plant taxa: algae (Atkinson, Gunning & John, 1972); bryophytes (liverworts and mosses) (Brown & Lemmon, 1990); pteridophytes (horse tails, ferns) (Lugardon, 1990); gymnosperms (conifers) (Kurmann, 1990) and angiosperms (Rowley, 1962; Rowley & Southworth, 1967; Dickinson & Heslop-Harrison, 1968). This very broad phylogenetic distribution suggests an ancient origin, perhaps in association with the spores of early land plants or the algal progenitors of embryophytes (Atkinson *et al.*, 1972). Irrespective of source, sporopollenin-associated lamellae are characteristically of 'unit-membrane' dimensions, consisting of two electron dense layers of 5–6 nm in thickness, separated by an electron lucent layer of 4–8 nm in thickness (Rowley, 1962; Rowley & Southworth, 1967). These dimensions are of the same order as those of cutin and suberin-associated lamellae. Moreover, in common with cutin and suberin, exine lamellae also stain well with osmium tetroxide.

In angiosperm exines, lamellae are most readily observed in the nexine II where they become compressed to form the final polylaminated layer of the exine. Figure 3C shows the polylamellate layer of a microspore from a genetically modified tobacco (Worrall *et al.*, 1992). The lamellae are particularly obvious here because one of the effects of the transgene on pollen wall ontogeny is to reduce their compression. The sequence of events at the cell surface that is responsible for the formation of the

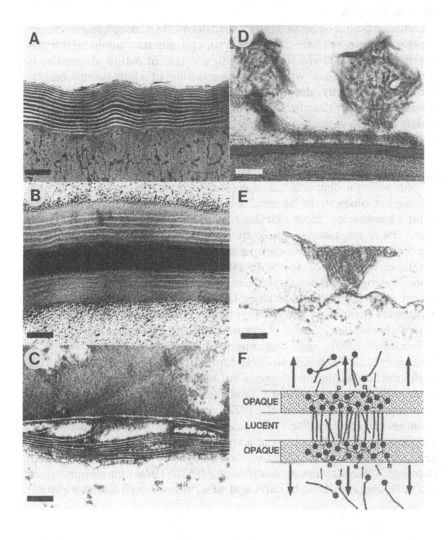

polylamellate layer is not well understood, but may begin with the formation of the lucent layer on the outside of the plasma membrane which then synthesises a layer of sporopollenin on both its the inner and outer faces. The completed lamella apparently moves away from the membrane and is included in the growing nexine II; a new lamella forms at the membrane, and the process is repeated 'treadmill-like' many times. However, there is no direct evidence that lamella formation is dependent on the plasma membrane, and some authors believe that exine precursors can polymerise some distance from the membrane (Horner, Lerston & Bowen, 1966), as occurs during the formation of the cuticle proper.

In certain primitive plants, such as the sporopollenous algae and liverworts, lamellae very obviously participate in sporopollenin deposition throughout the entire exine (Blackmore & Barnes, 1987, 1990). In liverworts, such as *Haplomitrium* (Fig. 3D), the spore wall consists of both a simple continuous layer adjacent to the protoplast (analogous to the nexine II of higher plants) and an elaborately sculptured outer layer, which shares features of the sophisticated sexines of angiosperms (Brown & Lemmon, 1986). At an immature stage, the spore wall is pre-patterned with a cellulosic material (primexine) within which variously oriented lamellae develop and subsequently accumulate sporopollenin; the lamellae remain obvious in the mature spore (Brown, Lemmon & Renzaglia, 1986). This contrasts with the situation in higher plants, where at maturity the sporopollenin of the sexine elements is apparently homogeneous and shows no obvious evidence of the participa-

---

Fig. 3. Ultrastructure of cutin, suberin and sporopollenin. A–E are transverse transmission electron micrographs (TEM) of thin sections stained with osmium tetroxide. A. Ultrastructure of cutin. Adaxial cuticle of an *Iris germanica* leaf showing prominent lamellae of the cuticle proper (photograph courtesy of P.J. Holloway, Long Ashton). B. Ultrastructure of suberin. Tuber periderm of *Solanum tuberosum* (photograph courtesy of P.J. Holloway.) C. Sporopollenin-associated lamellae in *Nicotiana tabacum* pollen wall. D. Mature spore wall of the liverwort *Haplomitrium hookeri* (photograph courtesy of B. Brown, University of Southwestern Louisiana). E. Longitudinal view of a probaculum in tetrad-stage microspore of *Lilium* (photograph courtesy of H.G. Dickinson, University of Oxford). F. Schematic representation of lamellae growth (adapted from Kolattukudy & Köller, 1983). Large arrows indicate direction of polymer growth; small arrows indicate movement of monomers on to the lamellae. Open boxes represent cellulose fibrils of the cell wall or primexine on to which some monomers are covalently attached. Magnification bar: 100 nm, A, B, C; 180 nm, D; 150 nm, E.

tion of lamellae. However, Rowley (1962) observed that, during early sexine formation in *Poa annua,* sporopollenin occurred in bundles of strands that occasionally interconnected to produce a system of lamellae. These structures disappeared as the sporopollenin became increasingly homogeneous as a result, suggested Rowley, of its progressive polymerisation owing to the availability of a 'comonomer'. Similarly, in *Lilium,* the immature sexine elements (probaculae) of tetrad-stage microspores have a lamellate appearance, as shown in Fig. 3E (Dickinson & Heslop-Harrison, 1968). According to these authors, the lamellae of the nexine II and sexine appear sufficiently homologous to suggest that a single mechanism is responsible for the formation of all parts of the exine.

Initially, the lamellae of the probaculae in both *Poa* and *Lilium* are apparently not composed of sporopollenin, since they are not resistant to acetolysis, the generally accepted microchemical test for sporopollenin (Erdtman, 1960). However, since acetolysis involves extended incubation at 100 °C in a mixture of acetic anhydride and sulphuric acid, and is therefore extremely aggressive, it is likely that only highly polymerised sporopollenin can survive: nascent sporopollenin may therefore not be detectable by this test. Thus, as suggested by Heslop-Harrison (1968c,d), probaculae might contain partially polymerised sporopollenin.

Thus sporopollenin polymerisation is probably universally associated with a trilamellate intermediate, which is more or less transitory in the case of the sexine, and persistent in both the nexine II and throughout the entire exine in many lower plants. The physical similarity of the sporopollenin lamellae to those of cutin and suberin is also evident and suggests that polymers with a high lipid content may naturally undergo self-assembly into a trilayered configuration (Fig. 3). In a model of suberin structure proposed by Kolattukudy and Köller (1983), the lucent layer is interpreted as being composed of the hydrophobic regions of long-chain fatty acids, whilst the hydrophilic ends of these molecules comprise the opaque layer in a complex with the phenolic component. A schematic representation is shown in Fig. 3F. According to this concept, the constant thickness of the lucent layer is determined by the chain length of the fatty acids. Cross-links between the molecules within both layers are assumed to form a rigid scaffold for the deposition of wax, since exhaustive extraction for waxes does not reduce the thickness of the lucent layer.

Sporopollenin biosynthesis genes – the spatial and temporal expression of the candidate pathways

The available biochemical evidence, particularly the potential for the biosynthesis of the putative sporopollenin monomers within the appropriate anther cell types, is also broadly supportive of an homology between cutin, suberin and sporopollenin. An obvious prerequisite of the model is the appropriate temporal and spatial activity of the fatty acid synthetase (FAS) complex and the phenylpropanoid pathway. In *Cucurbita pepo* anthers, sporopollenin accumulation commences during meiosis, continues throughout microspore development and ceases toward pollen maturation (Prahl *et al.*, 1985). It is likely that, in most species, the majority of the pollen wall sporopollenin is contributed by the tapetum (Heslop-Harrison, 1962) and therefore the pattern of sporopollenin accumulation largely reflects the activity of tapetally located enzymes. In contrast, the microspores contribute relatively little to the final mass of mature sporopollenin and this is reflected in the fact that the phase of sporopollenin biosynthesis by the microspores is short-lived and may occur entirely whilst the microspores are held within the post-meiotic tetrad (Heslop-Harrison, 1968c).

Long chain fatty acids, which may constitute the major monomeric component of sporopollenin (Guilford *et al.*, 1988), accumulate to high levels in both the tapetum and in pollen (Evans *et al.*, 1992). Cell-specific isoforms of acyl carrier protein (ACP), which channels fatty acid precursors into the FAS complex and is regarded as diagnostic of the activity of the complex (Ohlrogge, 1987), are associated with the tapetum and microspores (Fig. 2). In *Brassica*, a low level of ACP activity was detectable in immediately post-meiotic microspores, and this rapidly peaked at about the mid-maturation stage, and declined towards anthesis (Evans *et al.*, 1992). Unfortunately, no similar systematic studies have been carried out on ACP activity within the tapetum. However, cytochemical studies indicate that substantial quantities of lipid accumulate within the tapetum by the mid-vacuolate stage of pollen development and continues until the tapetum ruptures at the mid-maturation stage (Evans *et al.*, 1992). This suggests that there is considerable lipid biosynthetic activity within the tapetum throughout pollen development, including the phase of sporopollenin biosynthesis.

In pollen, the majority of the lipid biosynthetic activity is involved in the provision of storage lipids (Evans *et al.*, 1992; Roberts *et al.*, 1993). Similarly, in the tapetum, most lipid accumulates in elaioplasts and is transferred from the tapetum late in pollen development to coat the surface of the pollen grains (Dickinson, 1973). Consequently, any involvement of the microspore and tapetum lipid biosynthesis pathways in the synthesis of sporopollenin lipid-monomers is obscured by these other activities. Nevertheless, the important point is that the biosynthetic capacity is present in the sporopollenin-synthesising cell types of the anther.

The polymerisation of cutin and suberin is invariably extracellular and therefore requires the movement of monomers from the site of synthesis through the plasma membrane and past the extracellular carbohydrate matrix into the proximity of the growing polymer. At least in the case of cutin, non-specific lipid transfer proteins (nsLTPs) have been implicated in this trafficking activity (Sterk et al., 1991; Thoma, Kaneko & Somerville, 1993). Recently, potential anther-specific nsLTP isoforms have been reported in tobacco (Koltunow et al., 1990) and Brassica (Foster et al., 1992). The Brassica nsLTP cDNA, E2, which shows 55–60% amino acid sequence similarity with spinach, castor bean and maize nsLTPs, hybridises specifically to abundant messages within both the tapetum and the developing microspores, and encodes a potential hydrophobic signal sequence. If LTPs are indeed indicative of cutin formation or at least extracellular lipid transport, what role do these proteins play in developing microspores where the cuticle is formed of sporopollenin, and in the tapetum for which there is no evidence of the presence of a leaf-like cuticle? One possibility is that E2 represents a class of specialised LTPs that transport the fatty acid precursors of sporopollenin.

The general phenylpropanoid pathway provides activated CoA esters of cinnamic acid derivatives for the biosynthesis of flavonoids, lignin and other phenolic compounds via specific branch pathways (Fig. 2). The first enzyme in the general pathway, phenylalanine ammonia-lyase (PAL), has been localised to the tapetum (Herdt, Sütfeld & Wiermann, 1978; Kehrel & Wiermann, 1985). The tapetum is also the site of flavonoid pigment synthesis (Herdt et al., 1978; Beerhues et al., 1989; van der Meer et al., 1992) indicating that the general pathway and the flavonoid branch are active in this tissue. This was also shown directly by the antisense RNA inhibition of chalcone synthase expression within the tapetum of petunia plants which prevented the synthesis of flavonoid pigments and caused male sterility due to an arrest in pollen development (van der Meer et al., 1992). Promoter–gus fusions of the Arabidopsis PAL1 gene (Ohl, 1990) and the parsley 4-coumarate:coenzyme A ligase-1 gene (Hauffe et al., 1991; Reinold, Hauffe & Douglas, 1993) have also been analysed in transgenic tobacco plants. However, the expression of these genes in the tapetum and microspores apparently failed to parallel sporopollenin accumulation. Currently, there is no information relating to the activity of the lignin branch enzymes cinnamyl coenzyme A reducatase (CCR) and cinnamyl alcohol dehydrogenase (CAD), within any anther tissue, although lignin may be deposited in secondary wall thickenings of endothecial cells late during anther development to aid pollen release (Esau, 1965). Similarly, there are no published accounts of assays for components of the general phenylpropanoid pathway in the developing microspores, although the demonstration of tetrahydroxychalcone in petunia pollen suggests that the pathway is operational (de Vlaming & Kho, 1975).

Brooks and Shaw (1968) demonstrated that a range of carotenoids are extractable from the anthers of *Lilium* and that the concentration of these compounds increases dramatically during exine deposition. At least a proportion of this material is presumably synthesised within the tapetum since the tryphine or Pollenkitt deposited on the microspore surface upon tapetal disruption contains carotenoids of tapetal origin (Heslop-Harrison, 1968b). However, to date, none of the enzymes involved in carotenoid biosynthesis has been localised within the anther. Figure 2 outlines the carotenoid biosynthetic pathway and draws mainly on information from biochemical and molecular-genetic studies in bacteria (Chamovitz *et al.*, 1992) and tomato (Bartley *et al.*, 1992).

The possible mechanism of sporopollenin polymerisation

The hypothesised relatedness of sporopollenin with cutin, suberin and lignin would suggest that monomer polymerisation is also mediated via a common mechanism. However, the mode of cutin and suberin/lignin polymerisation is different (Fig. 2). Cutin polymerisation is catalysed by a hydroxyacyl-CoA: cutin transacylase, which transfers activated cutin monomers to the growing polymer (Kolattukudy, 1980; Kolattukudy & Espelie, 1985). Although there is little direct evidence, it is envisaged that suberin is polymerised in much the same way as lignin which involves the oxidation of phenolic alcohols to yield mesomeric phenoxy radicals which rapidly polymerise together, and also form covalent bonds with cell wall polysaccharides. The process is catalysed by cell wall-bound peroxidase isozymes and produces multiple ether and carbon–carbon bonds between the monomers, which accounts for the great resistance of lignin to hydrolytic enzymes. These types of bond are also found in suberin and sporopollenin.

The hydrogen peroxide (H_2O_2) needed for the peroxidase-catalysed polymerisation process is provided by a cell wall-bound malate dehydrogenase which generates NADH to drive peroxide generation by the peroxidase. The extent of lignification is tightly restricted and is always associated with the carbohydrate matrix of the cell wall. One explanation for this is that the large size of the peroxidase protein restricts its movement through the cell wall, and therefore superoxide radicals are formed close to the plasma membrane, from where they diffuse through the matrix. Superoxide radicals are stabilised in the presence of Ca^{2+} located in cell walls and polymerisation is therefore only possible close to or within the wall matrix.

Suberinisation is envisaged as an essentially analogous process, except that in addition to phenolics, fatty alcohols are also incorporated into the polymer, possibly also by peroxidase-catalysed polymerisation. The fatty alcohols may be esterified to the phenolics prior to polymerisation.

Linskens (1966) reported the activity of a peroxidase enzyme in anthers that appeared at the tetrad stage and persisted through the entire phase of sporopollenin deposition. Whether this activity is responsible for sporopollenin polymerisation is not known. However, in common with lignin and presumably suberin, the patterned sexine (tectum, baculae and foot-layer) is also formed within a cellulosic matrix, called the primexine by Heslop-Harrison (1968d). Synthesis of the primexine commences soon after the completion of meiosis II and initially forms a homogeneous layer between the plasma membrane of the new microspore and the inner face of the callose wall. Patterned arrays of multilamellate probaculae (Fig. 3E) then condense within the primexine and these subsequently develop into the mature wall elements by the accumulation of sporopollenin. This sequence of events gave rise to the idea that the primexine contains receptors that promote sporopollenin polymerisation. Since the similarity between sporopollenin and suberin polymerisation suggests a common mechanism, this receptor concept may be restated in a manner consistent with a suberin-like polymerisation mechanism as follows:

1. The primexine acts as a loose scaffold on to which sporopollenin monomers (fatty acids and phenolics) are covalently attached by the localised action of superoxide radicals generated at the plasma membrane.
2. Initially, the extent of polymerisation is low, resulting in acetolysis sensitive multilamellate probaculae.
3. Subsequently, the probaculae act as conduits for the cross-linking agent allowing more extensive polymerisation and eventually the acquisition of acetolysis resistance leading to the formation of baculae proper.

It is noteworthy that the oxidative polymerisation of carotenoids was unsuccessful using free radical catalysts such as those responsible for suberin and lignin polymerisation (Shaw, 1971).

The specification and generation of exine architecture and patterning

Pollen wall ontogeny involves the specification of three main architectural features: 1. generally uniform numbers of precisely positioned colpi; 2. patterned arrays of baculae; 3. exine stratification. In the majority of species, the position of the colpi is consistent and genetically determined. For example, in the lily tetrad, which has a two-by-two arrangement common to monocotyledons, the single aperture of the spore always develops at a specific position toward the perimeter of the cell. In the majority of dicotyledons, including *Brassica* and tobacco, the tetrad of microspores is arranged in a tetrahedron, and each spore forms the three radially symmetrical equatorial colpi. In these species, as first pointed out by Wodehouse (1935), this arrangement is determined by the positions of maximum contact with the other three spores within the tetrad.

How then is this most obvious architectural feature of the pollen grain determined? Significantly, primexine is not deposited in areas destined to become colpi or germinal apertures, which emphasises the importance of this matrix to sporopollenin polymerisation. In several species, including lily, the plasma membrane in the primexine-less areas of the future colpi is invariably associated with an underlying plate of endoplasmic reticulum – the colpal shield. Heslop-Harrison (1963*a*) proposed that the apposition of the colpal shield to the plasma-membrane blocks deposition of primexine, which in turn prevents sexine growth. Subsequently, experiments conducted by Dover (1972) in wheat implicated the meiotic cytoskeleton in the positioning of the single distal aperture, since the apertures form in positions close to the poles of the first meiotic division. Furthermore, complete disruption of the spindle by centrifugation prevented both division and colpal formation, whereas partial disruption usually resulted in colpus formation.

Sheldon and Dickinson (1986) suggested that the transduction mechanism by which spindle position influences colpus position operates via the spindle microtubule organising centre (MTOC) which forces cytoplasmic components, including smooth endoplasmic reticulum (SER) against the plasma membrane, thereby forming the colpal shield. Thus, at least in plants with a single distal aperture, the specification of aperture position is associated with the mode of microspore partitioning and displays symmetry related to that of the meiotic spindle. Although experimental data is lacking for tricolpate pollen grains, it is likely that the meiotic spindle plays a very similar role in aperture positioning.

Despite the clear diversity in pollen wall patterning, a basic feature shared by many species is the reticulate arrangement of baculae. This is very obvious in mature pollen of *Lilium* (Fig. 4A) and *Brassica* (Fig. 4B, right panel), where the basic pattern unit is very often the hexagon. Note that each line of baculae is common to two pattern elements, indicating a large degree of interdependency between elements, rather than an independent origin for each element. As shown in Fig. 4B (left panel), the positioning of the baculae into the final configuration is complete at

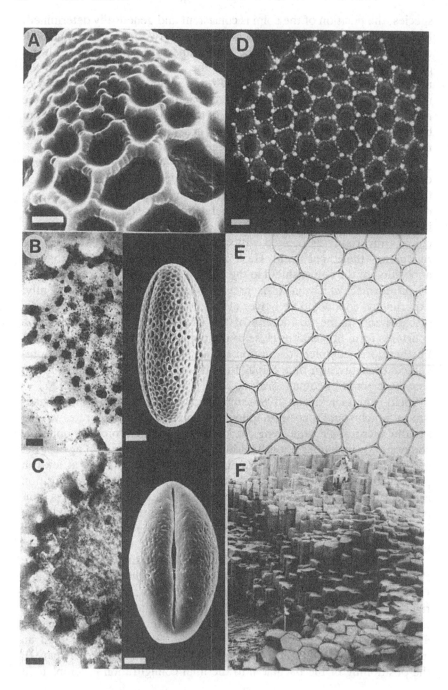

the tetrad stage. Tobacco pollen is tectate-perforate and therefore in SEMs the baculae are obscured from view by the tectum (Fig. 4C, right panel). Nevertheless, at the tetrad stage the arrangement of the baculae is quite similar to that of *Brassica* (Fig. 4B and 4C, left panels). This indicates that development of a tectate exine also involves the expression of pattern information. This is to be expected since, as stated earlier, the obviously patterned semitectate and intectate exine types are probably derivative of the fully tectate condition.

A particularly striking example of the hexagonal arrangement of baculae is provided by the immature pollen of *Ipomea purpurea* (Fig. 4D). In this species, the reticulate pattern of the tetrad stage microspore is remarkably consistent; of the 54 complete elements that make up the spore face depicted in Fig. 4D, 48 are regular hexagons. However, it is noteworthy that the pattern is not quite perfect since six of the constituent elements are pentagonal. Another interesting feature of *Ipomea* is that almost every three-way intersection (triple junction) between the polygonal elements carries a spine. These examples demonstrate that pattern specification involves the precise positioning of wall elements, principally baculae, and that the unit of pattern is very often a regular hexagon.

Several lines of evidence suggest that the genes responsible for determining the species-specific exine pattern are transcribed within the diploid nucleus of the pre-meiotic sporocyte, and that the pattern information is inherited by the microspores (Heslop-Harrison, 1963a, 1968a; Rogers & Harris, 1969). Unfortunately, since there is no evidence for the participation of any subcellular organelle in this process, the colpal-shield model cannot be extended to explain how this information is utilised to

---

Fig. 4. Reticulation in pollen walls, bubbles and basalt. A. Scanning electron micrograph (SEM) of *Lilium* pollen wall showing reticulate pattern formed by the baculae. The nexine I is visible through the lacunae. B. Reticulation in the semitectate exine wall of *Brassica napus*. *Left panel:* TEM of a glancing section through the surface of a tetrad-stage microspore. *Right panel:* SEM of a mature grain. C. Reticulation in the tectate-perforate pollen wall of *Nicotiana tabacum*. *Left and Right panels:* technical details as in 5B. D. Light micrograph of a tetrad-stage microspore of *Ipomea purpurea* stained with primuline to show the position of the young baculae (small dots) and spines (larger dots at triple junctions) (photograph, courtesy of L. Waterkeyn, Université Catholique de Louvain). E. Monolayer of soap bubbles floating over water. F. Basalt columns of the Giant's Causeway, Co. Antrim, Ireland. Magnification bar: 5 μm, A, B [right], C [right]; 28 nm, B [left], C [left]; 0.6 μm, D.

direct baculae positioning. However, some progress has been made toward formulating a satisfactory model (Sheldon & Dickinson, 1983, 1986; Dickinson & Sheldon, 1986). A series of elegant experiments involving the centrifugation of microsporocytesin *in situ* established that the reticulate patterning of the exine was prone to disruption as early as meiotic prophase, but as meiosis proceeded was only modified within the new cross walls of the tetrad (Sheldon & Dickinson, 1983). This indicated that the agent(s) responsible for imposing pattern on the primexine must appear in the cytoplasm at the beginning of meiosis, where it is sensitive to centrifugation, and is then progressively inserted into the plasma membrane during meiosis. However, since the colpal shield subsequently modifies the reticulate pattern, the pattern inducing material must remain latent, and malleable to a certain extent, until after colpus specification. To date, the only component of the prophase cytoplasm that constitutes a good candidate for the determinant of pattern formation are so-called coated vesicles (CVs) (Sheldon & Dickinson, 1983). These small protein-coated vesicles are formed from prophase onwards from SER, and migrate to the cell surface where they fuse with the plasma membrane. The implication is that the CVs insert protein into the plasma membrane which somehow influences the position of the probaculae.

Pollen wall patterning – bubbles and basalt

How might protein inserted in the plasma membrane specify pattern? Heslop-Harrison made the following observation:

> Sometimes when we contemplate biological pattern it is difficult to imagine how great a part of it could arise from physical causes, and correspondingly easy to slip into accepting that genetical control extends down to all detail. Yet this is obviously not so . . . (Heslop-Harrison, 1972).

Sheldon and Dickinson (1983) also acknowledged the role of purely physical phenomena in the specification of exine pattern when they pointed out that it is difficult to envisage a mechanism whereby CVs are inserted into the plasma membrane at positions subsequently occupied by probaculae. Instead, they proposed the conceptually simpler hypothesis that the vesicles fuse randomly with the plasma membrane, and the deposited protein undergoes self-assembly into pattern-specifying units. Sheldon and Dickinson (1983) offered two models that might explain how this happens, both of which are capable of accounting for the common hexagonal pattern of baculae.

The first mechanism is best illustrated by following the behaviour of soap bubbles within a foam (Fig. 4E). Soap bubbles in isolation are invariably circular in cross-section, but when aggregated in a foam with bubbles of a similar size a high proportion become hexagonal. This occurs in order to satisfy the dual constraints of maximum space-filling and minimum free energy. During the creation of a foam, bubbles adjust their shape to minimise the total area of interface and achieve a stable, low free energy, state. However, all such adjustments must conform to the geometrical constraints of space filling. For example, the original spherical shape has a minimum interfacial energy, but an aggregate of spheres does not fill space efficiently. Careful examination of Fig. 4E shows that, in two dimensions, all bubbles meet in threes along lines to form what are known as triple junctions. All the triple junction angles are equal and therefore 120°. Hexagons are usually formed since this geometrical shape combines the smallest number of triple-junctions (minimum free energy) with the ability to tessellate (maximum space-filling). As in *Ipomea* the pattern is not perfect: of the 31 complete elements, 23 are hexagonal, 5 septagonal and 3 are pentagonal. Here, pattern formation is entirely due to physical forces, and yet the parallel with pollen wall pattern is inescapable.

With respect to events at the plasma membrane, the model proposes that the CVs insert material into the membrane which is lipophobic and therefore initially tends to aggregate into circular plates that float within the fluid membrane. This is shown schematically in Fig. 5A. Elements of approximately equal area will be formed, provided the contribution of each CV is small relative to the final size of the plate. The behaviour of the plates as they grow and become juxtaposed is seen as analogous to that of bubbles, so that the final shape of the plates is most often hexagonal. The baculae would then form along lines specified by the interfaces between the plates; the boundary of each element is therefore specified by the interaction of two protein plates, which is consistent with real situation exemplified by *Ipomea* (Fig. 4D). Thus pattern is imprinted or stencilled on to the membrane.

How does this model account for the fact that the subsequent arrival of the colpal shield early in the tetrad stage prevents wall formation in this region and reduces the size of the lacunae of the reticulum around its edge? One possibility is that the spindle, which later determines the position of the colpus, interferes with the passage of CVs to the cell surface before the arrival of the colpal shield, resulting in the creation of smaller protein plates.

The second model is again entirely physical. Wodehouse (1935) in considering several aspects of pollen formation, pointed out that a reticulate pattern is a common theme in nature, and is formed where even shrinkage occurs within a uniform matrix. He chose the cracking of drying mud to

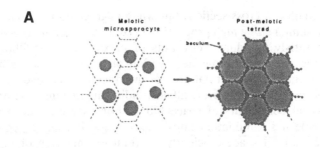

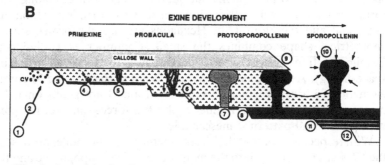

Fig. 5. A model of pollen wall ontogeny. A. Creation of exine pattern via the 'bubble' mechanism. The Figure depicts a surface view of a small section of the plasma membrane as the protein plates grow and eventually meet. B. Pollen wall ontogeny based on *Lilium*. Events begin before meiosis and run left to right until the mature wall is formed. 1. Transcription of 'pattern' genes in pre-meiotic diploid nucleus. 2. Insertion of labile pattern information into plasma membrane (via CVs) and eventual tessellation to produce a negative stencil composed of (protein) plates. 3. First phase of primexine synthesis over entire microspore surface except in areas destined to become colpi. 4. Conversion of primexine to a sporopollenin-receptive state through activity of factor(s) secreted from sites between the (protein) plates. 5. Second phase of primexine synthesis, more rapid than the first, resulting in limited primexine conversion and specification of baculae. 6. Wall elements apparent as lamellated probaculae which lack sporopollenin proper. 7. Consolidation of probaculae by appearance of protosporopollenin on receptive surfaces. 8. Final phase of primexine synthesis during which the pattern stencil dissipates or is circumvented (as in 4) to produce the nexine I. 9. Tetrad wall dissolved, exposing microspores to locular fluid. Remnants of primexine apparent between wall elements. 10. Wall elements further consolidated by tapetally derived sporopollenin. 11. Nexine II synthesised without participation of the primexine. 12. Intine synthesis initiated.

illustrate this point. Nature, though, offers far more striking examples, not least of which are the spectacular basalt columns of the Giant's Causeway, Ireland (Fig. 4F), and the Devil's Postpile, North America (see Plummer & McGeary, 1991). Geological formations of this type were created by sub-terranean volcanic eruptions which formed a flat pancake of larva. This then cooled at a uniform rate over a large area; the resultant even shrinkage gradually created correspondingly uniform lines of stress, which eventually caused the basalt to fracture into an array of regular polygons, usually hexagons. The physics of this process is analogous to that of foam formation as outlined above. However, such is the precision and uniformity of these formations that mythological creatures, rather than Nature, were credited with the necessary creative force.

In this case, the material delivered to the plasma membrane by the CVs is seen as possessing much less affinity for itself and spreads out to form a uniform layer, rather than distinct circular plates. Somehow the layer undergoes fracture to create a reticulate imprint within the membrane and baculae subsequently form along the fracture lines. Perhaps significantly, when tetrad-stage lily microspores were subjected to freeze–fracture, the procedure caused some shrinkage of the plasma membrane and the membrane fractures into a reticulate pattern (see Fig. 10, Dickinson & Sheldon, 1986). The shape and dimensions of the fragments reflect those of the exine reticulum. However, this observation is also consistent with the first model.

Pollen wall ontogeny

The next stage in the growth of the pollen wall involves synthesis of wall elements at positions specified by the latent pattern information. Due to the large size and accessibility of the microspores, this process is best documented in *Lilium* (Heslop-Harrison, 1968*c,d*; Heslop-Harrison, 1971; Dickinson, 1970; Sheldon & Dickinson, 1983; Dickinson & Sheldon, 1986). The process is depicted in Fig. 5B. The first readily observable event is the synthesis of the primexine between the plasma membrane and the inner face of the callose wall following the completion of meiosis II. Individual masses of 5–10 tubular protrusions of membrane (tubular accumulations: TAs) then originate from the plasma membrane at random positions along lines later occupied by the probaculae (Dickinson, 1970; Dickinson & Sheldon, 1986).

Next, the plasma membrane retreats except where it is apparently anchored within the primexine by the TAs. The TAs then act as foci for the condensation of the primexine which forms into a mass of lamellae, now termed a probaculum, orientated perpendicular to the cell surface.

The growth of the lamellae forces or facilitates the full retreat of the plasma membrane. Dickinson and Sheldon (1986) suggested that the TAs provide substances (peroxidase, sporopollenin precursors?) that promote condensation. This secretory model is particularly attractive since it could explain how in some species, such as *Helleborus* (Echlin & Godwin, 1968), the probaculae appear to condense at the surface of the plasma membrane. In these cases, the polymerising agent could simply be secreted from the plasma membrane without the intervention of TAs.

Initially, the probaculae are not resistant to acetolysis and therefore do not contain highly polymerised sporopollenin. As the tetrad stage progresses, the elements become increasingly resistant to acetolysis owing to the accumulation of a sporopollenin precursor material termed protosporopollenin by Heslop-Harrison (1968*d*). This material, which is synthesised solely by the microspore protoplast, is apparently not chemically identical to sporopollenin, but could simply represent sporopollenin in a state of partial polymerisation. Protosporopollenin only accumulates over the probaculae and therefore some mechanism must exist that restricts polymerisation to these sites. The model of sporopollenin polymerisation presented earlier, that postulates that the localised accumulation of sporopollenin is achieved by the action of a diffusible cross-linking agent (e.g. peroxide radicals), is consistent with this sequence of events.

How then does the pattern information restrict the supply of the cross-linking agent? One possibility is that the passage of at least one component of the polymerisation machinery (cross-linking agent, monomers) into the primexine is severely restricted by the protein plates of the pattern stencil. In the two models of pattern specification outlined above, polymerisation is accordingly only possible either between the protein plates or along the fracture lines.

Following microspore release, protosporopollenin is rapidly converted to sporopollenin, and the elements of the exine are thickened by sporopollenin of tapetal origin. Again this new material polymerises only over areas already covered by protosporopollenin or newly polymerised sporopollenin, presumably because of a high concentration of polymerisation sites in these areas. Hence, the principal features of exine pattern are determined within the tetrad and, following tetrad dissolution, are merely amplified by the addition of tapetally derived sporopollenin.

Now you see it, now you don't – specification of the tectum and foot-layer

The pollen wall of angiosperms, including tobacco, is stratified into tectum, bacula, foot-layer and nexine. The synthesis of these strata is centripetal – the tectum forms first adjacent to the callose wall, followed by the bacula and finally the foot-layer. Thus, the patterned baculate layer is sandwiched between two patternless layers. Since the pattern information is present within the plasma membrane prior to primexine synthesis, apparently the tectal layer, and possibly the foot-layer, escape the influence of the pattern information.

How is this possible? A clue is that like the wall elements, primexine synthesis is centrapetal, and therefore has the potential for stratification at the molecular level. Therefore, tectum specification might occur where some limiting component of the polymerisation machinery is able to diffuse laterally from the sites of secretion at the plasma membrane throughout most of the primexine, including over the areas otherwise blocked by the elements of the pattern stencil. Bacula, on the other hand, are specified when the extent of this diffusion is restricted. The modulation of the extent of this lateral diffusion could be achieved by changes in the rate of either primexine synthesis or secretion of the polymerisation machinery.

Recently, we showed that removal of the callose wall early in microsporogenesis has a profound affect on pollen wall formation (Worrall *et al.*, 1992). In tobacco, which normally forms a near complete tectum, this resulted in microspores that fail to produce a tectum. Bacula, on the other hand, are formed, but these are longer than those of the wild type. This suggests that the callose wall may provide a solid surface against which the tectum forms, rather than the alternative, that the low porosity of callose prevents the diffusion of polymerisation components out of the tetrad. The callose wall may also aid the lateral deflection of growing sporopollenous lamellae.

The final primexine-associated layer to form is the foot-layer or nexine I. Like the tectum, this layer is continuous, and therefore its synthesis marks a resumption in the capacity of the entire primexine to accumulate sporopollenin. This might involve either the loss of the pattern stencil or the evocation of a tectum-like mechanism.

In members of the Compositae, where the pollen form elaborate spines and ridges that rise above the tectum, the primexine may also function as a pathway for the diffusion of callose wall degrading enzymes. If the channelling were site specific this could account for how these supratectal features apparently form within the callose wall. In *Ipomea purpurea,* the elaborate reticulate exine is apparently formed by the deposition of primexine within a negative mould formed within the inner face of the callose wall (Waterkeyn & Bienfait, 1970). However, an alternative interpretation is that the callose wall does not act as a primexine mould *per se,* but instead pattern is created within the callose wall by the action of callose wall degrading enzymes held within the primexine.

Future directions

Despite the foregoing, the composition and structure of sporopollenin remain enigmatic. Undoubtedly, chemical analysis will continue and may prove successful. However, given the historical context of such analyses, more emphasis should be placed on biochemical and molecular genetic approaches. For example, sporopollenin may contain a lignin component and therefore the sporopollenin-synthesising tissues, i.e. the tapetum and young microspores, should express the lignin-specific enzymes CCR and CAD. Assays are available for both enzymes, and various molecular probes have been developed for CAD expression in tobacco (Knight, Halpin & Schuch, 1992). These would enable the temporal and spatial analysis of these enzymes in developing anthers.

By performing antisense experiments with appropriate promoters (Scott et al., 1991) the opportunity also exists to judge the effect of inhibiting any tapetally located CAD expression on pollen wall synthesis. Although potentially more difficult to interpret, components of the general phenylpropanoid pathway, such as PAL and 4CL, might be also be targeted in antisense experiments. In the event that the extensive suppression of the supply of phenolics had little or no effect on pollen wall synthesis, such compounds could be eliminated from any involvement in sporopollenin biosynthesis.

Similarly, a molecular genetic approach might be used to investigate the role of carotenoids in sporopollenin synthesis. cDNAs encoding several key enzymes in carotenoid biosynthesis pathway have been cloned from plants. These include a bean phytoene desaturase (Bartley et al., 1991) and the tomato fruit phytoene synthase Pys1 (Bartley et al., 1992) which was first identified in a ripening fruit-specific cDNA library and designated pTOM5 (Bird et al., 1991). Antisense experiments in tomato using a CaMV-pTOM5 antisense construct resulted in transgenic plants with significantly reduced levels of fruit carotenoids, and an associated accumulation of GGPP, but normal levels of leaf carotenoids (Bird et al., 1991). An antisense experiment could shed light on the role of carotenoids in pollen wall biosynthesis provided that promoters with activity within at least the tapetum are employed.

The specification of pollen wall patterning is also amenable to investigation via molecular genetics. Several approaches are feasible. Firstly, since the pattern is determined by genes expressed within the diploid sporocytes, recessive mutations in the pattern genes generated in an M1 population should show normal inheritance into the M2. Consequently, M2 populations of Arabidopsis, for example, could be screened for mutations affecting pollen wall pattern. Secondly, the specification of pattern might be manipulated using promoters from genes active in the microsporocyte during meiosis to insert proteins in the plasma membrane that might interfere with the formation of the putative protein plates. Unfortunately, such promoters are not available as yet. Whatever the approach, unravelling the mystery of pollen wall pattern specification remains a significant and surely a worthwhile challenge in plant biology.

Acknowledgements

I would like to thank Stefan C. Hyman for preparing TEM sections, and George McTurk for SEM. I am particularly indebted to Rachel Hodge and Wyatt Paul for their invaluable support. I am also grateful to Hugh Dickinson for numerous discussions and to Nickerson Biocem, Cambridge, UK, for financial support.

References

Atkinson, A.W., Gunning, B.E.S. & John, P.C.L. (1972). Sporopollenin in the cell wall of *Chlorella* and other algae: ultrastructure, chemistry, and incorporation of ^{14}C-acetate, studied in synchronous cultures. *Planta*, **107**, 1–32.

Bartley, G.N., Viitanen, P.V., Pecker, I., Chamovitz, D., Hirschberg, J. & Scolnik, P.A. (1991). Molecular cloning and expression in photosynthetic bacteria of a soybean cDNA coding for phytoene desaturase, an enzyme of the carotenoid biosynthesis pathway. *Proceedings of the National Academy of Sciences, USA,* **88**, 6532–6.

Bartley, G.N., Viitanen, P.V., Bacot, K.O. & Scolnik, P.A. (1992). A tomato gene expressed during fruit ripening encodes an enzyme of the carotenoid biosynthesis pathway. *Journal of Biological Chemistry*, **267**, 5036–9.

Beerhues, L., Forkmann, G., Schöpker, H., Stotz, G. & Wiermann, R. (1989). Flavonone 3-hydroxylase and dihydroflavonol oxygenase activities in anthers of *Tulipa*. The significance of the tapetum fraction in flavonoid metabolism. *Journal of Plant Physiology*, **133**, 743–6.

Bird, C.R., Ray, J.A., Fletcher, J.D., Boniwell, J.M., Bird, A.S., Teulieres, C., Blain, I., Bramley, P.M. & Schuch, W. (1991). Using antisense RNA to study gene function: inhibition of carotenoid biosynthesis in transgenic tomatoes. *Bio/Technology*, **9**, 635–9.

Blackmore, S. & Barnes, S.H. (1987). Embryophyte spore walls: origin, developmemt and homologies. *Cladistics*, **3, 185–95**.

Blackmore, S. & Barnes, S.H. (1990). Pollen wall development in angiosperms. In *Microspores: Evolution and Ontogeny*, ed. S. Blackmore & R.B. Knox, pp. 173–192. London: Academic Press.

Borg-Olivier, O. & Monties, B. (1993). Lignin, suberin, phenolic acids and tyramine in the suberized, wound-induced potato periderm. *Phytochemistry*, **32**, 601–6.

Brooks, J. & Shaw, G. (1968). Chemical structure of the exine of pollen walls and a new function for carotenoids in nature. *Nature*, **219**, 532–3.

Brooks, J. & Shaw, G. (1969). Evidence for extraterrestrial life: identity of sporopollenin with the insoluble organic matter present in the Orgueil and Murray meteorites and also in some terrestrial microfossils. *Nature*, **223**, 754–6.

Brown, R.C. & Lemmon, B.E. (1985). Spore wall development in the liverwort, *Haplomitrium hookeri*. *Canadian Journal of Botany*, **64**, 1174–82.

Brown, R.C. & Lemmon, B.E. (1990). Sporogenesis in bryophytes. In *Microspores: Evolution and Ontogeny,* ed. S. Blackmore & R.B. Knox, pp. 55–94. London: Academic Press.

Brown, R.C., Lemmon, B.E. & Renzaglia, K.S. (1986). Sporocytic control of spore wall pattern in liverworts. *American Journal of Botany*, **73**, 593–6.

Burczyk, J. & Hesse, M. (1981). The ultrastructure of the outer wall-layer of Chlorella mutants with and without sporopollenin. *Plant Systematics and Evolution*, **138**, 121–7.

Cecich, R.A. (1979). Development of vacuoles and lipid bodies in apical meristems of *Pinus banksiana*. *American Journal of Botany*, **66**, 895–901.

Chaloner, W.G. (1976). The evolution of adaptive features in fossil exines. In *The Evolutionary Significance of the Exine,* ed. I.K. Ferguson & J. Muller, pp. 11–14. London: Academic Press.

Chamovitz, D., Misawa, N., Sandmann, G. & Hirschberg, J. (1992). Molecular cloning and expression in *Escherichia coli* of a cyanobacterial gene coding for phytoene synthase, a carotenoid biosynthesis enzyme. *FEBS Letters*, **296**, 305–10.

Delwiche, C.F., Graham, L.E. & Thomson, N. (1989). Lignin-like compounds and sporopollenin in *Coleochaete*, an algal model for land plant ancestry. *Science*, **245**, 399–401.

Dickinson, H.G. (1970). Ultrastructural aspects of primexine formation in the microspore tetrad of *Lilium longiflorum*. *Cytobiology*, **1**, 437–49.

Dickinson, H.G. (1973). The role of plastids in the formation of pollen grain coatings. *Cytobios*, **8**, 25–40.

Dickinson, H.G. & Heslop-Harrison, J. (1968). Common mode of deposition for the sporopollenin of sexine and nexine. *Nature*, **220**, 926–7.

Dickinson, H.G. & Sheldon, J.M. (1986). The generation of patterning at the plasma membrane of the young microspore of *Lilium*. In *Pollen and Spores: Form and Function,* ed. S. Blackmore & I.K. Ferguson, pp. 1–17. New York: Academic Press.

Dover, G.A. (1972). The organisation and polarity of pollen mother cells of *Triticum aestivum*. *Journal of Cell Science,* **11,** 699–711.

Echlin, P. & Godwin, H. (1968). The ultrastucture and ontogeny of pollen in *Helleborus foetidus*. L. II. Pollen grain development through the callose special wall stage. *Journal of Cell Science,* **3,** 175–86.

Erdtman, G. (1960). The acetolysis method. *Svensk Botanisk Tidskrift,* **54,** 561–4.

Erdtman, G. (1969). *Handbook of Palynology*. Copenhagen: Hafner.

Esau, K. (1965). *Plant Anatomy*. New York: Wiley.

Evans, D.E., Taylor, P.E., Singh, M.B. & Knox, R.B. (1992). The interrelationship between the accumulation of lipids, protein and the level of acyl carrier protein during the development of *Brassica napus* L. pollen. *Planta,* **186,** 343–54.

Foster, G.D., Robinson, S.W., Blundell, R.P., Roberts, M.R., Hodge, R., Draper, J. & Scott, R.J. (1992). A *Brassica napus* mRNA encoding a protein homologous to phospholipid transfer proteins, is expressed specifically in the tapetum and developing microspores. *Plant Science,* **84,** 187–92.

Guilford, W.J., Schneider, D.M., Labovitz, J. & Opella, S.J. (1988). High resolution solid state ^{13}C NMR spectroscopy of sporopollenins from different taxa. *Plant Physiology,* **86,** 134–6.

Hauffe, K.D., Paszkowski, U., Schulze-Lefert, P., Hahlbrock, K., Dangl, J.L. & Douglas, C.J. (1991). A parsley 4CL-1 promoter fragment specifies complex expression patterns in transgenic tobacco. *The Plant Cell,* **3,** 435–43.

Herdt, E., Sütfeld, R. & Wiermann, R. (1978). The occurrence of enzymes involved in phenylpropanoid metabolism in the tapetum fraction of anthers. *European Journal of Cell Biology,* **17,** 433–41.

Herminghaus, S., Gubatz, S., Arendt, S. & Wiermann, R. (1988). The occurance of phenols as degradation products of natural sporopollenin – a comparison with 'synthetic sporopollenin'. *Zeitschrift für Naturforschung,* **43c,** 491–500.

Heslop-Harrison, J. (1962). Origin of exine. *Nature,* **195,** 1069–71.

Heslop-Harrison, J. (1963a). An ultrastructural study of pollen wall ontogeny in *Silene pendula*. *Grana Palynologica,* **4,** 1–24.

Heslop-Harrison, J. (1963b). Ultrastructural aspects of differentiation in sporogenous tissue. *Symposium of the Society of Experimental Biology,* **17,** 315–40.

Heslop-Harrison, J. (1968a). Tapetal origin of pollen-coat substances in *Lilium*. *The New Phytologist,* **67,** 779–86.

Heslop-Harrison, J. (1968b). Anther carotenoids and the synthesis of sporopollenin. *Nature,* **220,** 605.

Heslop-Harrison, J. (1968c). Pollen wall development. *Science,* **16,** 230–7.

Heslop-Harrison, J. (1968d). Wall development within the microspore tetrad of *Lilium longiflorum*. *Canadian Journal of Botany,* **46,** 1195–2.

Heslop-Harrison, J. (1971). The chemistry of sporopollenin. In *Sporopollenin,* ed. J. Brooks, P.R. Grant, M. Muir, P. van Gijzel & G. Shaw, pp. 1–30. London: Academic Press.

Heslop-Harrison, J. (1972). Pattern in plant cell walls: morphogenesis in miniature. *Proceedings of the Royal Institute of Great Britain,* **45,** 335–51.

Heslop-Harrison, J. (1976). The adaptive significance of the exine. In *The Evolutionary Significance of the Exine,* ed. I.K. Ferguson & J. Muller, pp. 27–37 London: Academic Press.

Higuchi, T. (1985). Biosynthesis of lignin. In *Biosynthesis and Biodegradation of Wood Components,* ed. T. Higuchi, pp, 141–60. New York: Academic Press.

Holloway, P.J. (1982). Structure and histochemistry of plant cuticular membranes: an overview. In *The Plant Cuticle,* ed. D.F. Cutler, K.L. Alvin & C.E. Price, pp. 1–32. London: Academic Press.

Holloway, P.J. & Wattendorf, J. (1987). Cutinised and suberised walls. *CRC Handbook of Plant Cytochemistry,* **11,** 1–35.

Horner, Jr. H.T., Lersten, N.R. & Bowen, C.C. (1966). Spore development in the liverwort *Riccardia pinguis. American Journal of Botany,* **53,** 1048–64.

Kehrel, B. & Wiermann, R. (1985). Immunochemical localisation of phenylalanine ammonia-lyase and chalcone synthase in anthers. *Planta,* **163,** 183–90.

Knight, M.E., Halpin, C. & Schuch, W. (1992). Identification and characterisation of cDNA clones encoding cinnamyl alcohol-dehydrogenase from tobacco. *Plant Molecular Biology,* **19,** 793–801.

Kolattukudy, P.E. (1980). Biopolyester membranes of plants: cutin and suberin. *Science,* **208,** 990–1000.

Kolattukudy, P.E. (1987). Lipid-derived defensive polymers and waxes and their role in plant–microbe interaction. In *The Biochemistry of Plants,* ed. P.K. Stumpf, pp. 291–314. New York: Academic Press.

Kolattukudy, P.E. & Espelie, K.E. (1985). Biosynthesis of cutin, suberin, and associated waxes. In *Biosynthesis and Biodegradation of Wood Components,* ed. T. Higuchi, pp. 161–207. New York: Academic Press.

Kolattukudy, P.E. & Köller, W. (1983). Fungal penetration of the first line defensive barriers of plants. In *Biochemical Plant Pathology,* ed. J.A. Callow, pp. 79–100. New York: Wiley.

Kurmann, M.H. (1990). Exine ontogeny in conifers. In *Microspores: Evolution and Ontogeny,* ed. S. Blackmore & R.B. Knox, pp. 157–172. London: Academic Press.

Koltunow, A.M., Truettner, J., Cox, K.H., Wallroth, M. & Goldberg, R.B. (1990). Different temporal and spatial gene expression patterns occur during anther development. *The Plant Cell,* **2,** 1201–24.

Lewis, N.G. & Yamamoto, E. (1990). Lignin: occurrence, biogenesis and biodegradation. *Annual Review of Plant Physiology and Plant Molecular Biology,* **41**, 455–96.

Linskens, H.F. (1966). Die Änderung des Protein- und Enzym-musters während der Pollenmeiose und Pollenentwicklung. *Planta,* **69**, 79–91.

Lugardon, B. (1990). Pteridophyte sporogenesis: a survey of spore wall ontogeny and fine structure in a polyphyletic plant group. In *Microspores: Evolution and Ontogeny,* ed. S. Blackmore & R.B. Knox, pp. 95–120. London: Academic Press.

Ohl, S., Hedrick, S.A., Chory, J. & Lamb, C.J. (1990). Functional properties of a phenylalanine ammonia-lyase promoter from *Arabidopsis. The Plant Cell,* **2**, 837–48.

Ohlrogge, J.B. (1987). Biochemistry of plant acyl carrier proteins. In *The Biochemistry of Plants, Vol. 9. Lipids: Structure and Function.* ed. P.K. Stumpf, pp. 137–157. London: Academic Press.

Plummer, C.C. & McGeary, D. (1991). *Physical Geology.* Dubuque, IA, USA: Wm.C. Brown.

Paul, W., Hodge, R., Smartt, S., Draper, J. & Scott, R.J. (1992). Isolation and characterisation of the tapetum–specific *Arabidopsis thaliana* A9 gene. *Plant Molecular Biology,* **19**, 611–22.

Prahl, A.K., Springstubbe, H., Grumbach, K. & Wiermann, R. (1985). Studies on sporopollenin biosynthesis: the effect of inhibitors of carotenoid biosynthesis on sporopollenin accumulation. *Zeitschrift für Naturforschung,* **40c**, 621–26.

Reinhold, S., Hauffe, K.D. & Douglas, C.J. (1993). Tobacco and parsley 4-coumarate:coenzyme A ligase genes are temporally and spatially regulated in a cell type-specific manner during tobacco flower development. *Plant Physiology,* **101**, 373–83.

Rittscher, M. & Wiermann, R. (1988). Studies on sporopollenin biosynthesis in *Tulipa* anthers II. Incorporation of precursors and degradation of the radiolabelled polymer. *Sexual Plant Reproduction,* **1**, 132–9.

Roberts, M.R., Hodge, R., Sorenson, A.-M., Ross, J., Murphy, D.J., Draper, J. & Scott, R. (1993). Characterisation of a new class of oleosins suggests a male gametophyte-specific lipid storage pathway. *The Plant Journal,* **3**, 629–36.

Rogers, C.M. & Harris, B.D. (1969). Pollen exine deposition, a clue to its control. *American Journal of Botany,* **56**, 101–6.

Rowley, J.R. (1962). Stranded arrangement of sporopollenin in the exine of microspores of *Poa annua.* Science, **137**, 526–8.

Rowley, J.R. & Southworth, D. (1967). Deposition of sporopollenin on lamellae of unit membrane dimensions. *Nature,* **213**, 703–4.

Scott, R., Hodge, R., Paul, W. & Draper, J. (1991). The molecular biology of anther differentiation. *Plant Science,* **80**, 167–91.

Schmidt, H.W. & Schönherr, J. (1982). Fine structure of isolated and non-isolated potato periderm. *Planta,* **154**, 76–80.

Schulze Osthoff, K. & Wiermann. (1987). Phenols as integrated compounds of sporopollenin form *Pinus* pollen. *Journal of Plant Physiology*, **131**, 5–15.

Sheldon, J.M. & Dickinson, H.G. (1983). Determination of patterning in the pollen wall of *Lilium henryi*. *Journal of Cell Science*, **63**, 191–208.

Sheldon, J.M. & Dickinson, H.G. (1986). Pollen wall formation in *Lilium:* the effect of chaotropic agents, and the organisation of the microtubular cytoskeleton during pattern development. *Planta*, **168**, 11–23.

Shaw, G. (1971). The chemistry of sporopollenin. In *Sporopollenin*, ed. J. Brooks, P.R. Grant, M. Muir, P. van Gijzel & G. Shaw, pp. 305–348. London: Academic Press.

Shaw, G. & Yeadon, A. (1966). Chemical studies on the constitution of some pollen and spore membranes. *Journal of the Chemical Society*, **C**, 16–22.

Sterk, P., Booij, H., Schellekens, G.A., Kammen, A.Van. & De Vries, S.C. (1991). Cell-specific expression of the carrot EP2 lipid transfer protein gene. *The Plant Cell*, **3**, 907–21.

Takahashi, M. & Skvarla, J.J. (1991). Exine pattern formation by plasma membrane in *Bougainvillea spectabilis* Willd. (Nyctaginaceae). *American Journal of Botany*, **78**, 1063–69.

Taylor, T.N. & Zavada, M.S. (1986). Developmental and functional aspects of fossil pollen. In *Pollen and Spores: Form and Function*, ed. S. Blackmore & I.K. Ferguson, pp. 165–178. New York: Academic Press.

Terras, F.R.G., Torrekens, S., Van Leuven, F., Osborn, R.W., Vanderleyden, J., Cammue, B.P.A. & Broekaert, W.F. (1993). A new family of basic cysteine-rich plant antifungal proteins from Brassicaceae species. *FEBS Letters*, **3**, 233–40.

Thoma, S., Kaneko, Y. & Somerville, C. (1993). A nonspecific lipid transfer protein from *Arabidopsis* is a cell-wall protein. *The Plant Journal*, **3**, 427–36.

van der Meer, I.M., Stam, M.E., van Tunen, A.J., Mol, J.N.M. & Stuitje, A.R. (1992). Antisense inhibition of flavonoid biosynthesis in petunia anthers results in male sterility. *The Plant Cell*, **4**, 253–62.

de Vlaming, P. & Kho, K.F.F. (1975). 2,2′,4′,6′-tetrahydroxychalcone in pollen of *Petunia hybrida*. *Phytochemistry*, **15**, 428–49.

Walker, J.W. (1976). Evolutionary significance of the exine in the pollen of primitive angiosperms. In *The Evolutionary Significance of the Exine*, ed. I.K. Ferguson & J. Muller, pp. 251–308. London: Academic Press.

Walker, J.W. & Skvarla, J.J. (1975). Primitively columellaless pollen: a new concept in the evolutionary morphology of angiosperms. *Science*, **187**, 445–7.

Wall, D. (1962). Evidence from recent plankton regarding the biological affinities of *Tasmanites* Newton, 1875 and *Leiosphaeridia*, Eisenack, 1958. *Geological Magazine*, **99**, 353–62.

Waterkeyn, L. & Beinfait, A. (1970). On a possible function of the callosic special wall in *Ipomoea purpurea* (L) Roth. *Grana*, **10**, 13–20.

Wattendorf, J. & Holloway, P.J. (1980). Studies on the ultrastructure and histochemistry of plant cuticles: the cuticular membranes of *Agave americana L. in situ*. *Annals of Botany*, **46**, 13–28.

Wehling, K., Niester, Ch., Boon, J.J., Willemse, M.T.M. & Wiermann, R. (1989). *p*-Coumaric acid – a monomer in the sporopollenin skeleton. *Planta*, **179**, 376–80.

Wodehouse, R.P. (1935). *Pollen Grains. Their Structure, Identification and Significance in Science and Medicine*. New York: McGraw-Hill.

Worrall, D., Hird, D.L., Hodge, R., Paul, W., Draper, J. & Scott, R. (1992). Premature dissolution of the microsporocyte callose wall causes male sterility in transgenic tobacco. *The Plant Cell*, **4**, 759–71.

D. TWELL

The diversity and regulation of gene expression in the pathway of male gametophyte development

Summary

The highly specialised angiosperm male gametophyte is both the site of production and the vehicle by which the male gametes are transported to the embryo sac to participate in fertilisation. Because of its relative simplicity compared with the sporophyte and accessibility for cytological and molecular analysis, the male gametophyte represents an excellent system in which to unravel the molecular basis of gene regulation and cellular differentiation in plants. The intent of this chapter is to review evidence for haploid gene expression in the developing male gametophyte, emphasising in particular the contribution of molecular cloning and transgenic approaches. cDNA cloning and RNA analysis has led to the characterisation of more than 30 microspore or pollen-expressed genes from 12 different plant species which are preferentially activated at particular stages of development. Two broadly defined groups of pollen genes are recognised; the 'early genes', which are activated prior to pollen mitosis I (PMI), and the 'late genes' which are first activated after PMI. Promoter studies have demonstrated differential gene expression between the vegetative and generative cells and insight into the significance of PMI for late gene activation. Detailed analyses of the promoters of several late pollen genes have led to the identification of several distinct cis-regulatory elements which control both the level and specificity of haploid gene expression. Some of these regulatory elements are functionally conserved among homologous promoters from diverse plant species and among different promoters within a single species. Recently, a gene encoding a pollen-specific DNA binding protein has been cloned, which may represent a transcriptional regulator of late gene transcription. Further analysis indicates that, in addition to transcriptional controls, certain members of the late pollen gene group are regulated

Society for Experimental Biology Seminar Series 55: *Molecular and Cellular Aspects of Plant Reproduction*, ed. R.J. Scott & A.D. Stead.
© Cambridge University Press, 1994, pp. 83–135.

post-transcriptionally by a mechanism involving their 5' untranslated leader sequences. The challenge for the future is to identify, with greater precision, the networks of structural and regulatory genes required for pollen development, and the development signals which interact with such regulatory networks.

Introduction

Sexual reproduction is strictly dependent on the successful production and union of the haploid male and female gametes. In angiosperm plants, the male gametes are produced and delivered to the female gametes by a discrete organism the male gametophyte or pollen grain. Unlike the free living multicellular and photosynthetic gametophytes of lower plants which produce motile male gametes, the angiosperm male gametophyte develops entirely within the sporophytic (anther loculus) tissues and, at maturity, consists of only three cells, one vegetative cell containing within it two non-motile sperm cells. Thus, although the primary function of the gametophyte, the production of the gametes, has been conserved during evolution, in higher plants the male gametophyte has become increasingly diminutive and dependent upon the sporophyte for its development and function.

Despite the apparent simplicity of angiosperm male gametophyte development and function when compared with the sporophyte, in common with other morphogenetic processes involving cellular differentiation, there is now overwhelming evidence that pollen development requires the strict control of gene expression from the haploid genome. The intent of this article is to review both early and more recent evidence for such haploid gene expression, and our current knowledge of the diversity of genes which constitute the haploid genetic programme. The relationship between this gene expression programme and the commitment to follow the gametophytic pathway is emphasised, together with likely areas of progress and approaches which are expected to be fruitful in elucidating the underlying mechanisms involved in pollen differentiation. Several reviews of gene expression and function in developing anthers, including sporophytic tissues, have been published in recent years, which reflect the rapid progress which is being made in this area (Bedinger, 1992; Ottaviano and Mulcahy, 1989; Mascarenhas, 1990, 1992; McCormick, 1991; Scott *et al.*, 1991*b*). Here I have attempted to synthesise the relevant aspects of these discussions, and review in detail our current understanding of the repertoire, developmental regulation and promoter structure of genes expressed within developing microspores and pollen grains.

The pathway of male gametophyte development

The completion of meiosis of the diploid microsporocytes (or pollen mother cells), within the anther loculus marks the initiation of a unique process of cellular differentiation which ultimately leads to the formation of a large number of male gametophytes which are shed from the plant as pollen grains. This series of events represents the haploid phase of development, and in the current context is collectively referred to as the pathway of male gametophyte development. In the majority (70%) of angiosperms, this pathway involves only a single mitotic cell division, pollen mitosis I (PMI), which serves as a developmental marker for the termination of microspore development and the initiation of pollen development. After pollen germination the generative cell undergoes a further mitotic division during pollen tube growth to form the two sperm cells. In the remaining 30% of angiosperm species, generative cell division occurs during pollen maturation prior to pollen shed. The phylogenetic distribution of species which shed pollen in a bicellular or tricellular condition demonstrates that tricellularity is a derived character which has evolved independently on numerous occasions (Brewbaker, 1967).

The two pathways of pollen development are illustrated in Fig. 1, for the model species *Nicotiana tabacum* (bicellular) and *Arabidopsis thaliana* (tricellular). It is interesting to note that, despite considerable divergence in floral morphology and pollen size, microspore and pollen development occur over similar time periods, and major cytological processes, such as vacuolation and the accumulation of metabolic reserves, are similar between the two pollen types (Schrauwen *et al.*, 1990; Regan & Moffat, 1990; Eady, Lindsey & Twell, 1994). Since the pathway of pollen development appears to be well conserved among angiosperms, it may be expected that many basic developmental processes, and the genes controlling them, are also shared.

Despite the absence of symplastic connections between the sporophytic tissues of the anther and the developing microspores there is considerable evidence that interactions between the sporophytic and gametophytic generations are fundamentally important for the completion of pollen development. For example, the tapetum actively secretes and controls the deposition of proteins, lipids, carbohydrates and secondary metabolites, which are required for membrane synthesis, energy requirements and the elaboration of the unique pollen wall (for review see Pacini, 1990). The tapetum and anther loculus thus provide essential metabolites and a suitable environment in which differentiation of the microspores can occur. However, at later stages of both microspore and pollen development apparently normal differentiation can be completed in *in vitro*

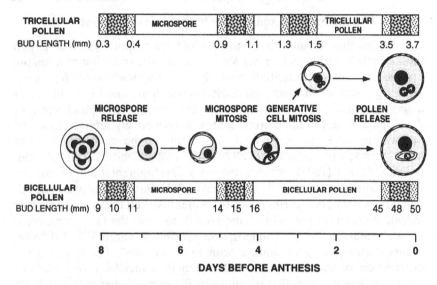

Fig. 1. Schematic diagram of the two pathways of pollen development in the model plant species *Nicotiana tabacum* (bicellular) and *Arabidopsis thaliana* (tricellular) shown in relation to flower bud length and time before anthesis. Stippling along bars represents the range over which particular developmental stages are observed.

culture if essential nutrients are provided (Tanaka & Ito, 1980, 1981; Benito Moreno *et al.*, 1988; Tupy, Rihova & Zarsky, 1991). Thus, the male gametophyte from its inception at PMI, and perhaps to a lesser extent the haploid microspore, is to a large extent in control of, and capable of, directing its own genetic programme to achieve maturity. This degree of autonomy presumably reflects the need to establish an independent genetic programme required for the highly specialised development, structure and function of the pollen grain, and in an evolutionary context reflects the independent genetic programme of phylogenetically less advanced gametophytes of lower plants.

Pollen mitosis I: a key developmental switch during pollen formation

One of the most striking features of pollen development is the highly asymmetric mitotic division of the microspore, pollen mitosis I (PMI), which results in the formation of two unequal cells, the vegetative and generative cells, which possess dramatically different structures and developmental fates. PMI by definition signifies the initiation of pollen development as opposed to microspore development and is a prerequisite for the correct maturation of the male gametophyte. Experimental evidence which supports the significant role of the asymmetry of this division has come from studies in which division asymmetry has been disrupted by the application of microtubule inhibitors (Tanaka & Ito, 1981; Zaki & Dickinson, 1990, 1991), or by physical means such as centrifugation (Terasaka & Niitsu, 1987). When division asymmetry is disrupted, two apparently identical cells are produced. These cells do not possess the nuclear characteristics of the generative cell, and superficially resemble vegetative cells. These results imply that division asymmetry is essential for the different behaviour of the generative nucleus after PMI. In particular, because the GC nucleus remains highly condensed following PMI, compared with the VC nucleus, this is evidence for a different nuclear architecture between the two cells. The extent to which these apparently equal cells possess characteristics of the vegetative cell is not known in any detail. This is one area where molecular markers such as gene-specific probes or promoter sequences may find application. For example, such probes may be used to determine whether vegetative cell-specific genes are activated in response to symmetrical division. Such detailed analysis and the identification of GC-specific markers are particularly important if we are to understand how such differential gene expression and cellular differentiation is maintained both during pollen maturation and during pollen tube growth in the style.

Evidence for haploid gene expression – the first 60 years

The most direct and conclusive evidence for post-meiotic transcription in the developing male gametophyte has come from molecular studies of the expression of specific genes and their associated regulatory sequences, which has taken place only within the past decade. Prior to such studies there was considerable interest in the genetic activity of the male gametophyte, and experimental evidence for haploid gene expression was largely derived from genetic studies, including observed deviations from mendelian segregation ratios and the identification of rare gametophytic mutants (for review see Ottaviano & Mulcahy, 1989). Subsequently new approaches including isozyme and RNA analysis were employed, such that prior to the earliest examples of the molecular cloning and analysis of genes expressed in the male gametophyte (Hamilton *et al*; 1989; Twell *et al.*, 1989a), there was considerable firm evidence for haploid transcription and translation. It is useful to examine this

evidence as several important concepts concerning the genetic programme of the male gametophyte were established on the strength of
this evidence.

Genetic evidence

Pollen mutants

The genetic analysis of gametophytic mutants provided the first evidence
for the post-meiotic function of genes in the male gametophyte. Parnell
(1921) identified phenotypic segregation following iodine staining in
pollen of rice plants heterozygous for the *glutinous* character of the endosperm. Similar observations were made by Brink and MacGillivray (1924)
for the corresponding *waxy* character of maize pollen and endosperm,
who further observed reduced pollen transmission of both *waxy* and *glutinous* alleles, and hypothesised, in referring to the growth of the pollen
tube, that 'this presumably is controlled by the action of certain factors
active in the tube nucleus'. However, is was not until 1940 that the first
gametophytically acting pollen lethal gene was described in *Zea mays*
(Singleton & Mangelsdorf, 1940). This gene, termed *small pollen*-1 (*sp*),
resulted in a phenotype in which two pollen sizes, large and small, were
produced in approximately equal numbers in plants heterozygous for *sp*.
The common observation of the lack of pollen transmission, and the
associated segregation for a similar *small pollen* phenotype in many chromosomal deficiencies in maize, indicates that important genes required
for male gametophyte development are both numerous and dispersed
throughout the genome (Stadler, 1933; Stadler & Roman, 1948). More
recently, the identification of the segregation of other biochemical deficiencies such as for alcohol dehydrogenase activity in maize pollen
(Freeling, 1976; Stinson & Mascarenhas, 1985) and β-galactosidase activity in *Brassica campestris* pollen (Singh, O'Neil & Knox, 1985) by histochemical methods has provided more direct evidence of gametophytic transcription and translation.

Distorted Mendelian segregation ratios which specifically affect the
transmission of particular alleles through pollen have been detected
repeatedly, and include examples where such genes affect pollen development, pollen germination or tube growth (for review see Ottaviano &
Mulcahy, 1989). Early examples were noted in *Melandrium*, *Datura* and
Hordeum (Jones, 1928). In *Medicago falcata* and lima beans, preferential
fertilisation was shown to be the consequence of unequal pollen tube
growth (Bemis, 1959). A pollen killing factor (*ki*) which caused microspore abortion was detected in wheat (Loegering & Sears, 1963), while
a distinct pollen killer factor was detected in *Nicotiana tabacum*, which

was carried by an alien chromosome of *N. plumbaginifolia* (Cameron & Moav, 1957). This factor, in contrast, caused abortion in pollen not carrying the alien chromosome. Deleterious alleles of genes which presumably affect haploid gene expression also include chlorophyll mutants in barley which showed distorted segregation ratios and reduced viability of microspores carrying mutants alleles (Doll, 1967). Other studies have shown reduced pollen transmission due to slower germination of pollen carrying the recessive alleles *waxy* (Sprague, 1933) and *opaque*-2 in maize (Sari Gorla & Rovida, 1980). Also in maize nine different gametophytic factors have been described which confer high competitive ability to pollen tubes carrying mutant Ga alleles (Bianchi & Lorenzoni, 1975). In contrast, a class of gametophytic mutants in maize, defective endosperm (de), negatively affect pollen tube growth and block endosperm development (Ottaviano, Petroni & Pe, 1987). A similar gametophytic/sporophytic overlap was identified for two embryo lethal mutants in *Arabidopsis* which show reduced gametophytic transmission (Meinke, 1982; Meinke & Baus, 1986). Further examples of gametophytic factors which influence pollen function are provided by gametophytic incompatibility systems which operate in many taxa and act at different stages during the progamic phase until fertilisation (Singh & Kao, 1992). The above examples illustrate that mutant screening and analysis may be a particularly valuable approach to identify genes which influence both the development and function of pollen in novel ways. Systematic screening for such gametophytic factors in model species such as *Arabidopsis* which are amenable to molecular genetic analysis is expected to gain importance in future studies.

Isozyme analysis

The first conclusive demonstration of haploid gene expression (transcription and translation) in the developing male gametophyte was obtained from the genetic analysis of dimeric enzymes such as alcohol dehydrogenase (ADH; Schwartz, 1971). Electrophoretically distinct variants of ADH encoded by different alleles of the *Adh1* locus (F and S) associate in sporophytic tissues to form FF and SS homodimers and F/S heterodimers in *Adh1*-F/*Adh1*-S heterozygotes. In contrast, only FF and SS homodimers are found in pollen extracts of heterozygotes as a result of their post-meiotic expression. Micro-electrophoretic analysis of heterodimeric isozymes such as phosphoglucoisomerase in total pollen extracts and sporophytic tissues of *Clarkia* extended the biochemical evidence for haploid gene expression (Weeden & Gottlieb, 1979). Mulcahy *et al.* (1979) further refined the technique of microelectrophoresis to single pollen grains of hybrid material, and showed segregation for specific protein bands. This technique was subsequently employed to demonstrate the segregation of isozymes of acid phosphatase in F1 hybrids of *Curcurbita* (Mulcahy *et al.*, 1981).

Molecular evidence

RNA synthesis

Prior to the isolation of pollen-expressed genes by molecular cloning, several studies involving *in vivo* RNA labelling and polysome profile analysis, demonstrated the synthesis of major RNA species in the developing male gametophyte. Early studies in *Tradescantia* and *Lilium* showed that different pollen species behaved similarly, with large amounts of rRNA synthesised prior to PMI, followed by a sharp decline before pollen shed (for review see Mascarenhas, 1975). In contrast, the major increase in ribosome and total RNA levels in tobacco pollen have been shown to occur after PMI (Tupy *et al.*, 1983). These data conclusively demonstrate haploid transcription of genes encoding major RNA species, but further suggest that pollen of different species have markedly different genetic programmes. Such differences presumably reflect the specialised requirement for different pollen types to develop and function in particular anther and stylar environments and are likely to contribute significantly to the observed bilateral sexual incompatibility among widely separated taxa.

Using improved extraction and fractionation methods, the synthesis and accumulation of both total and poly A+RNA were determined quantitatively during pollen development of tobacco (Schrauwen *et al.*, 1990). These studies showed a rapid increase in the synthesis of both total and mRNA soon after PMI, which was directly correlated with an increase in the synthesis and accumulation of total cell protein. During final maturation, the rate of protein accumulation in the pollen grain declines which may be associated with the dehydration phase (Zarsky *et al.*, 1985). Thus, the developmental marker, PMI, is correlated with the onset of a dramatically increased level of gene expression in the newly formed pollen grain. Detailed analysis of *in vitro* translation profiles of mRNA purified from isolated microspore and pollen populations by two-dimensional gel electrophoresis in tobacco (Schrauwen *et al.*, 1990) and maize (Bedlinger & Edgerton, 1990) supports this conclusion, but further demonstrates a major shift in the translatable mRNA population following PMI. Similarly 2D electrophoresis of newly synthesised proteins in staged isolated microspores and pollen in maize showed a dramatic decrease in the synthesis of a number of polypeptides in mid- to late bicellular pollen relative to uninucleate microspores, and the synthesis of a number of new ones, in particular very basic polypeptides (Mandaron *et al.*, 1990). Studies by Vergné and Dumas (1988) and Delvallée and Dumas (1988) further demonstrate a new set of proteins synthesised at the time of generative cell mitosis (pollen mitosis II; PMII). Taken together, these studies indicate that the pathway of pollen development may be considered most simply to comprise two distinct developmental phases, microspore and pollen development. However, because of the unique processes which occur during each of these phases, such as vacuolation, intine wall synthesis, starch synthesis and preparation for dehydration/germination, microspore and pollen development appear to demand the expression of different sets of combinations of genes from the haploid genome at discrete stages of differentiation. With the advent of molecular cloning and detection technology, it has become possible to investigate in some detail the division of labour between these two phases in terms of gene expression, and how such gene expression relates to distinct differentiation events.

Number of genes

Quantitative analysis of the numbers of different genes expressed in the male gametophyte provide a perspective of the complexity of gene expression and synthetic capacity associated with the haploid genome. Estimates based upon the kinetics of hybridisation of ^{3}H-cDNA with poly(A) RNA in *Tradescantia* and maize (Willing & Mascarenhas, 1984; Willing, Bashe & Mascarenhas, 1988), indicate that some 20 000 to 24 000 unique mRNAs are present in the pollen grain at maturity. This represents a significant reduction in complexity compared with the sporophyte, which synthesises approximately 30 000 unique mRNAs, and most likely reflects the simple structure and restricted activity of the gametophyte. Analysis of the abundance classes within pollen and shoot RNAs showed that pollen mRNAs in all classes are approximately 100-fold more abundant than their corresponding classes in shoot mRNA (Willing & Mascarenhas, 1984; Willing *et al.*, 1988). Such high levels of mRNA in the mature pollen grain may result from a need to accumulate and store mRNAs prior to their rapid translation during pollen germination. Indeed, the pattern of proteins synthesised during pollen germination *in vivo* is almost identical with that observed from *in vitro* translation of mRNA isolated from mature pollen (Mascarenhas *et al.*, 1984), and translation is an absolute requirement for pollen germination and/or early tube growth (for review see Mascarenhas, 1975). Taken together, these studies show that large numbers of genes are actively transcribed from the haploid genome of the male gametophyte late during development and many of their corresponding transcripts are stored and translated during pollen germination.

Similar quantitative analyses of microspore mRNA populations have not been carried out such that information on the diversity of gene expression in the microspore is rather limited. The available data are derived from two dimensional gel electrophoretic analysis of *in vitro* translation products synthesised from microspore mRNA in maize (Bedlinger & Edgerton, 1990) and tobacco (Schrauwen *et al.*, 1990), from *in vivo* labelling of newly synthesised proteins in maize microspores (Mandaron *et al.*, 1990), and from the molecular cloning and promoter analysis of only relatively few microspore expressed genes (see discussion below). Clearly, more work is needed in this area to complement the significant progress which has been made in understanding the repertoire of genes expressed in the maturing pollen grain. The identification and characterisation of such microspore-expressed or microspore-specific genes is important since they may be involved with the determinative influence of the microspore cytoplasm on PMI and the subsequent commitment to the gametophytic pathway (Raghavan, 1986; Bedlinger, 1992; Dickinson, 1993).

Molecular cloning of microspore and pollen-expressed genes

Progress in the molecular cloning of pollen-expressed genes has come from many laboratories with diverse interests; those interested in gene regulation and function, those interested in the manipulation breeding systems and the improvement of pollen characteristics, and those interested in the molecular structure of pollen allergens. Most laboratories have taken the direct and effective strategy of the construction of cDNA libraries from mRNA isolated from mature pollen, anthers or whole buds at different developmental stages. Where the intention was to clone abundant mRNAs specifically expressed in the developing male gametophyte, standard differential screening techniques were used. An alternative approach, termed 'cold plaque' screening, has led to the cloning of mRNAs present at low abundance, and involves picking non-hybridising plaques at random, after probing a cDNA library with labelled cDNA from the source tissue (Hodge *et al.*, 1991).

The differential screening approach has been particularly successful in identifying mRNAs which are specifically, or preferentially, expressed in the male gametophyte and has led to the cloning of more than 30 unique mRNAs from 12 different species (Table 1). Sequence analysis has led to the identification of sequence similarities with known genes and has provided important clues as to their functions. For example, clones encoding proteins with similarity to pectin degrading enzymes, including polygalacturonase (Brown & Crouch, 1990) and pectate lyase (Wing *et al.*, 1989), are abundantly expressed in the mature pollen grain of several species. Such enzymes are implicated in the control of pollen germination and/or pollen tube growth through the pistil (Wing *et al.*, 1989). Similarly, the related *lat*52 (Twell *et al.*, 1989*a*) and *Zm*13 (Hanson *et al.*, 1989) gene products, from tomato and maize respectively, show similarities to the kunitz trypsin inhibitor family (McCormick, 1991) and may be involved in providing resistance to insect pollen predation. Where antibodies to specific proteins were available, cDNA clones encoding the corresponding antigenic proteins have been cloned by immunoscreening of cDNA expression libraries. This approach, often involving screening libraries with serum from patients with specific allergen sensitivities, has led to the cloning of a rapidly increasing number of mRNAs encoding important pollen allergens (Table 1). Although detailed discussion of the functional analysis of cloned pollen-expressed genes is outside the scope of this review, with further effort these materials may be exploited to provide insight into a variety of important cellular processes which take place during pollen development and function such as cell wall synthesis, protein secretion and the nature of pollen–pistil interactions.

Developmental regulation of haploid gene expression

Much emphasis is often placed on the degree of specificity determined for many pollen-expressed genes, in particular because the expression of genes with absolute specificity strongly implies an exclusive role for the encoded gene product in male gametophyte development. The mechanisms which so tightly control pollen-specific gene expression are also of considerable interest. However, because many overlapping basic cellular functions exist between sporophytic cells and gametophytic cells, genes which encode important shared functions are often expressed in both generations. Such genes may therefore be independently regulated in each generation, and it is also of considerable interest to establish whether different developmental signals regulate their expression in different cell types. Thus, to obtain a more balanced view of the gametophytic gene expression programme, it is important to consider the precise patterns of accumulation and decay of both microspore and pollen-specific mRNAs (and their protein products) and those which exhibit gametophytic/sporophytic overlap. Furthermore, even where the function of isolated genes is unknown, which is often the case, detailed

Table 1. Cloned mRNAs expressed In developing microspores and/or pollen

Clone	Species	Function/Sequence/Similarity	Expression	References
AmbaI.1 (+2)	Ambrosia artemisiifolia	allergen (Amb a I), pectate lyases LAT56, LAT59, TP10, Zm58	P	Rafnar et al., 1991
AmbtV	Ambrosia trifida	allergen (Amb t V)	P	Ghosh et al., 1991
TUAI	Arabidopsis thaliana	α-tubulin	LP	Carpenter et al., 1992
BetvI	Betula verrucosa	allergen (Bet v I), bean elicitor induced proteins (PvPR1/2), potato/parsley PR proteins (STH2 and PR1-1), pea disease response gene (149), Asparagus wound induced gene (AoPRI)	P	Breiteneder et al., 1989 / Walter et al., 1990
BetvII		allergen (Bet v II), profilin, BMP1, LMP131A	P	Valenta et al., 1991
Bcp1 (−9)	Brassica campestris	Bgp1	Lp[a]	Theerakulpisut et al., 1991
Bgp1		Bcp1	P	Xu et al., 1993
Bp10	Brassica napus	ascorbate oxidase, Nt303, LAT51	LP	Albani et al., 1990
Bp19		pectin methyl esterase	EM-TP	Albani et al., 1991
I3		oleosin	EM-TP*	Roberts et al., 1991
E2		phospholipid transfer protein	EM-TP*	Foster et al., 1992
BA42		chalcone synthase	MS	Shen & Hsu, 1992
BMP1		profilin, LMP131A	P	Kim et al., 1993
CEX1/6		apg	EM-TP*	Roberts et al., 1993a
Bp4 A (+2)		—	EM-TP	Albani et al., 1990
BA54, BA73		—	EM-TP*	Shen & Hsu, 1992
Cor aI5 (+3)	Corylus avellana	allergen (Cor a I), BetvI, CarbI, AlngI	P	Breiteneder et al., 1993
7.8 (+2)	Dactylis glomerata	allergen	P	Walsh et al., 1989
SF3	Helianthus annus	LIM domain/zinc finger protein Eryf and GF-1 transcription factors	LP	Baltz et al., 1992 a,b
LMP131A	Easter lily	profilin, BMPI	P	Kim et al., 1993
1A,5A	Lolium perenne	allergen (Lol p Ia), Lol p II, Lol p III	P	Perez et al., 1990
13R		allergen (Lol p Ia), Lol p II, Lol p III	P	Griffith et al., 1991
12R		allergen (Lol p Ib), PhI pV	P	Singh et al., 1991
LAT51	Lycopersicon esculentum	ascorbate oxidase, Bp10, Nt303	LP	McCormick, 1991
LAT52		kunitz trypsin inhibitor, Zm13	LP	Twell et al., 1989a
LAT56		pectate lyases, LAT59, TP10, Zm58	LP	Wing et al., 1989
		Ambrosia allergens		

Clone	Species	Gene product / homologous clones	Stage	Reference
LAT59		pectate lyases, LAT56, TP10, Zm58, Ambrosia allergens	LP	Wing et al., 1989
LAT58	Nicotiana tabacum	—	LP	Ursin et al., 1989
Nt303		ascorbate oxidase, Bp10,	LP	Weterings et al., 1992
Tac25		actin	P	Thangavelu et al., 1993
TP10		pectate lyase, LAT56, LAT59, Zm58	LP	Rogers et al., 1992
Npg1		polygalacturonase, P1, 3C12	LP	Rogers & Lonsdale, 1992; Tebbut & Lonsdale, unpub.
P1 (+5)	Oenothera organensis	polygalacturonases, Npg1, 3C12	LP	Brown & Crouch, 1990
P3		—	LP	Brown & Crouch, 1990
P6 (late)		—	LP[b]	Brown & Crouch, 1990
Act1	Oryza sativa	actin	LP	Tsuchiya et al., 1992
Histone H3		histone H3	LP	Raghavan, 1989
Osc4, Osc6		—	MS-LP	Tsuchiya et al., 1992
Adh-1	Petunia hybrida	alcohol dehydrogenase	MS-LP	Gregerson et al., 1991
chiA		chalcone flavanone isomerase	LP	van Tunen et al., 1989
KBG7.2 (+3)	Poa pratensis	allergen (Poa p IX), Lol p II, Lol p III, AmbaV, PhlpV	P	Silvanovich et al., 1991; Olsen et al., 1991
Wx1.3	Solanum tuberosum*	granule bound starch synthase	T/EM-LP	Olmedilla et al., 1991
Tpc44 (+2)	Tradescantia paludosa	—	LP	Stinson et al., 1987
Adh-l	Zea mays	alcohol dehydrogenase	T-TP	Stinson & Mascarenhas, 1985
Zmc13		kunitz trypsin inhibitor, LAT52	LP	Stinson et al., 1987
3C12 (+2), W2247 (+3)		polygalacturonase, P1, Npg1	LP	Allen & Lonsdale, 1992
*PGc1, PGg6 (+7)		polygalacturonase, P1, 3C12, Npg1	LP	Niogret et al., 1991
Zm58		pectate lyase, LAT56, LAT59, TP10, Amb a I allergens	LP	Barakate et al., 1993; Mascarenhas, 1990
Zmc26 (+2)		—	LP	Stinson et al., 1987

All examples listed above represent clones which have been demonstrated to be expressed in microspores or pollen by direct RNA analysis or by cDNA cloning directly from pollen-derived cDNA libraries. Haploid expression of corresponding transcripts is listed according to stage at which they were first detected or the range of their presence where available: T/EM = tetrad/early microspore; MS = microspore; LP = transcripts first detectable after PMI accumulating until maturity; TP = transcripts decline before PMII (tricellular pollen); P = transcripts detected in mature pollen but no developmental analysis performed; * = examples where RNA analysis used whole buds or anthers and not isolated spores. Lp* = Bcp1 homologous transcripts detectable at low levels in microspores by in situ hybridisation; LP[b] = P6 transcripts were detectable significantly later than other Oenothera late pollen transcripts.

analysis of their patterns of expression can provide considerable insight into critical regulatory points in the pathway of pollen development.

Methodology

Two well-established approaches have been used to determine the developmental regulation of specific genes transcribed from the haploid genome. In the first approach, gene-specific probes have been used to determine mRNA levels in total or mRNA purified from anthers or isolated spores by northern blot analysis. This approach is important but is hampered by the difficulty in the isolation of RNA from precisely staged populations of spores, such that whole anthers or flower buds have often been used as a source of RNA. This has limited both sensitivity and the ability to make precise statements concerning haploid gene expression. The complementary approach of *in situ* hybridisation to tissue sections provides further essential information about cell specificity, but is often not particularly quantitative.

A more precise and informative approach involves the fusion of putative regulatory sequences to the reporter gene *uid*A (*gus*), encoding *E. coli* β-glucuronidase followed by plant transformation (Jefferson, Kavanagh & Bevan, 1987). This approach allows the sensitive fluorimetric determination of gene expression in low numbers of spores, and together with histochemical analysis provides essential data on the cell-specificity of gene expression within the anther. This approach also provides the best approximation of the transcriptional activity of particular regulatory sequences *in planta*. Although GUS fusion analysis is clearly superior to direct RNA analysis alone, it is important to use a combination of methods if misinterpretations as indicated above are to be avoided. The value of such a dual approach is further emphasised by a recent example of a lack of correspondence between *gus* mRNA and native gene transcript levels (Uknes *et al.*, 1993). Native tobacco PR1A transcripts are undetectable in pollen of untransformed tobacco plants, but the introduction of a PR1A promoter-*gus* fusion gene led to the expression of *gus* mRNA and enzyme in pollen of transgenic plants. There are several possible explanations for this discrepancy including the relative instability of the native PR1A transcript in pollen, cryptic promoter elements within the *gus* coding DNA, or position effect variation. Although combinations of these possibilities may be involved, the contention by the authors that cryptic promoter elements reside within the *gus* gene seems rather unlikely, since several independent studies show that truncated promoters linked to the *gus* gene are only rarely activated in pollen (Twell *et al.*, 1991; Topping, Wei & Lindsey, 1991). Thus, RNA analysis should be performed initially to determine the presence and broad developmental pattern of native transcript levels, and second, cell-specificity and detailed quantitative analysis of promoter-*gus* fusion expression should be performed in transgenic plants. Future work should take these factors into consideration if we are to gain further insight into the true repertoire and regulation of gene expression in the pathway of pollen development.

Genes activated in microspores and developing pollen

A compilation of promoters which has been shown to be transcriptionally activated in developing microspores and/or pollen currently includes the promoters of more than 30 different native plant genes isolated from 12 different species (Table 2). This expanding data base of regulatory sequences is increasingly useful in building up an understanding of the networks of genes which are coordinately regulated in pollen, and the *cis*-regulatory elements which control their expression (see discussion below). However, this listing includes many promoters for which the presence of the corresponding native mRNA in pollen has not yet been demonstrated (Table 2). Thus, more detailed analyses are required, firstly to assess the significance of the observed activity of particular promoters, and secondly to establish with some precision their developmental regulation. Systematic and detailed analysis should extend the broad division of genes expressed in the pathway of pollen development as either 'early' or 'late' genes as defined by Stinson *et al.* (1987), depending upon whether they are activated prior to, or following, PMI.

Genes activated in the developing microspore

The first conclusive evidence for the activation of specific genes in the developing microspore (early genes) came from detailed quantitative cytochemical analysis of ADH and β-galactosidase enzyme activities in maize (Stinson & Mascarenhas, 1985) and *Brassica campestris* (Singh *et al.*, 1985) respectively. In plants segregating for null alleles at the *Adh1* and *Gal* loci, enzyme activity was detectable in tetrads (ADH) and in young microspores (β-GAL) respectively. Both enzymes showed a similar biphasic pattern of activity, gradually increasing in the developing microspore until PMI, followed by a sharp increase until generative cell division (pollen mitosis II; PMII), after which enzyme activities remained constant or decreased during final maturation. The first molecular evidence to support these findings came from the analysis of actin mRNA levels in maize, which showed that actin transcripts were present in young microspores, increased dramatically in mid- to late pollen grains and declined at maturity (Stinson *et al.*, 1987). Similar transcript abundance

Table 2. Native plant gene promoters active in developing microspores and/or pollen

Species	Gene	Promoter	Species[b]	Activity	Reference
Arabidopsis thaliana	*AHA3	[1]~4 kb	At, Nt	P	DeWitt et al., 1991
Arabidopsis thaliana	apg	[2]−1501 to +6	At, Nt	MS	Roberts et al., 1993a
Arabidopsis thaliana	*acidic chitinase	1129 bp	At	P	Samac & Shah, 1991
Arabidopsis thaliana	*FedA	−1850 to +76	At	P	Casper & Quail, 1993
Arabidopsis thaliana	*HSP18.2	[2]−917 to +77	At	P	Takahashi et al., 1992
Arabidopsis thaliana	*PAL1	[1]−1816	At	MS–LP	Ohl et al., 1990
Arabidopsis thaliana	TUA1	−1500 to +56	At	MS–LP	Carpenter et al., 1992
Arabidopsis thaliana	*TUA2	[1]630 bp	At	P	Carpenter et al., 1993
Arabidopsis thaliana	*ACP A1	−900 to +74	Nt	P	Baerson & Lamppa, 1993
Asparagus officinalis	*AoPR1	−982 to +48	Nt	P	Warner et al., 1992
Brassica campestris	Bgpl	−767 to +100	At, Nt	LP	Xu et al., 1993
Brassica napus	Bp4 A	−235 to +5565	Nt	MS	Albani et al., 1990
Brassica napus	Bpl0	−399 to +64	Nt	LP	Albani et al., 1992
Brassica oleracea	*SLG-13	−3650 to −8	Nt	LP	Thorsness et al., 1991
Brassica oleracea	*S63 SLR1	[2]−1500 to +1	Nt	LP	Hacket et al., 1992
Craterostigma plantagineum	*CDeT2745	−1900 to +104	Nt	P	Michel et al., 1993
Glycine max	*GS-15	−3500 to +30	Lc	P	Marsolier et al., 1993
Lycopersicon esculentum	lat52	−492 to +110	At, Le, Nt	LP	Twell et al., 1990
Lycopersicon esculentum	lat52	−492 to +110	Ng	P	van der Leede-Plegt et al., 1992
Lycopersicon esculentum	lat52	−492 to +110	Ll, Nt, Nr, Pl	P	Nishihara et al., 1993
Lycopersicon esculentum	lat56	−1153 to +191	Nt	P	Twell et al., 1991
Lycopersicon esculentum	lat59	−1305 to +91	At, Le, Nt	LP	Twell et al., 1990

Species	Gene	Coordinates	Species[b]	Activity	Reference
Nicotiana tabacum	G10	−937 to +217	Nt	LP	Lonsdale et al., unpublished
Nicotiana tabacum	Npgl	−744 to +81	Nt	LP	Lonsdale et al., unpublished
Nicotiana tabacum	Nt303	−603 to +174	Nt	LP	Weterings et al., unpublished
Nicotiana tabacum	*Osmotin	[1]−1800	Nt	LP	Kononowicz et al., 1992
Nicotiana tabacum	*PRla	−903 to +29	Nt	P	Ukness et al., 1993
Oryza sativa	Actl		Os	P	Zhang et al., 1991
Oryza sativa	*RCH10	−1512 to +76	Nt	P	Zhu et al., 1993
Petunia hybrida	chiA P_{A2}	−437 to +3	Nt, Ph	LP	van Tunen et al., 1990
Petunia hybrida	chiA P_{A2}	−437 to +3	Ng	P	van der Leede-Plegt et al., 1992
Petunia hybrida	*chiB	−1700 to +62	Nt, Ph	LP	van Tunen et al., 1990
Petunia hybrida	*EPSPS	−1800 to −285	Ph, Nt		
Petroselinum crispum	*4CL-1	−597 to +17	Nt	P	Hauffe et al., 1993
Solanum tuberosum	*pgT16	−3300 to +13	St	P	Liu et al., 1991
Solanum tuberosum	*BG24	−1883 to +17	St	P	Liu et al., 1991
Triticum aestivum	Histone H3	−1711 to +57	Os	P	Guerrero et al., 1990
Zea mays	pZmcl3	−314 to +68	Nt	LP	Allen & Lonsdale, 1993
Zea mays	W2247	−2680 to −12	Nt	LP	

Activity column represents developmental stage where initial activity of promoter observed by histochemical staining for GUS activity or RNA analysis. MS = microspore; LP = post PMI; P = activity detected in mature pollen where developmental expression not analys in detail. Species[b] = plant species in which analysis of function has been performed: At = *Arabidopsis thaliana*; Lc = *Lotus corniculatum*; Le = *Lycopersicon esculentum*; Li = *Lilium longiflorum*; Ng = *Nicotiana glutinosa*; Nr = *Nicotiana rustica*; Nt = *Nicotiana tabacum*; Os = *Oryza sativa*; Pl = *Paeonia lactiflora*; Ph = *Petunia hybrida*; St = *Solanum tuberosum*; Zm = *Zea mays*. * = examples where direct evidence has not been obtained for the presence of the native mRNA in RNA isolated from pollen. [1] = precise coordinates of flanking regions used not reported.
[2] = promoter coordinates relative to translational initiation codon, where mapping of transcription start site 1 not been performed.

profiles have been reported for other genes in several plant species, including those with bicellular and tricellular pollen at dehiscence: histone H3 in rice by *in situ* hybridisation (Raghavan, 1989); potato granule bound starch synthase in RNA from isolated spores of tobacco (Olmedilla, Schrauwen & Wullems, 1991); *Petunia Adh*1 in RNA isolated from anthers (Gregerson *et al.*, 1991); rice actin in RNA isolated from anthers (Tsuchiya *et al.*, 1992).

Where anther-specific cDNA clones have been isolated, several appear to belong to the early group as defined above. Transcripts corresponding to two anther-specific genes from *Brassica napus* (Bp4 and Bp19) were shown, by RNA analysis from isolated spores, to be present in uninucleate microspores, to accumulate until PMII and decrease in mature pollen (Albani *et al.*, 1990, 1991). Two cloned anther-specific rice transcripts (Osc4 and Osc6) showed a similar early pattern of accumulation and decay, although RNA analysis was only performed in whole anthers (Tsuchiya *et al.*, 1992). Several additional anther-specific cDNA clones have been isolated from *B. napus* (No. 17, E2 and F2S), the transcripts of which were reported to be microspore-specific (Scott *et al.*, 1991a; Roberts *et al.*, 1991). These transcripts showed overlapping patterns of accumulation appearing first in buds containing uninucleate microspores and decreasing in mature pollen grains. However, since RNA analysis was performed using RNA isolated from whole buds their specificity and precise patterns of accumulation in developing spores remain unclear. Recent analysis by *in situ* hybridisation has, in fact, demonstrated that the E2 cDNA hybridises to mRNA present in both the tapetum and developing microspores (Foster *et al.*, 1992). Whether the promoters of any of the above cloned 'early' anther-specific genes are activated in tetrads or young microspores and/or in sporophytic tissues of the anther remains to be established.

Recently a gene (*apg*) has been isolated from *Arabidopsis* which is expressed both in the tapetum and developing microspores and encodes a proline rich protein which may be secreted to the microspore and pollen grain wall (Roberts *et al.*, 1993a). As shown in Fig. 2, detailed histochemical analysis of transgenic plants containing an *apg* promoter-*gus* fusion, demonstrates that the *apg* promoter is first activated in the developing microspore prior to the characteristic equatorial migration of the microspore nucleus (stage 1; Twell *et al.*, 1993). Quantitative analysis of GUS activity in isolated spores showed that GUS activity per mg protein declined dramatically between the microspore and pollen stages (Fig. 2A). However, because protein content increases dramatically after PMI, GUS activity per spore reached a maximum during mid-pollen development and declined in mature pollen (Fig. 2B). This pattern is consistent

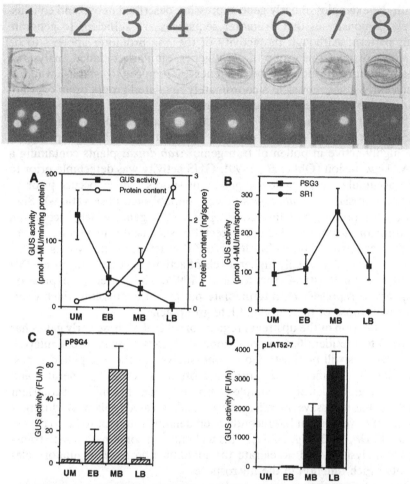

Fig. 2. Micrographs demonstrating the activation and developmental regulation of GUS activity directed by the *A. thaliana apg* promoter in tobacco during microspore development. *Top panel*: light micrographs showing representative microspores from *apg-gus* transformed tobacco at developmental stages 1–8 following incubation with X-glu. *Bottom panel*: the same microspores counter-stained with DAPI and viewed under UV epi-illumination. A and B. GUS activity and protein content measured in isolated spores of transgenic tobacco plants containing a −530 bp *apg* promoter-gus gene fusion. Developmental stages: UM = uninucleate; EB = early bicellular; MB = mid-bicellular; LB = late bicellular. SR1 = untransformed tobacco. C and D. Transient gene expression (GUS activity in fluorescence units per h/10⁶ spores) measured in isolated spores at different developmental stages after bombardment with −530 *apg-gus* (pPSG4; C) and –492 *lat52–gus* (pLAT52-7; D) plasmid DNA.

with the examples of early gene expression described above and conclusively demonstrates that promoter sequences are sufficient to generate this pattern. Although the activity of the *apg* promoter appears to be characteristic of this group of early genes, similarly detailed analysis of many other microspore-expressed genes are required to establish whether such early genes represent a coordinately regulated group, or an overlapping group of sequentially activated genes.

Another promoter which appears to be activated during microspore development is the *Arabidopsis PAL1* promoter, which was reported to be highly active in pollen of transgenic *Arabidopsis* plants containing a PAL1-*gus* fusion (Ohl *et al.*, 1990). GUS activity was detectable prior to the accumulation of yellow pigment in spores, which occurs prior to pollen mitosis I in *Arabidopsis* (Twell, unpublished observations). Similarly, the promoter of the *Arabidopsis TUA1* gene, which encodes an isoform of α-tubulin which is detectable specifically in mature flower buds, was shown to direct low levels of GUS activity first in young microspores, with substantially higher levels of activity accumulating after PMI until pollen maturity (Carpenter *et al.*, 1992). This pattern of expression appears to represent an intermediate pattern of expression which overlaps with those of the early and late group genes.

Examination of the upstream regions of several of these early genes has not led to the identification of extended regions of sequence similarity. However, as will be illustrated for the analysis of the late pollen genes, functionally significant similarities are often short and degenerate, and thus difficult to identify by simple comparison. The feasibility of transient biolistic-based assays in microspores and isolated pollen at different stages of development has recently been demonstrated for tobacco (Twell *et al.*, 1989b, 1993; Fig. 2C,D). This technique in combination with transgenic analysis should accelerate the identification of *cis*-regulatory elements which function in the microspore.

In addition to the transcriptional activation of early genes in the microspore, their corresponding transcript and/or protein levels accumulate rapidly in the young pollen grain. Thus, transcriptional activation of these genes appears to be maintained following PMI. Because the expression pattern of such early genes overlaps with that of the late genes, do distinct regulatory factors control early gene expression in the microspore and developing pollen grain? Detailed promoter analysis should allow the identification of regulatory elements which function in both phases of development and whether these are identical. A further characteristic of the early pattern of expression is the sharp decline in transcript levels at pollen maturity. Thus, early gene promoters may either be transcriptionally silenced during mid- to late pollen development and/or their transcripts may be rapidly turned over. Recent transient expression experiments, in which isolated microspores and pollen were bombarded with early (*apg–gus*) and late (*lat52–gus*) promoter constructs, showed that, while *lat52–gus* expression continued to increase until maturity, *apg–gus* expression showed a dramatic decline in mature pollen (Fig. 2C,D; Twell *et al.*, 1993). These data strongly suggest that early gene promoters are transcriptionally silenced in the latter stages of pollen development. Whether, in addition, specific mechanisms exist to enhance the turnover of early gene transcripts in maturing pollen remains an open question.

Genes expressed in the developing pollen grain

The late pollen genes represent the largest characterised group of genes expressed in the pathway of pollen development. Stinson *et al.* (1987) first reported the isolation of pollen-specific cDNA clones from mature pollen libraries of *Tradescantia paludosa* (pTpc44 and pTpc70) and maize (pZmc13 and pZmc30). Northern blot hybridisations using RNA isolated from purified populations of spores showed that transcripts were first detectable in young pollen soon after PMI and increased in concentration until pollen shed. This characteristic late pattern of transcript accumulation was subsequently described for five anther-specific cDNA clones from tomato, the LAT or late anther tomato clones (Twell *et al.*, 1989a; Ursin *et al.*, 1989). Very similar patterns of transcript accumulation in developing pollen have since been described using anther and pollen-specific cDNA clones isolated from *Oenothera organensis* (Brown & Crouch, 1990), tobacco (Weterings *et al.*, 1992; Rogers, Harvey & Lonsdale, 1992) and *B. campestris* (Theerakulpisut *et al.*, 1991). More detailed analysis may reveal subtle differences in their patterns of expression, but currently the isolated late pollen genes form a fairly homogeneous, coordinately regulated group. One exception to the above is provided by the pollen-specific P6 transcript from *Oenothera* which appears significantly later than other late pollen transcripts (Brown & Crouch, 1990). This indicates that further detailed studies of additional late pollen-expressed genes may reveal differential regulation among genes activated after PMI.

Developmental regulation of late pollen gene promoters

To gain further insight into the mechanisms regulating late pollen gene expression putative promoter regions of several of these genes have been linked to the *gus* reporter gene and their activities analysed in transgenic plants. Such studies have allowed the time of activation and developmental regulation of these genes to be established in far greater detail than by RNA analysis alone. Although this approach has provided convincing evidence for the transcriptional activation of late pollen genes only after PMI, very few studies have fully exploited the sensitivity of this approach to define developmental regulation.

The earliest examples of such analyses showed that the promoters of the tomato *lat52* and *lat59* genes were coordinately activated in developing pollen grains in close association with PMI (Twell, Yamaguchi & McCormick, 1990). Furthermore, the patterns of accumulation of GUS activity in developing pollen were very similar when identical promoter fusions were introduced into tomato and tobacco (Twell *et al.*, 1990). Recently, more detailed histochemical analysis of transgenic tobacco plants showed that the *lat52* and *lat59* promoters were first activated soon after PMI prior to the migration of the generative cell into the interior of the vegetative cell (Fig. 3; Eady, Lindsey & Twell, 1994). Detailed developmental analysis of the promoter of the maize homologue of the *lat52* gene, *Zm*13, in transgenic tobacco are in strict accord with these data (Guerrero *et al.*, 1990), which suggests that regulatory mechanisms for late pollen promoter activation are conserved among species which shed pollen in a bicellular or tricellular condition. These studies further indicate that there is a critical developmental switch in transcriptional activity associated with PMI. This tight link between PMI and late gene activation may indicate that passage of the microspore nucleus through PMI is a prerequisite for late gene activation. In this regard, recent analysis of the regulation of the *lat52* promoter in *Arabidopsis* shows that the *lat52* promoter is activated prematurely in late uninucleate microspores prior to PMI (Fig. 3; Eady *et al.*, 1994), but during pollen development shows a pattern of activity very similar to that observed in tobacco

---

Fig. 3. *Top*: A–F. Micrographs demonstrating the activation of the *lat52* promoter in tobacco (A–C) and *Arabidopsis* (D–F) containing the gene fusion −492 bp *lat52* promoter-*gus*. The top (light) panels show spores incubated with X-glu at late uninucleate (A, D, E), early bicellular (B, F) and early to mid-bicellular (C) stages. The corresponding bottom (dark) panels represent the same spores counter-stained with DAPI and viewed under uv epi-illumination. *Bottom*: Graphs showing the developmental accumulation of GUS activity in isolated spores of transgenic tobacco (open circles) and *Arabidopsis* (closed circles) containing the −492 bp *lat52*–*gus* construct. Total protein content per grain in tobacco is shown as open squares. B. Shows detail of values during the transition from microspore to pollen grain (pollen mitosis I), which occurs over the period indicated by the shaded bar.

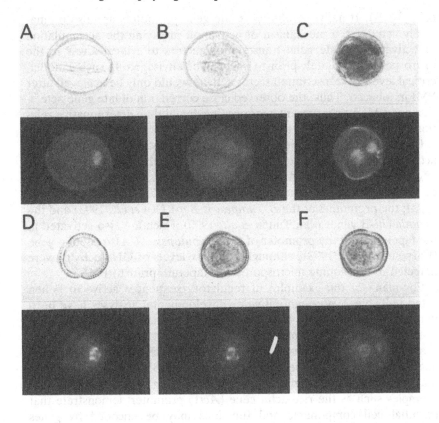

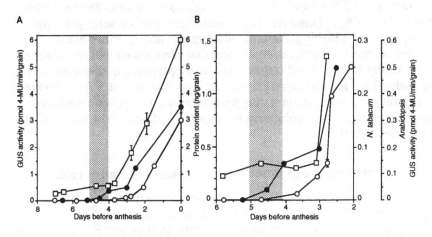

(Fig. 3A,B). If native *Arabidopsis* late gene promoters are also prematurely activated, a mechanism of activation involving the accumulation or activation of late gene transcription factors to critical levels in the microspore immediately prior to PMI can be envisaged. In such a model, critical levels of transcription factor activity would only be achieved after PMI in tobacco. Thus, the observed close correlation of late gene activation and PMI may represent a more coincidental than determinative relationship depending upon the plant species.

Other transgenic studies which have demonstrated the high level activation of late pollen genes after PMI include: the upstream promoter (P_{A2}) of the petunia *ChiA* gene (van Tunen *et al.*, 1990); the promoter of the maize exo-polygalacturonase gene W2247 (Allen & Lonsdale, 1993); the promoters of the *B. campestris Bgp*1 (Xu *et al.*, 1993) and the *Petunia chi*-B genes (van Tunen *et al.*, 1990) which are also activated in the tapetum; and the promoter of the *Arabidopsis TUA*1 α-tubulin gene (Carpenter *et al.*, 1992), although here low levels of GUS activity were detected in developing microspores and tapetum prior to PMI.

For many of the examples of regulatory sequences active in pollen listed in Table 2, no quantitative or developmental analyses have been performed because they have been studied primarily because of their predicted sporophytic patterns of expression. Despite this, these examples illustrate the diversity of promoters which are utilised in pollen and further emphasise the high degree of overlap which exists between the gametophytic/sporophytic gene expression profiles. Particular examples such as the rice actin gene (*Act*1) promoter demonstrate that essential cell components and functions may be encoded by genes expressed and differentially regulated in both generations (Zhang, McElroy & Wu, 1991). However, the recent cloning of an actin gene from tobacco which is highly preferentially expressed during pollen maturation (Thangavelu *et al.*, 1993), demonstrates that certain members of the actin gene family have evolved distinct regulatory properties, having become specialised for gametophytic expression. This may reflect a requirement for increased levels of actin during particular stages of pollen germination and tube growth or a requirement for a specialised isoform of actin within the pollen cytoskeleton.

Gene expression in the vegetative and generative cells

The large vegetative cell (VC) and diminutive generative cell (GC) have very different morphologies and fates during development. The extreme dimorphism in cytoplasmic volume, organelle, protein and RNA content clearly suggests that these cells have very different synthetic and genetic capacities. This is supported by ultrastructural analyses, cytochemical RNA staining and the analysis of protein and histone profiles of partially purified vegetative and GCs (for review see Mascarenhas, 1975; Sunderland & Huang, 1987). However, there is currently no information on the degree of overlap of gene expression between the vegetative and generative cells. Indeed, to date, no generative cell expressed sequences have been isolated. This has most likely resulted from the predominance of the vegetative cell cytoplasm over that of the GC in combination with the differential screening approach which has been used to identify pollen-expressed genes.

Direct evidence for the presence of specific transcripts within the vegetative cytoplasm has been provided by the *in situ* hybridisation of tomato (Ursin *et al.*, 1989), maize (Hanson *et al.*, 1989), *Oenothera* (Brown & Crouch, 1990), *B. campestris* (Theerakulpisut *et al.*, 1991) and tobacco (Reijnen *et al.*, 1991) cDNA clones to mature or germinated pollen grains. However, in these studies it was not possible to determine whether transcripts were present in the generative or sperm cell cytoplasma. Recently, a novel approach to determine the cell-specificity of regulatory sequences within the pollen grain has been devised which has been used to directly demonstrate differential gene expression between the vegetative and GCs (Twell, 1992). By linking the promoter of the tomato *lat52* gene to a GUS fusion protein containing the tobacco etch virus NIa nuclear-targeting signal (Restrepo *et al.*, 1990), the *lat52* promoter was shown to be activated specifically from the vegetative nucleus, and not in the GC during pollen development or tube growth (Fig. 4A,B; Twell, 1992). Using this approach it has also been demonstrated that the VC-specific activity of the *lat52* promoter is maintained in *Arabidopsis* (Fig. 4C,D; Eady *et al.*, 1994). Together these data provide direct evidence for the differential gene expression programmes of the VC and GC and strongly suggest that this occurs at the level of transcription.

It is likely that many other cloned pollen-expressed genes and their associated promoters are preferentially if not specifically activated in the VC based upon their level of expression and observed patterns of transcript accumulation. Efforts are currently under way in several laboratories to identify proteins and mRNAs which are expressed in GCs and in sperm cells with the aim to identify genes which have functions which are specifically associated with gamete differentiation. The isolation of such genes would provide important molecular markers with which to investigate the differential genetic determination of the vegetative, generative and sperm cells.

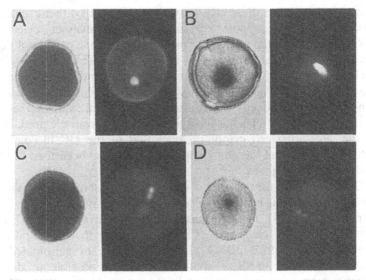

Fig. 4. Histochemical localisation of *lat*52 promoter activity to the veget- ative cell of tobacco and *Arabidopsis*. Light panels represent micro- graphs of X-glu treated pollen from transgenic tobacco (A and B) and *Arabidopsis* (C and D) containing an unmodified *lat*52 promoter–*gus* fusion (A and C) or a construct containing the *lat*52 promoter driving a GUS fusion protein containing a nuclear localisation signal (B and D). Adjacent dark panels represent the same pollen grains counter- stained with DAPI and viewed under uv by epi-illumination to visualise vegetative and generative or sperm cell nuclei.

Identification of *cis*-regulatory elements which function in pollen

An important approach in investigating the molecular basis of both stage and cell-specific and coordinate gene expression during pollen develop- ment is the identification of *cis*-regulatory DNA sequences which func- tion in the promoters of pollen-expressed genes. Functional analysis of identified shared sequence similarities among pollen-expressed pro- moters is essential if we are not to be misled by the apparently coincid- ental similarities which have been identified between the increasing number of promoter sequences available. Although detailed functional analyses of pollen-expressed promoters may not be feasible on a large scale using a transgenic approach, a rapid transient expression assay for gene expression in pollen has been developed involving microprojectile bombardment (Twell *et al.*, 1989*b*, 1991). Using both transgenic analyses and, in particular, the rapid transient expression assay in combination with site-directed mutagenesis, it is now possible to functionally investigate the role and precise identity of particular regulatory elements. Using both approaches, the structure and organisation of regulatory elements has been defined in considerable detail for three coordinately regulated LAT promoters from tomato (Twell *et al.*, 1991).

The promoter of the tomato *lat*52 gene has been most extensively analysed and provides some insight into the complexity and organisation of regulatory elements which control cell and stage-specific gene expression in the pollen grain. A functional map of the *lat*52 promoter is presented in Fig. 5A. Sequences within the *lat*52 core promoter region −71 to +110 (52-C) are sufficient to activate gene expression in the pollen grain after PMI and at similar levels in the developing endosperm (Twell *et al.*, 1991). Two upstream regions of the promoter (52-A and 52-B), referred to as the upstream activation domain, strongly enhance expression in pollen, but not in endosperm. This upstream activation domain together with the core domain (52-C) constitutes a regulatory unit which is highly preferentially active in the developing pollen grain. Experiments in which the upstream activation region 52-A was fused to the heterologous CaMV35S core (−90 bp) promoter demonstrated that the 52-A activation function in pollen is not strictly dependent on *lat*52 core sequences (Twell *et al.*, 1991). Whether such upstream sequences are sufficient to mediate the specificity of the *lat*52 promoter remains to be established.

Similar deletion analyses have been carried out in varying degrees of detail for several other late pollen promoters (Table 3). The results of these analyses show that *cis*-regulatory elements sufficient to direct transcription in developing pollen are always located relatively close to the transcription start site, within 70 to 150 bp, and multiple upstream promoter elements serve to increase transcriptional activity of core promoter regions. Evidence for negative regulatory elements has been obtained for particular upstream regions of the tomato *lat*59 and maize *Zm*13 promoters (Twell *et al.*, 1991; Guerrero *et al.*, 1990) and for the *B. campestris Bgp*1 promoter (Xu *et al.*, 1993). Such elements may serve to modulate the degree of transcriptional activation to achieve the correct levels of accumulation of particular late gene products in the mature pollen grain.

The identities of two distinct *cis*-regulatory elements within the *lat*52 upstream activation domain have been defined using the transient expression assay. The reiterated PB core motif (TGTGGTT) occurs 3 times (PBI, PBII, PBIII) within the upstream domain (Fig. 5A). The PBI motif has directly been shown to represent a positive regulatory element by

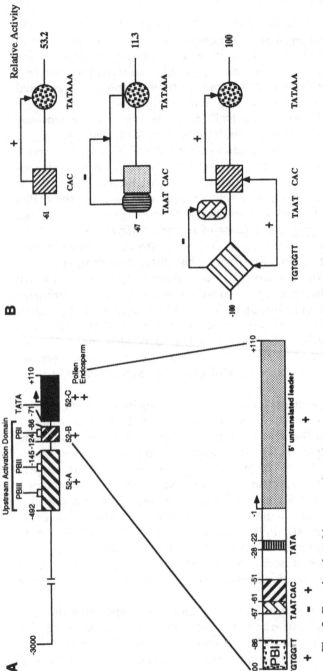

Fig. 5. Functional architecture of regulatory elements within the *lat52* promoter. A. *Top:* Functional map of the *lat52* promoter determined by 5' deletion analysis in transgenic tomato. The upstream activation domain contains two distinct activator regions (52-A and 52-B) which increase expression from the core promoter (52-C) specifically in pollen. 52-C alone is sufficient for low level expression specifically in pollen and endosperm. PBI-III represents the positions of the reiterated PB motif (TGTGGTT). *Bottom:* Shows a more detailed map of regulatory elements within the region −100 to +110 which have positive (+) or negative (−) effects on promoter activity from transient expression studies in tobacco pollen. B. Results of promoter deletion analysis within the core promoter interpreted to illustrate putative regulatory interaction between *trans*-acting factors binding at regulatory elements.

targeted mutation of the GG core, while a 20 bp oligonucleotide containing the PBII motif was shown to enhance the activity of the −90 CaMV35S promoter in pollen (Twell *et al.*, 1991). Recent targeted analysis of the PBIII motif, which is also found in the tobacco pollen-specific *Nt*303 promoter at position −340 bp (12/13 bp identity) did not lead to significant effects of *lat*52 promoter activity (Weterings *et al.*, 1992; Weterings & Twell, unpublished observations). Thus, although only the PBI and PBII motifs are functional in the transient expression assay, it is possible that the PBIII motif may exert an additional influence only in its native chromosomal configuration.

Fine structure analysis of the *lat*52 TATA proximal promoter region −100 to +110 bp has been carried out by 2 bp substitution analysis of the PBI element, further 5′ deletion analysis, and by modified linker scanning mutogenesis (Bate & Twell, unpublished observations). These data showed that both the GT and GG base pairs within the PB core motif (GTGG) are essential for maximum levels of activity. Substitution of GG with CA led to a fivefold decrease in promoter activity. However, substitution of the AT rich sequences flanking both sides of the PBI motif led to a two- to threefold stimulation of core promoter activity. This may indicate the involvement of protein factors which bind at the PBI core motif and additional factors which bind at flanking sites and modulate the activity of PBF. Structural interpretations of the derepression of the *lat*52 promoter by PBI flanking mutations appear less likely since both upstream and downstream mutations stimulate promoter activity.

Further deletion analysis of the core promoter indicates the presence of additional positive (TCCACCAT) and negative regulatory elements (ATAATAG) positioned between PBI and the TATA box. Targeted mutation of the TCCACCAT motif (CACC Box) demonstrate that both PBI and the CACC Box are required for maximal levels of core promoter activity (Fig. 5B). Based upon these analyses a model of the regulatory interactions which appear to operate to control *lat*52 core promoter activity is presented in Fig. 5B. In this model, positive regulatory factors bind at the core PBI and CACC elements and act to stimulate the efficiency of transcription initiation by stimulating the assembly or stability of proteins within the TATA complex. Protein factors which bind to the negative regulatory element TAAT adjacent to the PBI and CACC elements may function to precisely modulate the level of *lat*52 activation and/or may contribute to the developmental regulation and cell-specific control of the *lat*52 promoter activity in pollen. Such negative regulatory factors may act in a dominant negative manner to block *lat*52 promoter activation in sporophytic tissues in a manner similar to the core promoter of the petunia *chs* gene (van der Meer *et al.*, 1992). Modification of such

Table 3. *Deletion analysis of promoters activated in microspores and/or pollen*

Gene	Species[1]		Promoter regions	Specificity/ Activity	References/ Motifs
Bgp1	Bc	At	**−168 to +100**	pollen, tapetum	Xu et al., 1993
			−767 to −580	+ve	
			−580 to −322	−ve	
			−168 to −116	+ve	
CDeT27-45	Cp	Nt	**−197 to +104**	pollen, embryo	Michel et al., 1993
			−197 to −158	+ve	
lat52	Le	Le	**−71 to +110**	pollen, endosperm	Twell et al., 1991 GTGG/CCAC GTGG
			−492 to −145	+ve	
			−124 to −86	+ve	
			−71 to −41*	+ve	TCCACCAT
lat56	Le	Nt	**−103 to +191***	pollen	Twell et al., 1991 GTGG
			−163 to −118*	+ve	
			−118 to −103*	+ve	
			−103 to −95*	+ve	
lat59	Le	Le	**−115 to +93**	pollen, root cap, testa endosperm	GTGA/CACCAT Twell et al., 1991
			−1305 to −806	+ve	
			−806 to −603	−ve	
			−115 to −97*	+ve	GTGA
			−97 to −45*	+ve	GTGA

Gene	Species[1]	Species	Promoter region	Specificity/activity	Motifs / Reference
Nt303	Nt	Nt	**−150 to +173**	pollen	Weterings et al., unpublished
			−107 to −89*	+ve	AAATGA
			−89 to −62*	+ve	TCATTT
PAL1	At	At	**−260 to BglII**	pollen, sporophytic	Ohl et al., 1990
			−1816 to −260	+ve	
			−260 to BglII	+ve	
TUA1	At	At	**−97 to +56**	pollen, ms, tapetum	Carpenter et al., 1992
			−1500 to −533	+ve	GTGA
			−271 to −217	+ve	GTGA
			−97 to −39	+ve	
Npg1	Nt	Nt	**−86 to +85***	pollen	Lonsdale et al., unpublished
			−182 to −86*	+ve	GTGG (X2)
			−86 to +85*	+ve	GTGA
Zm13	Zm	Nt	−314 to +61	pollen	Guerrero et al., 1990
			−260 to −100*	+ve	Hamilton et al., 1992
			−100 to −54*	+ve	GTGG/TCAC/CCAC
					CCAC

All regulatory promoter regions listed represent those identified by transgenic analysis with the exception of those highlighted with an asterix (*) which represent regulatory regions defined in transient expression assays. Promoter positions shown in bold represent minimal promoter regions which function in pollen. Species[1] = first column is the species from which the gene was isolated; second column is the species in which functional analysis was performed. Motifs represent occurrences of the GTGG,GTGA and CACC motifs shown to be functional in the *lat* promoters. Specificity/activity column indicates site of expression where expression in diverse sporophytic tissues is listed as sporophytic. ms = microspore. Npg1 is a pollen-specific polygalacturonase gene from tobacco (Lonsdale et al., unpublished).

negative factors in the vegetative cell after PMI may subsequently allow positive regulatory factors to bind at PBI and CACC. Since promoter deletions down to −71 bp are actively expressed in pollen and not derepressed in sporophytic tissues (Twell *et al.*, 1991), the putative negative element TAAT may be sufficient to control specificity in the above model. Experiments involving analysis of the effect of further targeted mutations on tissue-specific expression are under way to investigate this model.

Shared regulatory elements among late pollen promoters

The *lat*52 PBI element discussed above also occurs within the *lat*56 promoter as part of a region of extended sequence identity termed the 52/56 Box (TGTGGTTATATA). Targeted mutation of this sequence within the *lat*52 promoter demonstrates a quantitative role for this element (Twell *et al.*, 1991). Furthermore, the *lat*56 promoter contains the sequence CACCAT at position −89 bp, which is identical with the *lat*52 CACC element, although the spacing between the 52/56 Box and CACC element is not conserved. A second shared regulatory element which has been shown to be functional in the transient expression assay is the 56/59 Box (GAAA/TTTGA). This element is present at similar positions in the *lat*56 (−97 bp) and *lat*59 (−108 bp) promoters (Fig. 6). Substitution mutation analysis has shown that only the GTGA motif within the 56/59 Box is essential for maximum activity (Twell *et al.*, 1991). The direct demonstration of the function of common DNA sequences within the promoters of three coordinately regulated late pollen genes strongly implies the involvement of shared transcription factors. Such factors may serve to coordinately activate different late pollen genes which are required for pollen maturation or during pollen germination or tube growth. Since essential nucleotides within the core motifs of the 52/56 and 56/59 boxes can be restricted to the rather similar 4 bp sequences, GTGG and GTGA, it is conceivable that a single common transcription factor may recognise both sequences. Such small target sequences are also likely to occur frequently in many other promoters, such that their position and/or spacing are likely to be important in determining the strength of functional interactions. Putative transcription factors which interact with such elements may be present specifically in developing pollen or may be expressed constitutively and recruited by particular sporophytic or gametophytically activated promoters.

Although no extensive sequence identities to either the 52/56 or 56/59 Boxes are apparent within the promoter of the maize Zm13 gene, the GTGG and GTGA and CCAC core motifs occur within −168 bp of the transcription start site (Fig. 6). Using the transient expression assay in *Tradescantia* pollen, Hamilton *et al.* (1992) localised quantitative elements required for high level *Zm*13 promoter activity to the region −260 to −100 bp which contains a single copy of each element (Table 3; Fig. 7). Additional sequences which were essential for promoter activity were located between −100 and −54 bp which includes GTGG in reverse orientation. It remains to be established whether these and/or additional regulatory elements are functionally conserved, which would suggest that transcription factors which regulate these promoters have been conserved between tomato and maize, species with bi- and tricellular pollen, respectively. Recent transient expression analysis of the tobacco *Npg*1 promoter provides further support for the role of the GTGG motif in other late pollen promoters (Tebbut & Lonsdale, unpublished observations). The *Npg*1 promoter region −182 to −86 bp which positively regulates *Npg*1 promoter activity contains two copies of the GTGG motif (Table 3; Fig. 6).

Comparison of the tomato *lat*56 promoter with its tobacco homologue *G*10 (Rogers, Harvey & Lonsdale, 1992) has revealed extensive blocks of sequence identity within the region from the initiator AUG up to position −98 bp (Twell & Lonsdale, unpublished observations). Blocks of sequence identity stop abruptly immediately upstream of the 56/59 Box core motif GTGA (Fig. 6). This finding strongly supports the important role of the GTGA motif within the 56/59 Box. It is of further interest that the spacing between the GTGA motif and the TATA Box is exactly conserved in the *lat*56 and *G*10 promoters and that the GTGA motif lies adjacent to the conserved sequence CACACCAT (Fig. 6). Two additional copies of the GTGA motif also flank the TATA box of both promoters. Regarding upstream activating sequences the *G*10 promoter also contains the GTGG motif in reverse orientation in a similar position to the functional GTGG motif in the *lat*56 promoter. Further mutational analysis is required to establish the function of such sequences and to investigate the significance of the conservation of spacing between them.

Deletion analysis of the *lat*59 promoter in transgenic tomato defined the minimal promoter region required for high level preferential expression in pollen to be −115 to +91 bp (Twell *et al.*, 1991). Sequences between −115 and −45 bp were absolutely required for promoter activity in pollen. Using the transient expression assay the GTGA motif within the 56/59 Box (−100 bp) was shown to account for 80% of the activity of the −115 bp promoter (Twell *et al.*, 1991). Recent analysis of deletion constructs which remove the 56/59 Box in transgenic tobacco confirm that this is indeed a quantitative element, but that sequences downstream

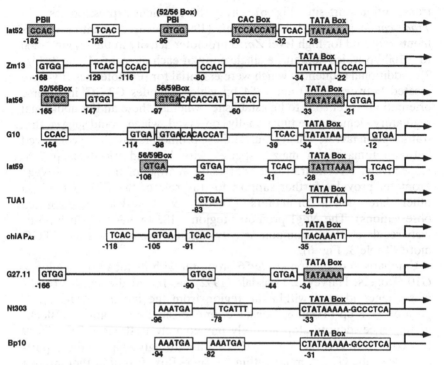

Fig. 6. Organisation of regulatory elements and the conservation of these elements within various late pollen gene promoters. Regulatory elements including the TATA elements, which have been precisely identified by functional analysis, are included within the shaded boxes. The length of the promoter regions shown reflect only those regions which have been shown to be functional by 5′ deletion analysis. All occurrences of the core regulatory motifs GTGG (reverse CCAC) and GTGA (reverse TCAC) within these functional promoter regions are shown, where GTGG represents the 52/56 Box core and GTGA represents the 56/59 Box core. Extended regions of sequence identity including conservation of sequence position between the *lat*56 and *G*10 promoters and between the *Nt*303 and *Bp*10 promoters are shown.

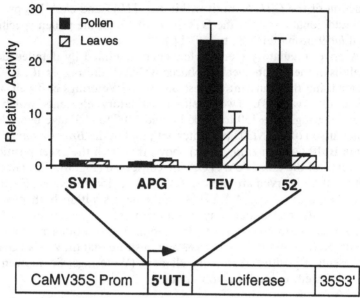

Fig. 7. *In vivo* assay of the ability of the *lat*52 5'-untranslated leader to act as a translational enhancer. The graph shows the effect of substituting different 5' untranslated leaders (5'-UTL) on luciferase expression driven by the CaMV35S promoter following bombardment into pollen and leaves of tobacco. The 5'-UTLs correspond to those of the microspore activated *apg* gene (APG), the known translational enhancer region from tobacco etch virus (TEV), the *lat*52 gene (52) and a synthetic polylinker (SYN). Activities in pollen and leaves were normalised to those obtained for the SYN 5'-UTL construct in each tissue such that the relative activity is a measure of the fold enhancement.

of position −97 bp remain sufficient to confer low levels of pollen-specific expression (Twell, Patel, Worrall & Lindsey, unpublished observations). This region contains an additional GTGA motif at position −82 bp which may be involved in minimal promoter function (Fig. 6). Deletion analysis of the *Arabidopsis TUA*1 promoter showed that the region −97 to −39 bp was essential for *TUA*1 promoter function in pollen of transgenic plants (Table 3; Carpenter *et al.*, 1992). A GTGA motif is present at position −83 bp in the *TUA*1 promoter and the spacing between this motif and the TATA Box is conserved between the *TUA*1 and *lat*59 promoters (Fig. 6). The upstream pollen-specific promoter (P$_{A2}$) of the *Petunia chi*A gene also contains three copies of the GTGA motif between −118 and −88 bp (Fig. 6). Further analysis is required to establish the function of the GTGA motif within the *TUA*1 and *chi*A P$_{A2}$ promoters and additional copies of the GTGA motif within the pollen-specific minimal *lat*59 promoter(−82 to +91 bp).

A novel regulatory element has been identified by detailed deletion analysis of the pollen-specific tobacco *Nt*303 (Weterings *et al.*, 1992) promoter using the transient expression assay (Weterings & Twell, unpublished observations). Two positive regulatory elements which occur within the regions −150 to −106 bp and −106 to −89 have been defined. Comparison of the *Nt*303 promoter with that of the *Brassica napus* homologue Bp10 (Albani *et al.*, 1992) shows that both the exact position and sequence of the motif AAATGA are conserved (Fig. 6). Extended similarity is also observed around the TATA Box (15/17 identity; Fig. 6), and a further copy of the AAATGA motif occurs within both promoters. Experiments are under way to directly test the function of these sequences and to investigate their relationship with other regulatory elements. Recent mutational analysis demonstrate that the AAATGA motif represents a positive regulatory element (Weterings, Schrauwen, Wullems & Twell, unpublished observations).

Transposon-mediated promoter analysis

An alternative approach to identify regulatory sequences within pollen-active promoters has been fruitful in maize (Chen *et al.*, 1987; Kloeckener-Gruissem, Vogel & Freeling, 1992). Characterisation of two transposable element *Mutator* (*Mu*) induced mutations at the *Adh*1 locus involving the *Adh*-1 promoter showed that duplication of the TATA Box occurred in one line, and in a second line, resulting from *Mu* excision, the TATA Box region was deleted. The insertion line carrying the TATA duplication showed reduced expression in roots but not in pollen sequences which determine pollen expression must reside at or downstream of the TATA Box, since the size of the *Mu*3 element at 1.85 kb is likely to disrupt interactions between upstream factors and TATA associated factors. In the TATA deletion line partial restoration of ADH activity occurred in roots but expression in pollen was reduced. The TATA deletion led to different aberrant transcription initiation sites in pollen and roots. Transcripts in pollen were present at wild type levels but were 50 to 100 nucleotides longer at their 5′ ends. The reduced levels of ADH activity in pollen of this line are most easily explained by a reduction in the efficiency of translation of the aberrant transcripts. Since no upstream AUG triplets or stable secondary structures were predicted within the extended transcripts the ability of the transcript to be recognised by translation initiation factors may be altered (Kloeckener-

Gruissem *et al.*, 1992). Similar analysis of revertant alleles of the unstable maize *Adh1–Fm335* mutation, caused by excision of the *Dissociation* (*Ds*1) transposable element from the *Adh*1 5′ untranslated leader, provides additional insight into mechanisms regulating pollen-specific expression (Dawe, Lachman-Singh & Freeling, 1993). Among 16 revertant alleles, one allele underexpressed pollen ADH1 activity and another overexpressed pollen ADH1 activity at 43% and 163% of wild-type levels respectively. These changes in ADH1 activity in pollen reflected changes in steady-state mRNA levels and identify the region between +45 and +52 bp as a region regulating transcript accumulation in pollen. The sequence motif GGGACTGA, duplicated by the insertion of *Ds*1 in *Adh1–Fm335*, is conserved among four other *Adh*1 alleles is thus implicated in promoting pollen-specific expression. As discussed by Dawe *et al.* (1993) this motif may represent an intragenic transcriptional promoter element or may serve to promote mRNA stability.

Transcription factors which regulate gene expression in pollen

The identity or properties of transcription factors which regulate gene expression in the male gametophyte are currently unknown, although several *cis*-regulatory promoter elements which are likely to be recognised by such factors have been defined. Work is under way in several laboratories to characterise such factors by gel retardation analysis using known *cis*-regulatory elements and by screening lambda-based cDNA expression libraries with multimerised *cis*-element probes. Despite such targeted efforts, a pollen-specific cDNA clone (SF3) has been isolated, by differential screening of a mature anther cDNA library of sunflower, which encodes a putative regulatory protein (Baltz, Domon & Steinmetz, 1992*a*). The SF3 transcript is specifically expressed in maturing pollen and the encoded protein shares significant sequence similarity to a group of regulatory proteins from animals collectively known as LIM proteins (Baltz *et al.*, 1992*b*). LIM proteins contain one or two copies of the conserved LIM domain which possess a zinc finger structure characteristic of many proteins which are able to bind DNA or RNA. The organisation of the zinc finger structure, including its proximity to a basic region within SF3 is similar to that of the chicken LIM domain transcription factor Eryf and its murine homologue GF-1 (Evans & Felsenfield, 1989; Orkin, 1990; Tsai *et al.*, 1989). Preliminary evidence suggests that the LIM domains alone are sufficient to bind DNA in *in vitro* assays and that an SF3 homologue is present in *Arabidopsis* (Baltz & Steinmetz, personal communication). Thus, SF3 may represent the first example of a regulatory gene which functions in the developing male gametophyte. If so, then it will be an important challenge to identify target genes which may be regulated by the SF3 protein. Given the appearance of SF3 transcripts in maturing pollen, a large number of candidate target promoters are already available, and experiments are under way to investigate their functional relationship with SF3.

Post-transcriptional regulation and the progamic phase

The hydration and germination of pollen grains on the stigma surface represents the initiation of the progamic phase of male gametophyte development. In most angiosperms, particularly the phylogenetically more advanced species with tricellular pollen, germination and pollen tube growth occur rapidly, within 2–60 min, after pollination. This is clearly a desirable trait where pollen competition occurs resulting from heavy pollen loads and limiting numbers of ovules (Mulcahy, 1979). A second advantage of rapid fertilisation is that it reduces the risk of inter-ference from adverse environmental conditions. To enable the rapid activation of synthetic and catabolic processes necessary for rapid growth, the pollen grain accumulates and stores large amounts of both protein and RNA. Thus, high rates of both transcription and translation are important during pollen maturation. In contrast, by the use of specific inhibitors, the germination and early tube growth of pollen is demon-strated to be largely independent of transcription but strictly dependent on translation. These results are consistent with the concept that the pollen grain synthesises all of the proteins and mRNAs necessary for germination and early tube growth during pollen maturation. However, species specific differences are apparent in the presence of polysomes in fresh pollen and the response to inhibitors of transcription and translation which emphasises differences in the importance of transcription and translation during the progamic phase among diverse pollen species (Mascarenhas, 1975; Hoekstra & Bruinsma, 1979).

Further insight into this concept was provided by experiments in which the pattern of proteins synthesised in *in vitro* grown pollen tubes was found to be very similar to that obtained from *in vitro* translation of mRNA isolated from mature pollen (Mascarenhas *et al.*, 1984). Thus, many diverse mRNA species appear to be stored and subsequently trans-lated during pollen germination. This strongly suggests a role for their encoded proteins during germination and early tube growth. Such stored transcripts include the late pollen transcripts several of which have been shown to persist in the growing pollen tube for up to 20 h after germina-tion *in vitro* (Brown & Crouch, 1990; Ursin *et al.*, 1989; Weterings *et al.*,

1992). In contrast Stinson *et al.* (1987) observed a rapid decline, within 30–60 minutes, of transcript levels corresponding to two late pollen-specific cDNAs and actin mRNA in maize pollen grown *in vitro*. Whether this rapid increase in transcript turnover is peculiar to maize pollen or is an artefact of the *in vitro* growth conditions, which are poor for maize pollen, requires further *in vivo* studies. Only one other detailed quantitative study of mRNA levels during pollen germination and tube growth has been performed. Weterings *et al.* (1992) showed that transcripts of the tobacco pollen-specific *Nt303* gene increased two–three-fold following germination *in vitro*, reached a maximum after 2 h and declined rapidly after 5 h to undetectable levels after 20 h. This increase was shown to be correlated with *de novo* transcription of the *Nt303* gene. This example serves to illustrate that transcriptional processes continue to play a role in the germinating pollen grain. However, since the majority of *Nt303* mRNA present in germinating pollen is presynthesised prior to dehydration, it further emphasises the importance of translational processes early in the progamic phase.

An important question regarding the storage of late pollen transcripts is whether they are translated prior to storage in the dehydrated pollen grain. Following the demonstration that pollen-specific polygalacturonase-related late pollen proteins were synthesised in the developing pollen grain of *Oenothera* (Brown & Crouch, 1990) several studies, mostly involving pollen allergens, have shown that late pollen encoded transcripts are also translated prior to dehydration (Table 1). Thus, typically late pollen transcripts appear to be translated concomitant with their appearance in the cytoplasm. However, it may be envisaged that transcripts produced later during pollen maturation which are destined for storage and subsequent translation, are either not translated and exist perhaps as ribonucleoprotein complexes, or are simply translated at lower efficiency during the onset of pollen dehydration. In the latter scenario, it may be an advantage for certain mRNAs to retain high rates of translation in an increasingly hostile cellular environment.

Concentration of the cytoplasm concurrent with pollen dehydration may result in an increase in cytoplasmic ion concentration which may increase the stability of secondary or tertiary mRNA structures. If structures are stabilised within 5'-untranslated leader (5'-UTL) regions this would progressively decrease the efficiency of translation. Studies of the translation of 5'-UTLs with different potentials for secondary structure in cell free extracts, in which the K+ ion concentration was varied, support this concept (for review see Sonenberg, Guertin & Lee, 1982). In order to maintain translational efficiency under such conditions, one strategy would be to maintain a structureless 5'-UTL sequence. In this regard, examination of the 5'-UTLs of a number of cloned late pollen transcripts revealed very low indices of secondary structure, despite the fact that these leaders range in size from 80 to 273 nt, which is considerably longer than the majority of other plant 5'-UTLs (30–50 nt; Joshi, 1987). An alternative strategy to maintain translation under conditions where secondary structures are stabilised may involve the binding of specific protein factors to the 5'-UTLs of particular leaders or an increase in the concentration of translational initiation factors such as eIF-4E, the cap-binding protein. Indeed, eIF-4E overexpression has been shown to increase the translational efficiency of mRNAs with stable secondary structure in mammalian cells (Koromilas, Lazaris-Karatzas & Sonenberg, 1992). Although no information is available on the activity or abundance of eIF-4E in pollen, a late pollen-specific gene encoding a protein very closely related to eIF-4A has been identified in tobacco (Owttrim, Hofman & Kuhlemeier, 1991; Brander & Kuhlemeier, personal communication). Since eIF-4A possesses ATP-dependent RNA helicase activity in association with the cap binding complex eIF-4F, increased levels of eIF-4A may contribute to an increased translational efficiency during pollen maturation by unwinding of 5'-UTLs.

With regard to the above, the role of the *lat*52 5'-UTL sequence in contributing to pollen-specific gene expression has been investigated. In particular because the *lat*52 5'-UTL shows some unusual features compared with the majority of eukaryotic 5'-UTLs, namely that it is unusually long (110 nt) and rich in adenylate residues (61%). In common with many other 5' -UTLs the *lat*52 5'-UTL is devoid of any predicted stable secondary structure (-7.5 kcal/mmol). The results of this analysis has led to the discovery that the *lat*52 5'-UTL functions as a translational enhancer in both wheat germ and rabbit reticulocyte in *in vitro* translation systems (Bate, Turner, Foster & Twell, unpublished observations). Translation of chimeric transcripts containing the *lat*52 5'-UTL linked to the firefly luciferase (*luc*) coding region is enhanced 3-to 5-fold above transcripts containing a 5'-UTL consisting of synthetic polylinker sequences. The level of enhancement mediated by the *lat*52 5'-UTL is comparable to the levels of enhancement observed for the potato virus S 5'-UTL (VTE), and other plant viral 5'-UTLs which have been shown to act as translational enhancers (Carrington & Freed, 1990; Gallie & Walbot, 1992).

Of particular interest are the results of *in vivo* experiments designed to investigate the relative levels of translational enhancement in different plant tissues. Using the microprojectile-bombardment-mediated transient expression assay (Twell *et al.*, 1989*b*), the activity of different 5'-UTL sequences was determined in both germinating pollen and in leaves. Constructs encoded chimeric transcripts differing only in their 5′-UTL sequence, which were inserted between the CaMV35S promoter and the initiator AUG of the *luc* coding region. Relative levels of enhancement in leaves were similar for constructs containing the *lat*52 or the VTE 5′-UTL, which in turn were similar to the levels of enhancement observed *in vitro* in leaves (Fig. 7). In pollen, however, the *lat*52 5′-UTL led to at least 20-fold higher levels of LUC activity relative to a synthetic polylinker 5′-UTL sequence. In contrast, the VTE 5′-UTL led to similar relative levels of enhancement in both leaves and pollen. These data strongly suggest that the *lat*52 5′-UTL acts as a translational enhancer *in vivo*, and further suggest that specific mechanisms operate to mediate the preferential activity of this 5′-UTL in germinating pollen. The nature of such mechanisms may involve *lat*52 5′-UTL-specific RNA-binding proteins and/or modulation of components of the basic translational machinery in pollen as discussed above. In future experiments, it is important to establish whether other pollen 5′-UTLs act as translational enhancers, to identify whether enhancement involves particular nucleotide sequences, and whether translation proceeds by conventional ribosome scanning or by alternative mechanisms involving internal ribosome entry used by some mammalian and plant viral transcripts (Hershey, 1991).

Concluding remarks

It is only within the past five years that significant progress has been made in understanding the repertoire of gene expression and the regulatory mechanisms which operate in developing microspores and pollen grains. Significantly, research effort in this area has been driven by both academic interest and the biotechnology industry whose interests lay primarily in the genetic manipulation of male fertility for hybrid seed production (Mariani *et al.*, 1990). This combined effort has had considerable benefit in the progress of research aimed at understanding male gametophyte development and function, in particular by providing molecular probes, improved technologies and most importantly through the stimulation of research interest and activity in this area. With the continuing effort which is being made in several laboratories, we can expect exciting new findings concerning the identity and regulation of protein factors which control haploid gene expression and the developmental signals which interact with such regulatory genes, which together determine the differentiation of the male gametophyte.

Acknowledgements

Work in the author's laboratory is supported by The Royal Society, AFRC, SERC and The Ministry of Agriculture, Food and Fisheries. I am particularly indebted to Neil Bate, Colin Eady, David Lonsdale and Koen Weterings for allowing their unpublished work to be discussed in this chapter.

References

Albani, D., Altosaar, I., Arnison, P.G. & Fabijanski, S.F. (1991). A gene showing sequence similarity to pectin esterase is specifically expressed in developing pollen of *Brassica napus*. Sequences in its 5′ flanking region are conserved in other pollen-specific promoters. *Plant Molecular Biology*, **16**, 501–13.

Albani, D., Robert, L.S., Donaldson, P.A., Altosaar, I., Arnison, P.G. & Fabijanski, S.F. (1990). Characterisation of a pollen-specific gene family from *Brassica napus* which is activated during early microspore development. *Plant Molecular Biology*, **15**, 605–22.

Albani, D., Sardana, R., Robert, L.S., Altosaar, I., Arnison, P.G. & Fabijanski, S.F. (1992). A *Brassica napus* gene family which shows sequence similarity to ascorbate oxidase is expressed in developing pollen – molecular characterization and analysis of promoter activity in transgenic plants. *The Plant Journal*, **2**, 331–42.

Allen, R.L. & Lonsdale, D.M. (1992). Sequence analysis of three members of the maize polygalacturonase gene family expressed during pollen development. *Plant Molecular Biology*, **20**, 343–5.

Allen, R.L. & Lonsdale, D.M. (1993). Molecular characterisation of one of the maize polygalacturonase gene family members which are expressed during late pollen development. *The Plant Journal*, **3**, 261–71.

Baerson, S.R. & Lamppa, G.K. (1993). Developmental regulation of an acyl carrier protein gene promoter in vegetative and reproductive tissues. *Plant Molecular Biology*, **22**, 255–67.

Baltz, R., Domon, C. & Steinmetz, A. (1992*a*). Characterization of a pollen-specific cDNA from sunflower encoding a zinc finger protein. *The Plant Journal*, **2**, 713–21.

Baltz, R., Evrard, J.L., Domon, C. & Steinmetz, A. (1992*b*). A LIM motif is present in a pollen-specific protein. *The Plant Cell*, **4**, 1465–6.

Barakate, A., Martin, W., Quigley, F. & Mache, R. (1993). Characterization of a multigene family encoding an exopolygalacturonase in maize. *Journal of Molecular Biology*, **229**, 797–801.

Bedinger, P. (1992). The remarkable biology of pollen. *The Plant Cell*, **4**, 879–87.

Bedinger, P. & Edgerton, M.D. (1990). Developmental staging of maize microspore proteins. *Plant Physiology*, **92**, 474–9.

Bemis, W.P. (1959). Selective fertilization in lima beans. *Genetics,* **44**, 555–62.

Benito Moreno, R.M., Macke, F., Alwen, A. & Heberle-Bors, E. (1988). *In-situ* seed production after pollination with *in vitro*-matured, isolated pollen. *Planta,* **176**, 145–8.

Bianchi, A. & Lorenzoni, C. (1975). Gametophytic factors in *Zea mays*. In *Gamete Competition in Plants and Animals,* ed. D.L. Mulcahy, pp. 65–82. Amsterdam: Elsevier.

Breiteneder, H., Ferreira, F., Hoffman-Sommergruber, K., Ebner, C., Breitenbach, M., Rumpold, H., Kraft, D. & Scheiner, O. (1993). Four recombinant isoforms of Cor a I, the major allergen of hazel pollen, show different IgE-binding properties. *European Journal of Biochemistry,* **212**, 355–62.

Breiteneder, H., Pettenburger, K., Bito, A., Valenta, R., Kraft, D., Rumpold, H., Scheiner, O. & Breitenbach, M. (1989). The gene coding for the major birch pollen allergen BetvI, is highly homologous to a pea disease resistance response gene. *EMBO Journal,* **8**, 1935–8.

Brewbaker, J.L. (1967). The distribution and phylogenetic significance of binucleate and trinucleate pollen grains in the Angiosperms. *American Journal of Botany,* **54**, 1069–83.

Brink, R.A. & MacGillivray, J.H. (1924). Segregation for the waxy character in maize pollen and differential development of the male gametophyte. *American Journal of Botany,* **11**, 465–9.

Brown, S.M. & Crouch, M.L. (1990). Characterization of a gene family abundantly expressed in *Oenothera organensis* pollen that shows sequence similarity to polygalacturonase. *The Plant Cell,* **2**, 263–74.

Cameron, D.R. & Moav, R.M. (1957). Inheritance in *Nicotiana tabacum*. XXVII. Pollen killer, an alien genetic locus inducing abortion of microspores not carrying it. *Genetics,* **42**, 326–35.

Carpenter, J.L., Kopczak, S.D., Snustad, D.P. & Silflow, C.D. (1993). Semi-constituitive expression of an *Arabidopsis thaliana* α-tubulin gene. *Plant Molecular Biology,* **21**, 937–42.

Carpenter, J.L., Ploense, S.E., Snustad, D.P. & Silflow, C.D. (1992). Preferential expression of an α-tubilin gene of *Arabidopsis* in pollen. *The Plant Cell,* **4**, 557–71.

Carrington, J.C. & Freed, D.D. (1990). Cap-independent enhancement of translation by a plant polyvirus 5' nontranslated region. *Journal of Virology,* **64**, 1590–7.

Casper, T. & Quail, P.H. (1993). Promoter and leader regions involved in the expression of the *Arabidopsis* ferredoxin A gene. *The Plant Journal,* **3**, 161–74.

Chen, C.-H., Oishi, K.K., Kloeckener-Gruissem, B. & Freeling, M. (1987). Organ-specific expression of maize *Adh1* is altered after *Mu* transposon insertion. *Genetics,* **116**, 469–77.

Dawe, R.K., Lachmansingh, A.R. & Freeling, M. (1993). Transposon-mediated mutations in the untranslated leader of maize *Adh1* that increase and decrease pollen-specific gene expression. *The Plant Cell,* **5**, 311–19.

Delvallée, I. & Dumas, C. (1988). Anther development in *Zea mays*: changes in proteins, peroxidase and esterase pattern. *Journal of Plant Physiology,* **132**, 210–17.

DeWitt, N., Harper, J.F. & Sussman, M.R. (1991). Evidence for a plasma membrane proton pump in phloem cells of higher plants. *The Plant Journal,* **1**, 121–8.

Dickinson, H.G. (1993). The regulation of sexual development in plants. *Philosophical Transactions of the Royal Society of London B,* **339**, 147–57.

Doll, H. (1967). Segregation frequencies of induced chlorophyll mutants in barley. *Hereditas,* **58**, 464–72.

Eady, C., Lindsey, K. & Twell, D. (1994). Differential activation and conserved vegetative cell-specific activity of a late pollen promoter in species with bi-and tricellular pollen. *The Plant Journal,* **5** 543–50.

Evans, T. & Felsenfield, G. (1989). The erythroid-specific transcription factor *EryF*1: a new finger protein. *Cell,* **58**, 877–85.

Foster, G.D., Robinson, S.W., Blundell, R.P., Roberts, M.R., Hodge, R., Draper, J. & Scott, R.J. (1992). A *Brassica napus* messenger RNA encoding a protein homologous to phospholipid transfer proteins, is expressed specifically in the tapetum and developing microspores. *Plant Science,* **84**, 187–92.

Freeling, M. (1976). Intragenic recombination methods in maize: pollen analysis methods and the effect of parental Adh+isoalleles. *Genetics,* **83**, 701–17.

Gallie, D.R. & Walbot, V. (1992). Identification of the motifs within the tobacco mosaic virus 5′-leader responsible for enhancing translation. *Nucleic Acids Research,* **20**, 4631–8.

Ghosh, B., Perry, M.P. & Marsh, D.G. (1991). Cloning the cDNA encoding the *Ambt*V allergen from giant ragweed (*Ambrosia trifida*) pollen. *Gene,* **101**, 231–8.

Gregerson, R., McLean, M., Beld, M., Gerats, A.G.M. & Strommer, J. (1991). Structure, expression, chromosomal location and product of the gene encoding ADH1 in *Petunia*. *Plant Molecular Biology,* **17**, 37–48.

Griffith, I.J., Smith, P.M., Pollock, J., Theerakulpisut, P., Avjioglu, A., Davies, S., Hough, T., Singh, M.B., Simpson, R.J., Ward, L.D. & Knox, R.B. (1991). Cloning and sequencing of *Lol p*I, the major allergenic protein of rye-grass pollen. *FEBS Letters,* **279**, 210–15.

Guerrero, F.D., Crossland, L., Smutzer, G.S., Hamilton, D.A. & Mascarenhas, J.P. (1990). Promoter sequences from a maize pollen-specific gene direct tissue-specific transcription in tobacco. *Molecular and General Genetics,* **224**, 161–68.

Hackett, R.M., Lawrence, M.J. & Franklin, F.C.H. (1992). A *Brassica* S-locus related gene promoter directs expression in both pollen and pistil of tobacco. *The Plant Journal,* 2, 613–17.

Hamilton, D.A., Bashe, D.M., Stinson, J.R. & Mascarenhas, J.P. (1989). Characterization of a pollen-specific genomic clone from maize. *Sexual Plant Reproduction,* 2, 208–12.

Hamilton, D.A., Roy, M., Rueda, J., Sindhu, R.K., Sanford, J. & Mascarenhas, J.P. (1992). Dissection of a pollen-specific promoter from maize by transient transformation assays. *Plant Molecular Biology,* 18, 211–18.

Hanson, D.D., Hamilton, D.A., Travis, J.I., Bashe, D.M. & Mascarenhas, J.P. ((1989). Characterization of a pollen-specific cDNA clone from *Zea mays* and its expression. *The Plant Cell,* 1, 173–9.

Hauffe, K.D., Lee, S.P., Subramaniam, R. & Douglas, C.J. (1993). Combinatorial interactions between positive and negative *cis*-acting elements control spatial patterns of *4CL-1* expression in transgenic tobacco. *The Plant Journal,* 4, 235–53.

Hershey, J.W.B. (1991). Translational control in mammalian cells. *Annual Review of Biochemistry,* 61, 717–55.

Hodge, R., Paul, W., Draper, J. & Scott, R. (1991). Cold-plaque screening: a simple technique for the isolation of low abundance, differentially expressed transcripts from conventional cDNA libraries. *The Plant Journal,* 2, 257–60.

Hoekstra, F.A. & Bruinsma, J. (1979). Protein synthesis of binucleate and trinucleate pollen and its relationship to tube emergence and growth. *Planta,* 146, 559–66.

Jefferson, R.A., Kavanagh, T.A. & Bevan, M.W. (1987). GUS fusions: β-glucuronidase as a sensitive and versatile gene fusion marker in higher plants. *EMBO Journal,* 6, 3901–7.

Jones, D.F. (1928). *Selective Fertilisation.* Chicago: University of Chicago Press.

Joshi, C.P. (1987). An inspection of the domain between putative TATA box and translation start site in 79 plant genes. *Nucleic Acids Research,* 15, 6643–53.

Kim, S.-R., Kim, Y. & An, G. (1993). Molecular cloning and characterization of anther-preferential cDNA encoding a putative actin depolymerising factor. *Plant Molecular Biology,* 21, 39–45.

Kloeckener-Gruissem, B., Vogel, J.M. & Freeling, M. (1992). The TATA box promoter region of maize *Adh*1 affects its organ-specific expression. *EMBO Journal,* 11, 157–66.

Kononowicz, A.K., Nelson, D.E., Singh, N.K., Hasegawa, P.M. & Bressan, R.A. (1992). Regulation of the osmotin gene promoter. *The Plant Cell,* 4, 513–24.

Koromilas, A.E., Lazaris-Karatzas, A. & Sonenberg, N. (1992). mRNAs containing extensive secondary structure in their 5′ non-coding region translate efficiency in cells overexpressing initiation factor eIF-4E. *EMBO Journal*, **11**, 4153–8.

Liu, X.-Y., Rocha-Sosa, M., Hummel, S., Wilmitzer, L. & Frommer, W.B.(1991). A detailed study of the regulation and evolution of the two classes of patatin genes in *Solanum tuberosum* L. *Plant Molecular Biology*, **17**, 1139–54.

Loegering, W.Q. & Sears, E.R. (1963). Distorted inheritance of stem-rust resistance of timstein wheat caused by a pollen-killing gene. *Canadian Journal of Genetics and Cytology*, **5**, 65–72.

McCormick, S. (1991). Molecular analysis of male gametogenesis in plants. *Trends in Genetics*, **7**, 298–303.

Mandaron, P., Niogret, M.F., Mache, R. & Monegar, F. (1990). *In vitro* protein synthesis in isolated microspores of *Zea mays* at several stages of development. *Theoretical and Applied Genetics*, **80**, 134–8.

Mariani, C., de Beuckeleer, M., Truettner, J., Leemans, J. & Goldberg, R.B. (1990). Induction of male sterility in plants by a chimeric ribonuclease gene. *Nature*, **347**, 737–41.

Marsolier, M.-C., Carrayol, E. & Hirel, B. (1993). Multiple functions of promoter sequences involved in organ-specific expression and ammonia regulation of a cytosolic soybean glutamine synthetase gene in transgenic *Lotus corniculatus*. *The Plant Journal*, **3**, 405–14.

Mascarenhas, J.P. (1975). The biochemistry of angiosperm pollen development. *Botanical Review*, **41**, 259–314.

Mascarenhas, J.P. (1990). Gene activity during pollen development. *Annual Review of Plant Physiology, Plant Molecular Biology*, **41**, 317–38.

Mascarenhas, J.P. (1992). Pollen gene expression. In *International Review of Cytology: Sexual Reproduction in Flowering Plants*, ed. S.D. Russell & C. Dumas, pp. 3–18. San Diego: Academic Press.

Mascarenhas, N.T., Bashe, D., Eisenberg, A., Willing, R.P., Xiao, C.M. & Mascarenhas, J.P. (1984). Messenger RNAs in corn pollen and protein synthesis during germination and pollen tube growth. *Theoretical and Applied Genetics*, **68**, 323–6.

Meinke, D.W. (1982). Embryo-lethal mutants of *Arabidopsis thaliana*: evidence for gametophytic expression of the mutant genes. *Theoretical and Applied Genetics*, **63**, 381–6.

Meinke, D.W. & Baus, A.D. (1986). Gametophytic gene expression in embryo-lethal mutants of *Arabidopsis thaliana*. In *Biotechnology and Ecology of Pollen*, ed. D.L. Mulcahy, G. Bergamini & E. Ottaviano, pp. 57–84. Berlin and New York: Springer-Verlag.

Michel, D., Salamini, F., Bartels, D., Dale, P., Baga, M. & Szalay, A. (1993). Analysis of a dessication and ABA-responsive promoter isolated from the resurrection plant *Craterostigma plantagineum*. *The Plant Journal*, **4**, 29–40.

Mulcahy, D.L. (1979). The rise of the angiosperms: a genecological factor. *Science*, **206**, 20–3.

Mulcahy, D.L., Mulcahy, G.B. & Robinson, R.W. (1979). Evidence for postmeiotic genetic activity in pollen of *Curcurbita* species. *Journal of Heredity*, **70**, 365–8.

Mulcahy, D.L., Robinson, R.W., Ihara, M. & Kesseli, R. (1981). Gametophytic transcription for acid phosphatases in pollen of *Curcurbita* species hybrids. *Journal of Heredity*, **72**, 353–4.

Niogret, M.F., Dubald, M., Mandaron, P. & Mache, R. (1991). Characterization of pollen polygalacturonase encode by several cDNA clones in maize. *Plant Molecular Biology*, **17**, 1155–64.

Nishihara, M., Ito, M., Tanaka, I., Kyo, M., Ono, K., Irifune, K. & Morikawa, H. (1993). Expression of the beta-glucuronidase gene in pollen of lily (*Lilium longiflorum*), tobacco (*Nicotiana tabacum*), *Nicotiana rustica* and peony (*Peonia lactiflora*) by particle bombardment. *Plant Physiology*, **102**, 357–61.

Ohl, S., Hedrick, S.A. Chory, J. & Lamb, C.J. (1990). Functional properties of a phenylalanine ammonia-lyase promoter from *Arabidopsis*. *The Plant Cell*, **2**, 837–48.

Olmedilla, A., Schrauwen, J.A.M. & Wullems, G.J. (1991). Visualization of starch-synthase expression by *in situ* hybridisation during pollen development. *Planta*, **184**, 182–6.

Olsen, E., Zhand, L., Hill, R.D., Kisil, F.T., Sehon, A.H. & Mohapatra, S.S. (1991). Identification and characterization of the *Poa p* IX group of basic allergens of Kentucky Bluegrass pollen. *Journal of Immunology*, **147**, 205–11.

Orkin, S.H. (1990). Globin gene regulation and switching: circa 1990. *Cell*, **63**, 665–72.

Ottaviano, E. & Mulcahy, D.L. (1989). Genetics of angiosperm pollen. *Advances in Genetics*, **26**, 1–64.

Ottaviano, E., Petroni, D. & Pe, M.E. (1987). Gametophytic expression of genes controlling endosperm development in maize. *Theoretical and Applied Genetics*, **75**, 252–8.

Owttrim, G.W., Hofman, S. & Kuhlemeier, C. (1991). Divergent genes for translation initiation factor eIF-4A are coordinately expressed in tobacco. *Nucleic Acids Research*, **19**, 5491–6.

Pacini, E. (1990). Tapetum and microspore function. In *Microspores: Evolution and Ontogeny*, ed. S. Blackmore & R.B. Knox, pp. 213–237. London: Academic Press.

Parnell, F.R. (1921). Note on the detection of segregation by examination of the pollen of rice. *Journal of Genetics*, **11**, 209–12.

Perez, M., Ishioka, G.Y., Walker, L.E. & Chesnut, R.W. (1990). cDNA cloning and immunological characteristics of the rye grass allergen *Lolp*I. *Journal of Biological Chemistry*, **265**, 16210–15.

Rafnar, T., Rogers, B.L., Bond, J.E., Kuo, M. & Klapper, D.G. (1991). Isolation of cDNA clones coding for the major short ragweed allergen *Amb a* I(antigen E.). *Journal of Biological Chemistry*, **266**, 1229–36.

Raghavan, V. (1986). *Embryogenesis in Angiosperms. A Developmental and Experimental Study*. Cambridge: Cambridge University Press.

Raghavan, V. (1989). mRNAs and a cloned histone gene are differentially expressed during anther and pollen development in rice (*Oryza sativa* L.). *Journal of Cell Science*, **92**, 217–29.

Regan, S.M. & Moffat, B.A. (1990). Cytochemical analysis of pollen development in wild-type *Arabidopsis* and a male-sterile mutant. *The Plant Cell*, **2**, 877–89.

Reijnen, W.H., van Herpen, M.M.A., de Groot, P.F.M., Olmedilla, A., Schrauwen, J.A.M., Weterings, K.A.P. & Wullems, G.J. (1991). Cellular localization of a pollen-specific mRNA by *in situ* hybridization and confocal laser scanning microscopy. *Sexual Plant Reproduction*, **4**, 254–7.

Restrepo, M.A., Freed, D.D. & Carrington, J.C. (1990). Nuclear transport of plant potyviral proteins. *The Plant Cell*, **2**, 987–8.

Roberts, M.R., Foster, G.D., Blundell, R.P., Robinson, S.W., Kumar, A., Draper, J. & Scott, R. (1993a). Gametophytic and sporophytic expression of an anther-specific *Arabidopsis thaliana* gene. *The Plant Journal*, **3**, 111–20.

Roberts, M.R., Hodge, R., Ross, J.H.E., Sorensen, A., Murphy, D.J., Draper, J. & Scott, R.(1993b). Characterization of a new class of oleosins suggests a male gametophyte-specific lipid storage pathway. *The Plant Journal*, **3**, 629–36.

Roberts, M.R., Robson, F., Foster, G.D., Draper, J. & Scott, R.J. (1991). A *Brassica napus* mRNA expressed specifically in developing microspores. *Plant Molecular Biology*, **17**, 295–9.

Rogers, H.J., Harvey, A. & Lonsdale, D.M. (1992). Isolation and characterization of a tobacco gene with hemology to pectate lyase which is specifically expressed during microsporogenesis. *Plant Molecular Biology*, **20**, 493–502.

Rogers, H.J. & Lonsdale, D.M. (1992). Genetic manipulation of male sterility for the production of hybrid seed. *Plant Growth Regulators*, **11**, 21–6.

Samac, D.A. & Shah, D.M. (1991). Developmental and pathogen-induced activation of the *Arabidopsis* acidic chitinase promoter. *The Plant Cell*, **3**, 1063–72.

Sari-Gorla, M. & Rovida, E. (1980). Competitive ability of maize pollen. Intergametophytic effects. *Theoretical and Applied Genetics*, **57**, 37–41.

Schrauwen, J.A.M., de Groot, P.F.M., van Herpen, M.M.A., van der Lee, T., Reynen, W.H., Weterings, K.A.P. & Wullems, G.J. (1990). State-related expression of mRNAs during pollen development in lily and tobacco. *Planta*, **182**, 298–304.

Schwartz, D. (1971). Genetic control of alcohol dehydrogenase – a competition model for regulation of gene action. *Genetics*, **67**, 411–25.

Scott, R., Dagless, E., Hodge, R., Paul, W., Soufleri, I. & Draper, J. (1991*a*). Patterns of gene expression in developing anthers of *Brassica napus*. *Plant Molecular Biology*, **17**, 195–207.

Scott, R., Hodge, R., Paul, W. & Draper, J. (1991*b*). The molecular biology of anther differentiation. *Plant Science*, **80**, 167–91.

Shen, J.B. & Hsu, F.C. (1992). *Brassica* anther-specific genes: characterization and *in situ* localisation of expression. *Molecular and General Genetics*, **234**, 379–89.

Silvanovich, A., Astwood, J., Zhang, L., Olsen, E., Kisil, F., Sehon, A., Mohapatra, S. & Hill, R. (1991). Nucleotide sequence analysis of three cDNAs coding for *Poa p*IX isoallergens of Kentucky bluegrass pollen. *Journal of Biological Chemistry*, **266**, 1204–10.

Singh, A. & Kao, T.-H. (1992). Gametophytic self-incompatibility: biochemical, molecular genetic and evolutionary aspects. In *International Review of Cytology: Sexual Reproduction in Flowering Plants*, ed. S.D. Russell & C. Dumas, pp. 449–483. San Diego: Academic Press.

Singh, M.B., Hough, T., Theerakulpisut, P., Avjioglu, A., Davies, S., Smith, P.M., Taylor, P., Simpson, R.J., Ward, L.D., McCluskey, J., Puy, R. & Knox, R.B. (1991). Isolation of a cDNA encoding a newly identified major allergenic protein of ryegrass pollen: intracellular targeting to the amyloplast. *Proceedings of the National Academy of Sciences, USA*, **88**, 1384–8.

Singh, M.B., O'Neil, P.M. & Knox, R.B. (1985). Initiation of postmeotic β-galactosidase synthesis during microsporogenesis in oilseed rape. *Plant Physiology*, **77**, 225–8.

Singleton, W.R. & Mangelsdorf, P.C. (1940). Gametic lethals on the fourth chromosome of maize. *Genetics*, **25**, 366–90.

Sonenberg, N., Guertin, D. & Lee, K.A.W. (1982). Capped mRNAs with reduced secondary structure can function in extracts from poliovirus-infected cells. *Molecular and Cell Biology*, **2**, 1633–8.

Sprague, G.F. (1933). Pollen tube establishment and deficiency of *waxy* seeds in certain maize crosses. *Proceedings of the National Academy of Sciences, USA*, **19**, 838–41.

Stadler, L.J. (1933). On the genetic nature of induced mutations in plants. II. A haploviable deficiency in maize. *Research Bulletin of the Missouri Agricultural Experimental Station*, **204**, 29.

Stadler, L.J. & Roman, H. (1948). The effect of X-rays upon mutation of a gene A in maize. *Genetics*, **33**, 273–303.

Stinson, J. & Mascarenhas, J.P. (1985). Onset of alcohol dehydrogenase synthesis during microsporogenesis in maize. *Plant Physiology*, **77**, 222–4.

Stinson, J.R., Eisenberg, A.J., Willing, R.P., Pe, M.P., Hanson, D.D. & Mascarenhas, J.P. (1987). Genes expressed in the male gametophyte of flowering plants and their isolation. *Plant Physiology*, **83**, 442–7.

Sunderland, N. & Huang, B. (1987). Ultrastructural aspects of pollen dimorphism. In *International Review of Cytology, Pollen: Cytology and Development*, ed. K.L. Giles & J. Prakash, pp. 175–220. London: Academic Press.

Takahashi, T., Naito, S. & Komeda, Y. (1992). The *Arabidopsis* HSP18.2 promoter/GUS gene fusion in transgenic *Arabidopsis* plants: a powerful tool for the isolation of regulatory mutants of the heat shock response. *The Plant Journal*, **2**, 751–61.

Tanaka, I. & Ito, M. (1980). Induction of typical cell division in isolated microspores of *Lilium longiflorum* and *Tulipa gesneriana*. *Plant Science Letters*, **17**, 279–85.

Tanaka, I. & Ito, M. (1981). Control of division patterns in explanted microspores of *Tulipa gesneriana*. *Protoplasma*, **108**, 329–40.

Terasaka, O. & Niitsu, T. (1987). Unequal cell division and chromatin differentiation in pollen grain cells. I. centrifugal, cold and caffeine treatments. *Botanical Magazine, Tokyo*, **100**, 205–16.

Thangavelu, M., Belostotsky, D., Bevan, M.W., Flavell, R.B., Rogers, H.J. & Lonsdale, D.M. (1993). Partial characterization of the *Nicotiana tabacum* actin gene family: evidence for pollen-specific expression of one of the gene family members. *Molecular and General Genetics*, **240**, 290–5.

Theerakulpisut, P., Xu, H., Singh, M.B., Pettitt, J.M. & Knox, R.B. (1991). Isolation and developmental expression of *Bcp*1, an anther-specific cDNA clone in *Brassica campestris*. *The Plant Cell*, **3**, 1073–84.

Thorsness, M.K., Kandasamy, M.K., Nasrallah, M.E. & Nasrallah, J.B. (1991). A *Brassica* S-locus gene promoter targets toxic gene expression and cell death to the pistil and pollen of transgenic *Nicotiana*. *Developmental Biology*. **143**, 173–84.

Topping, J.F., Wei, W., Lindsey, K. (1991). Functional tagging of regulatory elements in the plant genome. *Development*, **112**, 1009–19.

Tsai, S.-F., Martin, D.I.K., Zon, L.I., D'Andrea, A.D., Wong, S. & Orkin, S.-H. (1989). Cloning of cDNA for the major DNA binding protein of the erythroid lineage through expression in mammalian cells. *Nature*, **339**, 446–51.

Tsuchiya, T., Toriyama, K., Nasrallah, M.E. & Ejiri, S. (1992). Isolation of genes abundantly expressed in rice anthers at the microspore stage. *Plant Molecular Biology*, **20**, 1189–93.

Tupy, J., Rihova, L. & Zarsky, V. (1991). Production of fertile tobacco pollen from microspores in suspension culture and its storage for *in situ* pollination. *Sexual Plant Reproduction*, **4**, 284–7.

Tupy, J., Suss, J., Hrabetova, E. & Rihova, L. (1983). Developmental changes in gene expression during pollen differentiation and maturation in *Nicotiana tabacum* L. *Biologia Plantarum*, **25**, 231–7.

Twell, D. (1992). Use of a nuclear-targeted β-glucuronidase fusion protein to demonstrate vegetative cell-specific gene expression in developing pollen. *The Plant Journal*, **2**, 887–92.

Twell, D., Klein, T.M., Fromm, M.E. & McCormick, S. (1989b). Transient expression of chimeric genes delivered into pollen by microprojectile bombardment. *Plant Physiology*, **91**, 1270–4.

Twell, D., Patel, S., Sorensen, A., Roberts, M., Scott, R., Draper, J. & Foster, G. (1993). Activation and developmental regulation of an *Arabidopsis* anther-specific promoter in microspores and pollen of *Nicotiana tabacum*. *Sexual Plant Reproduction*, **6**, 217–24.

Twell, D., Wing, R., Yamaguchi, J. & McCormick, S. (1989a). Isolation and expression of an anther-specific gene from tomato. *Molecular and General Genetics*, **217**, 240–5.

Twell, D., Yamaguchi, J. & McCormick, S. (1990). Pollen-specific gene expression in transgenic plants: Coordinate regulation of two different tomato gene promoters during microsporogenesis. *Development*, **109**, 705–13.

Twell, D., Yamaguchi, J., Wing, R.A., Ushiba, J. & McCormick, S. (1991). Promoter analysis of three genes that are coordinately expressed during pollen development reveals pollen-specific enhancer sequences and shared regulatory elements. *Genes and Development*, **5**, 496–507.

Uknes, S., Dincher, S., Freidrich, L., Negrotto, D., Williams, S., Thompson-Taylor, H., Potter, S., Ward, E. & Ryals, J. (1993). Regulation of pathogenesis-related protein-1a gene expression in tobacco. *The Plant Cell*, **5**, 159–69.

Ursin, V.M., Yamagushi, J. & McCormick, S. (1989). Gametophytic and sporophytic expression of anther-specific genes in developing tomato anthers. *The Plant Cell*, **1**, 727–36.

Valenta, R., Duchene, M., Pettenburger, K., Sillaber, C., Valent, P., Bettelheim, P., Breitenbach, M., Rumpold, H., Kraft, D. & Scheiner, O. (1991). Identification of profilin as a novel pollen allergen: IgE autoreativity in sensitised individuals. *Science*, **253**, 557–60.

van der Leede-Plegt, L.M., van de Ven, B.C.E., Bino, R.J., van der Salm, T.P.M. & van Tunen, A.J. (1992). Introduction and differential use of various promoters in pollen grains of *Nicotiana glutinosa* and *Lilium longiflorum*. *The Plant Cell Reporter*, **11**, 20–4.

van der Meer, I.M., Brouwer, M., Spelt, C.E., Mol, J.N.M. & Stuitje, A.R. (1992). The TACPyAT repeats in the chalcone synthase promoter of *Petunia hybrida* act as a dominant negative *cig*-acting module in the control of organ-specific expression. *The Plant Journal*, **2**, 525–35.

van Tunen, A.J., Hartman, S.A., Mur, L.A. & Mol, J.N.M. (1989). Regulation of chalcone flavanone isomerase (CHI) gene expression in *Petunia hybrida*: The use of alternative promoters in corolla, anthers and pollen. *Plant Molecular Biology*, **12**, 539–51.

van Tunen, A.J., Mur, L.A., Brouns, G.S., Rienstra, J.-D., Koes, R.E. & Mol, J.N.M. (1990). Pollen- and anther-specific promoters from petunia: tandem promoter regulation of the *chiA* gene. *The Plant Cell*, **2**, 393–401.

Vergné, P. & Dumas, C. (1988). Isolation of viable wheat male gameto-phytes of different stages of development and variation in their protein patterns. *Plant Physiology*, **75**, 865–8.

Walsh, D.J., Matthews, J.A., Denmeade, R. & Walker, M.R. (1989). Cloning of cDNA coding for an allergen of Cocksfoot grass (*Dactylis glomerata*) pollen. *International Archives of Allergy and Applied Immunology*, **90**, 78–83.

Walter, M.H., Liu, J.-W., Grand, C., Lamb, C.J. & Hess, D. (1990). Bean pathogenesis-related (PR) proteins deduced from elicitor-induced transcripts are members of a ubiquitous new class of conserved PR proteins including pollen allergens. *Molecular and General Genetics*, **222**, 353–60.

Warner, S.A.J., Scott, R. & Draper, J. (1992). Isolation of an asparagus intracellular PR gene (AoPR1) wound responsive promoter by the inverse polymerase chain reaction and its characterization in transgenic tobacco. *The Plant Journal*, **3**, 191–201.

Weeden, N.F. & Gottlieb, L.D. (1979). Distinguishing allozymes and isozymes of phosphoglucoisomerases by electrophoretic comparisons of pollen and somatic tissues. *Biochemical Genetics*, **17**, 287–96.

Weterings, K., Reijnen, W., van Aarssen, R., Kortstee, A., Spijkers, J., van Herpen, M., Schrauwen, J. & Wullems, G. (1992). Characterisation of a pollen-specific cDNA clone from *Nicotiana tabacum* expressed during microgametogenesis and germination. *Plant Molecular Biology*, **18**, 1101–11.

Willing, R.P., Bashe, D. & Mascarenhas, J.P. (1988). An analysis of the quantity and diversity of messenger RNAs from pollen and shoots of *Zea mays*. *Theoretical and Applied Genetics*, **75**, 751–3.

Willing, R.P. & Mascarenhas, J.P. (1984). Analysis of the complexity and diversity of mRNAs from pollen and shoots of *Tradescantia*. *Plant Physiology*, **75**, 865–8.

Wing, R.A., Yamaguchi, J., Larabell, S.K., Ursin, V.M. & McCormick, S. (1989). Molecular and genetic characterization of two pollen-expressed genes that have sequence similarity to pectate lyases of the plant pathogen *Erwinia*. *Plant Molecular Biology*, **14**, 17–28.

Xu, H.L., Davies, S.P., Kwan, B.Y.H., O'Brien, A.P., Singh, M. & Knox, R.B. (1993). Haploid and diploid expression of a *Brassica campestris* anther-specific gene promoter in arabidopsis and tobacco. *Molecular and General Genetics*, **239**, 58–65.

Zaki, M.A.M. & Dickinson, H.G. (1990). Structural changes during the first divisions of embryos resulting from anther and microspore culture in *Brassica napus*. *Protoplasma*, **156**, 149–62.

Zaki, M.A.M. & Dickinson, H.G. (1991). Microspore-derived embryos in *Brassica*: the significance of division asymmetry in pollen mitosis I to embryogenic development. *Sexual Plant Reproduction*, **4**, 48–55.

Zarsky, V., Capkova, E., Hrabetova, E. & Tupy, J. (1985). Protein changes during pollen development in *Nicotiana tabacum. Biologia Plantarum,* **27**, 438–41.

Zhang, W., McElroy, D. & Wu, R. (1991). Analysis of rice Act1 5′ region activity in transgenic rice plants. *The Plant Cell,* **3**, 1155–65.

Zhu, Q., Doerner, P.W. & Lamb, C.J. (1993). Stress induction and developmental regulation of a rice chitinase promoter in transgenic tobacco. *The Plant Journal,* **3**, 203–12.

D.L. HIRD, D. WORRALL, R. HODGE,
S. SMARTT, W. PAUL and R. SCOTT

Characterisation of *Arabidopsis thaliana* anther-specific gene which shares sequence similarity with β-1,3-glucanases

Introduction

Callase is a complex of β-1,3-glucanase activities which play a crucial role in microsporogenesis. During microsporogenesis, archaesporial cells in the anther give rise to microsporocytes and tapetal cells (Fig. 1). In almost all higher plants a thick wall of callose, a β-1,3-glucan polymer, is deposited between the cell membrane and the primary cell wall of the microsporocyte. As meiosis progresses, callose is also deposited along the cellular plates formed during cytokinesis, until each individual microspore is completely encased in callose. The tapetum forms a single layer of cells surrounding the anther locule. The tapetum is an extremely metabolically active tissue and is thought to play a nutritive role in microspore development (Chapman, 1987). Another critical function of the tapetum is the synthesis of callase which is required for the dissolution of the callose walls of the tetrad (Frankel, Izhar & Nitsan, 1969). After the primary and callosic walls of the tetrad have been degraded, the individual microspores are freed into the locule and continue their development into mature pollen grains. The developmental importance of callase activity is illustrated by the occurrence of mutants in petunia and sorghum where callase activity is premature (Frankel *et al.*, 1969; Warmke & Overman, 1972) or delayed (Izhar & Frankel, 1971). These plants are male sterile because the inappropriate expression of callase results in microspore abortion. More recently, work by Worrall *et al.* (1992) demonstrated that premature secretion of an engineered β-1,3-glucanase from the tapetum of tobacco resulted in partial or total male sterility. Taken together, these data show that the timing of callase activity is extremely important in microspore development.

β-1,3-Glucanases are hydrolytic enzymes most frequently associated with a role in plant defence (van Loon & van Kammen, 1970). Basic

Society for Experimental Biology Seminar Series 55: *Molecular and Cellular Aspects of Plant Reproduction*, ed. R.J. Scott & A.D. Stead.
© Cambridge University Press, 1994, pp. 137–158.

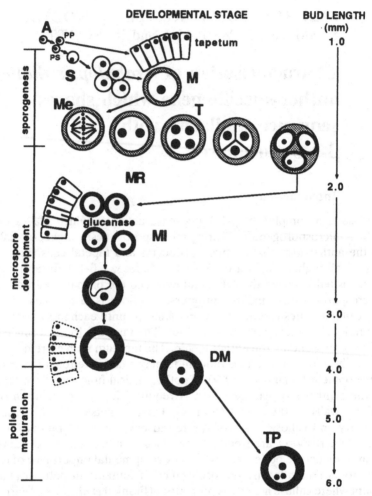

Fig. 1. *Brassica* microspore development (taken from Scott *et al.*, 1991). Diagram showing the cytological changes that occur during microsporogenesis. The relationship between bud length and the developmental stage of the anther is shown. The lightly shaded areas indicate regions of callose, the darkly shaded areas, sporopollenin. *Abbreviations*: A, archaesporial cell; PP, primary parietal cell; PS, primary sporogenous cell; S, sporocyte; M, meiocyte; Me, meiosis; T, tetrad; MR, microspore release; MI, microspore interphase; DM, dinucleate microspore; TP, trinucleate pollen.

forms of the enzyme are usually found in the vacuole (Boller & Vögeli, 1984) whereas acidic forms are secreted into the extracellular space (Parent & Asselin, 1984). Assays for β-1,3-glucanase activity are based on the nature of enzymatic action on the substrate, β-1,3-glucan polymers. Endoglucanases cleave the substrate internally yielding short chain reducing sugars. Exoglucanase hydrolysis releases single glucose units only. These different end products are the basis of assays which distinguish β-1,3-glucanase activities. Recently, in addition to their role in plant defence, another class of β-1,3-glucanases has been identified which is developmentally expressed in healthy plant tissue (Lotan Ori & Fluhr, 1989). Developmentally regulated β-1,3-glucanases can be detected in the stigma and style of tobacco flowers and are thought to be unrelated to the defence response (Lotan *et al.*, 1989). At present, no definitive role has been associated with these enzymes. Callase is the exception, as its role in microspore development has been clearly defined (Frankel *et al.*, 1969; Stieglitz & Stern, 1973) and, moreover, is critical for successful pollen production.

Work by Stieglitz (1977) in lily, demonstrated that callase consists of two components; a 62 kilodalton (kD) exoglucanase and a 32 kD endoglucanase. The graph in Fig. 2, adapted from her paper (1977) shows the dual peaks of glucanase activity in the anther at the time of microspore release. *In vitro* experiments by Stieglitz (1977) showed that semi-purified endoglucanase activity was sufficient for callose wall dissolution, and that the exoglucanase alone had no discernible effect. Stieglitz therefore suggested that the role of the exoglucanase is to degrade the oligosaccharides, produced as a result of endoglucanase activity, into readily metabolisable glucose for use by the developing microspores.

Sequence analysis of a *Brassica napus* cDNA and *Arabidopsis thaliana* genomic clone

In our work on anther development, a series of anther-specific *Brassica napus* cDNAs were isolated from a cDNA library encompassing the period of microsporogenesis (Scott *et al.*, 1991). Initial sequencing of one of these, A6, revealed sequence similarity to β-1,3-glucanases. The A6 cDNA was used as a probe to isolate two genomic clones G61 and G62 from an *Arabidopsis thaliana* genomic library (Roberts *et al.*, 1993). Initial sequencing of these showed that they were 96–98% identical to each other in both the coding and non-coding regions. Work was therefore concentrated on G62, and a 2.7 kb contiguous sequence constructed. The resulting deduced amino acid (aa) sequence was aligned with the A6 sequence, several published plant endo-β-1,3-glucanase sequences and a

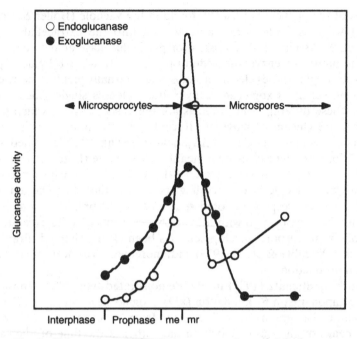

Fig. 2. Changes in endo-β-1,3-glucanase and exo-β-1,3-glucanase activity during microsporogenesis in *Lilium* anthers (cultivar Sonata) (adapted from Stieglitz, 1977). Glucanase activity in units per bud is plotted against anther development. Abbreviations: me, meiosis; mr, microspore release.

yeast endo-β-1,3-glucanase (Fig. 3). No plant exo-β-1,3-glucanase sequences have been published, and yeast exoglucanase sequences (Vazquez de Aldana *et al.*, 1991; Chambers *et al.*, 1993; Muthukumar *et al.*, 1993) display no similarity to plant or yeast endoglucanase sequences despite utilising the same substrate.

The A6 gene contains an open reading frame encoding 478 amino acids. This gives a protein of predicted molecular mass 53 kD and a calculated isoelectric point of 9.2. The polypeptide contains a putative hydrophobic signal sequence of 21 amino acids which conforms to the criteria set out by von Heijne (1983). The A6 protein also has a long C-terminal tail which extends well beyond those of other β-1,3-glucanases including vacuolar glucanases which contain a carboxy-terminal targeting sequence. A recent review by Meins *et al.* (1991) identified regions thought to be important in glucanase function, these are overlined in Fig. 3*a*. Work by various groups has also identified specific residues essential for catalytic activity (Moore & Stone, 1972; Høj *et al.*, 1989b; Macgregor & Ballance, 1991). These residues are highlighted in bold type in Fig. 3*a* and are conserved in the A6 sequences. This suggests that A6 has the essential residues necessary to function as a glucanase enzyme. Dendrogram analyses (Fig. 3*b*) of the alignment illustrates that the A6 sequences are significantly diverged from the other members of the alignment. This, in addition to the presence of a unique 114 amino acid C-terminal tail, suggests that A6 may be a member of a distinct group of plant glucanases. The alignment also suggests that A6 may be an endoglucanase since although plant and yeast endoglucanases are similar, yeast exoglucanases show no similarity to plant or yeast endoglucanases.

A Southern blot of Brassica DNA probed with the A6 cDNA shows a complex banding pattern, with 6–10 hybridising bands present in both *B. napus* and the *B. oleracea* parent (Fig. 4). A similarly complex banding pattern is seen in the *B. campestris* parent (data not shown). This suggests that A6 is a member of a large gene family in brassica. In support of this theory, three partial cDNAs also isolated from the *B. napus* cDNA library (Scott *et al.*, 1991) (A11, A20 and A28 shown in Fig. 3), show a very high degree of similarity to the A6 sequence, and are probably also β-1,3-glucanases.

The spatial and temporal expression of the *A. thaliana* A6 gene

To learn more about the spatial and temporal expression of the A6 gene, promoter fusions were made. An 885 base pair promoter fragment was linked to the reporter gene *gus* (Bevan, 1984) and the RNase *barnase* (Paul *et al.*, 1992). These constructs were transformed into tobacco and *B. napus*. Histochemical staining of anthers from A6-*gus* tobacco transformants showed tapetal expression of GUS (data not shown). GUS expression was also seen in the mature microspores and pollen of the tobacco, but not in the mature microspores and pollen of the *B. napus* A6-*gus* transformant. Fluorometric data (Fig. 5) from both tobacco anthers (a) and *B.napus* buds (b) showed a temporal pattern of GUS activity which peaked at the time of microspore release in both species. This developmental pattern is similar to that of callase as described by Stieglitz (see Fig. 2). The A6-*barnase* transformed tobacco were male sterile, female fertile and otherwise phenotypically normal. Upon dehiscence, the anther locules were completely devoid of pollen. Tapetal expression of the RNase *barnase* disrupts pollen development causing male sterility. Fig. 6 shows anther sections from a representative tobacco

(a)

```
At 2                                                      MLRDARRYLAKS
Np basic                                         LQMAAIILLGLLVSSTEIVGAQS
Pv ß13                                                              Q
Nt PR-Q                                        QFLFSLQMAHLIVTLLLLSVLTLATLDFTGAQ
Nt acidic                                        MTLCIKNGFLAAALVLVGLLICSIQMIGAQS
Hv ß1314             MAGQGVASMLALALLLGAFASIPQSMGEFFNFLLWQWNSATTLPGVES
Hv ß13                                          MARKDVASMFAAALFIGAFAAVPTSVQS
At A6                                     MSLLAFFLFTILVFSSSCCSATRFQGHRYMQRKTMLDLASK
Bn A6                                         FFLFTLVVFSSTSCSAVGFQHPHRYIQKKTMLELASK
Yeast endo                                               MRFSTTLATAATALFFTASQVSA


At 1           VG-VCYGRNGNNLPSPAETIALFKQKNIQRVRLYSPDHDVLAALRGSNIEVTLGLPNSYL
At 3           IG-VCYGRNGNNLRPASEVVALYQQRNIRRMRLYDPNQETLNALRGSNIELVLDVPNPDL
At 2           IG-RC-----GSLQTTKHPANA----------LYGPDPGALAALRGSDIELILDVPSSDL
Np basic       VG-VCYGMLGNNLPPASQVVQLYKSKNIRRMRLYDPNQAALQALRGSNIEVMLGVPNSDL
Pv ß13         IG-VCYGMMGNNLPSANEVINLYRSNNIRRMRLYDPNGAALGALRNSGIELILGVPNSDL
Nt PR-Q        AG-VCYGRQGNGLPSPADVVSLCNRRNNIRRMRIYDPDQPTLEALRGSNIELMLGVPNPDL
Nt acidic      IG-VCYGKHANNLPSDQDVINLYNANGIRKMRIYNPDTNVFNALRGSNIEIILDVPLQDL
Hv ß1314       IG-VCYGMSANNLPAASTVVNMFKSNGINSMRLYAPDQAALQAVGGTGVNVVVGAPNDVL
Hv ß13         IG-VCYGVIGNNLPSRSDVVQLYRSKGINGMRIYFADGQALSALRNSGIGLILDIGNDQL
At A6          IG-INYGRRGNNLPSPYQSINFIKSIKAGHVKLYDADPESLTLLSQTNLYVTITVPNHQI
Bn A6          IG-INYGRQGNNLPSPYQSINFIKLIKAGHVKLYDADPESLTLLSQTNLYVTIAVPTHQI
Yeast endo     IGELAFNLGVKNNDGTCKSTSDYET-ELQALKSYT-----------STVKVYAASDCNTL
                *                                         .   .  .


At 1           QSVASSQSQANAWVQTYVMNYANGVRFRYISVGNEVK---I---SDSYAQFLVPAMENID
At 3           QRLASSQAEADTWVRNNVRNYAN-VTFRYISVGNEVQ---P---SDQAASFVLPAMQNIE
At 2           ERLASSQTEADKWVQENVQSYRDGVRFRYINVGNEVK---P---S--VGGFLLQAMQNIE
Np basic       QNIAANPSNANNWVQRNVRNFWPAVKFRYIAVGNEVS---PVTGTSSLTRYLLPAMRNIR
Pv ß13         QGLATNADTARQWVQRNVLNFWPSVKIKYIAVGNEVS---PVGGSSWYAQYVLPAVQNVY
Nt PR-Q        ENVAASQANADTWVQNNVRNY-GNVKFRYIAVGNEVS---PLNENSKYVPVLLNAMRNIQ
Nt acidic      QSL-TDPSRANGWVQDNIINHFPDVKFKYIAVGNEVS---P-GNNGQYAPFVAPAMQNVY
Hv ß1314       SNLAASPAAAASWVRNSNIQAY-PKVSFRYVCVGNEVS---GGATQN-----LVPAMKNVQ
Hv ß13         ANIAASTSNAASWVQNNVRPYYPAVNIKYIAAGNEVQ---GGATQS-----ILPAMRNLN
At A6          TALSSNQTIADEWVRTNILPYYPQTQIRFVLVGNEIL---SYNSGNVSVN-LVPAMRKIV
Bn A6          TSLSANQTTAEDWVKTNILPYYPQTQIRFVLVGNEIL---SVKDRNITGN-VVPAMRKIV
Yeast endo     QNLGPAAEAEGF---TIFVGVWPTDDSHYAEKAALQTYLPKIKESTVAGFLVGSEALYR
               .   .   .


At 1           RAVLAAGLGGRIKVSTSVDMGVLGESYPPSKGSFRGDVM---VIMEPIIRFLVSKNSPLL
At 3           RAV--SSLG--IKVSTAIDTRGIS-GFPPSSGTFTPEFR---SFIAPVISFLSSKQSPLL
At 2           NAVSGAGLE--VKVSTAIATDTTTDTSPPSQGRFRDEYK---SFLEPVIGFLASKQSPLL
Np basic       NAISSAGLQNNIKVSSSVDMTLIGNSFPPSQGSFRNDVR---SFIDPIIGFVRRINSPLL
Pv ß13         GAVRAQGLHDGIKVSTAIDMTLIGNSYPPSQGSFRGDVR---SYLDPIIGYLLYASAPLH
Nt PR-Q        TAISGAGLGNQIKVSTAIETGLTTDTSPPSNGRFKDDVR---QFIEPIINFLVTNRAPLL
Nt acidic      NALAAAGLQDQIKVSTATYSGILANTYPPKDSIFRGEFN---SFINPIIQFLVQHNLPLL
Hv ß1314       GALASAGLGH-IKVTTSVSQAILGVYSPPSAGSFTGEAD---AFMGPVVQFLARTGAPLM
Hv ß13         AALSAAGLGA-IKVSTSIRFDEVANSFPPSAGVFK---N---AYMTDVARLLASTGAPLL
At A6          NSLRLHGIHN-IKVGTPLAMDSLRSSFPRSNGTFREEITG--PVMLPLLKFLNGTNSYFF
Bn A6          NSLRAHGIHN-IKVGTPLAMDSLRSTFPPSNSTFRGDIAL--PLMLPLLKFLNGTNSYFF
Yeast endo     NDLTASQLSDKINDVRSVVADISDSDGKSYSGKQVGTVDSWNVLVAGYNSAVIEASDFVN
                .   .   .    .


At 1           LNLYT**Y**FSYAGNIGQIRLDYALFTAPSGIVS-DPPRSYQNLFDAMLDAMYSALEKFGGAS
At 3           VNNYP**Y**FSYTGNMRDIRLDYILFTAPSTVVN-DGQNQYRNLFHAILDTVYASLEKAGGGS
At 2           VNLYP**Y**FSYMGDTANIHLDYALFTAQSTVDN-DPGYSYQNLFDANLDSVYAALEKSGGGS
Np basic       VNIYP**Y**FSYAGNPRDISLPYALFTAPNVVVQ-DGSLGYRNLFDAMSDAVYAALSRAGGGS
Pv ß13         VNVYP**Y**FSYSGNPRDISLPYALFTSPNVVVR-DGQYGYQNLFDMLDSVHAAIDNTRIGY
Nt PR-Q        VNLYP**Y**FAIANNA-DIKLEYALFTSSEVVVN-DNGRGYRNLFDAILDATYSALEKASGSS
Nt acidic      ANVYP**Y**FGHIFNTADVPLSYALFTQQEA-----NPAGYQNLFDALLDSMYFAVEKAGGQN
Hv ß1314       ANIYP**Y**LAWAYNP**S**AMDMSYALFTASGTVV-QDGSYGYQNLFDTTVDAFYTAMAKHGGSN
Hv ß13         ANVYP**Y**FAYRDNPGSISLNYATFQPGTTVRDQNNGLTYTSLFDAMVDAVYAALEKAGAPA
At A6          LNVHP**Y**FRWSRNPMNTSLDFALFQGHSTYTDPQTGLVYRNLLDQMLDSVLFAMTKLGYPH
Bn A6          INLQP**Y**FRWSRNPNHTTLDFALFQGNSTYTDPHTGLVYHNLVDQMLDSVIFAMTKLGYPY
Yeast endo     ANAFS**Y**WQ--------------------GQTNQNASY-SFFDDINQALQVIQSTKGSTD
                *  .*                     *  .. ..   .
```

A6-*barnase* transformant (A, C, E) and a wild-type tobacco (B, D, F). The first developmental differences occur after initial callose deposition by the microsporocytes and after meiosis. In the wild-type (B and D) the tetrads are surrounded by a characteristic callose wall which is then degraded releasing the microspores into the locule (F). In the A6-*barnase* anther (A and C) there is little evidence of further callosic deposition on the cellular plates between the individual microspores. The callose that is present is not broken down by callase activity and the microspores remain in tetrads. At a later developmental stage the tapetum, still clearly visible in the wild-type (F) as a darkly staining layer, has completely degenerated in the A6-*barnase* anther (E) and microspore development has been curtailed. Figure 6 demonstrates the catastrophic consequences of RNase expression in the single cell layer of the tapetum. It also con-

```
                      ____a____                              ____b____
At 1        LEIVVAETGWPTGGGVD-TNIE--NARIYNNNLIKHVKNG----TPKRPGKEIETYLFAI
At 3        LEIVVSESGWPTAGGAA-TGVD--NARTYVNNLIQTVKNG----SPRRPGRATETYIFAM
At 2        LEIVVSETGWPTEGAVG-TSVE--NAKTYVNNLIQHVKNG----SPRRPGKAIETYIFAM
Np basic    IEIVVSESGWPSAGAFA-ATTN--NAATYYKNLIQHVKRG----SPRRPNKVIETYLFAM
Pv ß13      VEVVVSESGWPSDGGFG-ATYD--NARVYLDNLVRRAGRG----SPRRPSKPTETYIFAM
Nt PR-Q     LEIVVSESGWPSAGAGQLTSID--NARTYNNNLISHVKGG----SPKRPSGPIETYVFAL
Nt acidic   VEIIVSESGWPSEGNSA-ATIE--NAQTYYENLINHVKSGA--GTPKKPGKAIETYLFAM
Hv ß1314    VKLVVSESGWPSAGGTA-ATPA--NARIYNQYLINHV--GR--GTPRHPG-AIETYVFSM
Hv ß13      VKVVVSESGWPSAGGFA-ASAG--NARTYNQGLINHV--GG--GTPKKRE-ALETYIFAM
At A6       MRLAISETGWPNFGDIDETGANILNAATYNRNLIKKMSASPPIGTPSRPGLPIPTFVFSL
Bn A6       IRIAISETGWPNSGDIDEIGANVFNAATYNRNLIKKMTATPPIGTPARPGSPIPTFVFSL
Yeast endo  ITFWVGETGWPTDGTNFESS-------YPS--VDNAKQFWKEGICSMRAWGVNVIVFEA
             .  ..*.***. *.      .        *        .        ..   . .*.
```

```
At 1        YDEN---QKPTPPYVEKFWGLFYPNKQPKYDINFN
At 3        FDEN---SKQGPE-TEKFWGLFLPNLQPKYVVNFN
At 2        FDEN---KKE-PTY-EKFWGLFHPDRQSKYEVNFN
Np basic    FDEN---NKNPE--LEKHFGLFSPNKQPKYPLSFGFSDRYWDISAENNATAASLISEM
Pv ß13      FDEN---QKSPE--IEKHFGLFKPSKEKKYPFGFGAQRMQRLLLMSSMQHIPLRVTCKLE
Nt PR-Q     FDED---QKDPE--IEKHFGLFSANMQPKYQISFN
Nt acidic   FDEN---NKEGDI-TEKHFGLFSPDQRAKYQLNFN
Hv ß1314    FNEN---QK-DNG-VEQNWGLFYPNMQHVYPISF
Hv ß13      FNEN---QKTGDA-TERSFGLFNPDKSPAYNIQF
At A6       FNEN---QKSGSG-TQRHWGIFDPDGSPIYDVDFTGQTPLTGFNPLPKPTNNVPYKGQVW
Bn A6       FNEN---KKPGSG-TQRHWGILHPDGTPIYDIDFTGQKPLTGFNPLPKPTNNVPYKGQVW
Bn A20                  PGSG-TERHWGILNPDGTQIYDIDLSGTRPVSSLGSLPKPNNNVPFKGNVW
Yeast endo  FDEDWKPNTSGTSDVEKHWGVFTSSDNLKYSLDCDFS
             ..*.         .   .. .*....*  *
```

```
Pv ß13      PSSQSLL
At A6       CVPVEGANETELEETLRMACAQSNTTCAALAPGRECYEPVSIYWHASYALNSYWAQFRNQ
Bn A6       CVPVEGANETELEEALRMACARSNTTCAALVPGRECYEPVSVYWHASYALNSYWAQFRSQ
Bn A20      CVAVEGANETELGQALDFACGRSNATCAALAPGRECYAPVSVTWHASYAFSSYWAQFRNQ
Bn A11                                                                RTE
```

```
At A6       SIQCFFNGLAHETTTNPGNDRCKFPSVTL
Bn A6       NVQCYFNGLAHETTTNPGNDRCKFPSVTL
Bn A20      SSQCYFNGLARETTTNPGNEQCKFPSVTL
Bn A11      NVRCYFNGLARMTTVNPGNDRCKFPGVTL
Bn A28            RMTTINPGNDRCKFPGVTL
```

(b)

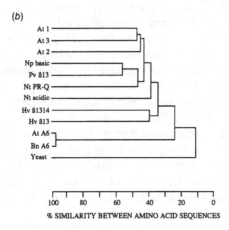

% SIMILARITY BETWEEN AMINO ACID SEQUENCES

Fig. 3. Alignment of the deduced amino acid sequences of the *B. napus* A6 cDNA and the *A. thaliana* G62 genomic clone with selected plant endo-β-1,3-glucanases, an endo-β-1,3;1,4-glucanase and a yeast endo-β-1,3-glucanase. (*a*) The sequences shown are – At1, At2 and At3; β-glucanases from *A. thaliana* (Dong *et al.*, 1991). Np basic; class I β-glucanase from *N. plumbaginifolia* (de Loose *et al.*, 1988). Pv β13; class I β-1,3-glucanase from *Phaseolus vulgaris* (Edington *et al.*, 1991). Nt PR-Q; class III β-1,3-glucanase from *N. tabacum* (Payne *et al.*, 1990). Nt acidic; class II β-1,3-glucanase from *N. tabacum* (Linthorst *et al.*, 1990). Hv β1314; class II 13,14-β-glucanohydrolase from *Hordeum vulgare* (Slakeski *et al.*, 1990). Hv β13; class II 1,3-β-glucan endohydrolase from *H. vulgare* (Høj *et al.*, 1989a). At A6; *A.thaliana* A6 sequence. Bn A6; *B. napus* A6 sequence. Yeast endo; class II endo-β-1,3-glucanase from *Saccharomyces cerevisiae* (Klebl & Tanner, 1989; Mrsa *et al.*, 1993). Regions a and b are thought to be involved in catalytic activity and important conserved residues are shown in bold type. These are conserved in the *B. napus* and *A. thaliana* A6 sequences. *represents a conserved residue, . represents a conservative substitution. (*b*) Dendrogram alignment of the sequences shown in Fig. 3(*a*). The A6 sequences are less similar to the other members of the alignment.

firms the anther-specific nature of A6 expression in tobacco, since the transformants are otherwise phenotypically normal and wild-type pollen crosses result in normal seed set.

Both the histochemical/fluorometric and RNase data demonstrate that A6 is active in the tapetum at the time of callase release. This supports the theory that A6 may be part of the callase complex of β-1,3-glucanase

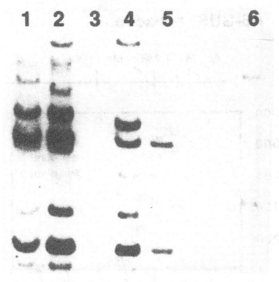

Fig. 4. The complexity of the A6 gene family in Brassica and *A. thaliana*. Southern blot of *Hind*III cut genomic DNA probed with [^{32}P]-labelled A6 cDNA. Lane 1, *B. oleracea* (Brussels sprout); lane 2, *B. oleracea* (cabbage); lane 4, *B. napus* (swede); lane 5, *B. napus* (oil seed rape); lane 6, *A. thaliana*.

activities described by Stieglitz. If this is the case, does A6 represent the exoglucanase or endoglucanase component of callase? As stated earlier, Stieglitz partially purified a 62 kD exoglucanase and a 32 kD endoglucanase from lily. In general, exoglucanases tend to be around 60 kD in size (a 63 kD carrot exoglucanase (Kurosaki, Tokitoh & Nishi, 1991), a 62 kD *Acacia* exoglucanase (Liénart, Comtat & Barnoud, 1986), a 68 kD soyabean exoglucanase (Cline & Albersheim, 1981) and a 40 kD maize exoglucanase (Labrador & Nevins, 1989), whereas endoglucanases are approximately 25–30 kD (Meins *et al.*, 1991). If the situation is conserved between lily and brassica, then, on the basis of size alone, A6 is likely to be an exoglucanase.

Using an anti-A6 antibody to determine the actual size of the A6 protein in *B. napus*

To investigate the size of the A6 protein in *B. napus* and to determine whether the protein undergoes any post-translational modification, a

(a) A6-GUS tobacco

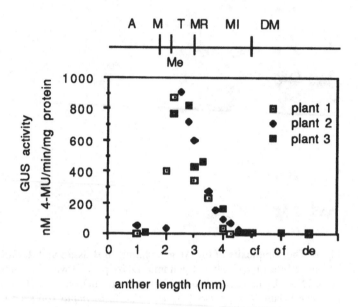

anther length (mm)

(b) A6-GUS *B. napus*

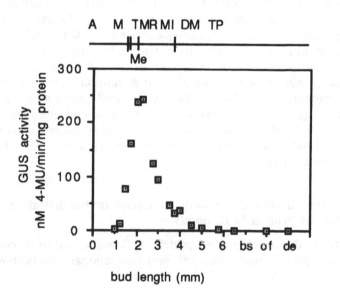

bud length (mm)

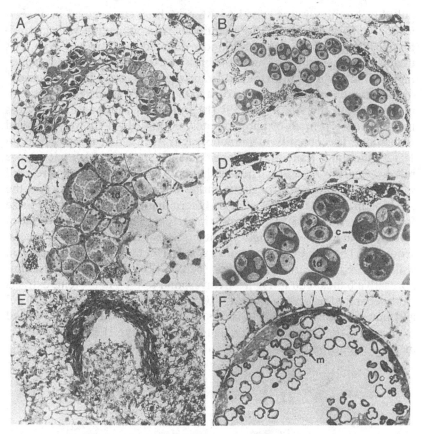

Fig. 6. Tapetal expression of barnase causes microspore abortion in A6-barnase tobacco transformants. Comparison of microsporogenesis in anthers of A6-barnase (A, C, E) and wild-type (B,D,F) tobacco. A, 2.4 mm *A6-barnase* anther 60X; B, 3.0 mm wild-type anther 60X; C, 2.5 mm *A6-barnase* anther 150X; D, 3.0 mm wild-type anther 150X; E, 3.5 mm *A6-barnase* anther 60X; F, 4.0 mm wild-type anther 60X. *Abbreviations:* c, callose wall; m, microspore; t, tapetum; td, tetrad.

Fig. 5. Fluorometric analyses of GUS activity in A6-GUS tobacco anthers (a) and *B. napus* buds (b). GUS activity (in nM 4-MU/minute/ mg protein) is plotted against anther length and developmental stage. *Abbreviations*: A, archaesporial cell; M, meiocyte; Me, meiosis; T, tetrad; MR, microspore release; MI, microspore interphase; DM, dinucleate microspore; TP, trinucleate pollen; cf, closed flower; bs, bud splitting; of, open flower; de, dehisced.

polyclonal antibody was raised to A6 protein overexpressed in *Escherichia coli*. The resulting anti-A6 antibody, in conjunction with an anti-tomato PR glucanase antibody (kindly donated by Pierre de Wit), was used to investigate the size and distribution of the A6 protein in *B. napus*. Figure 7 shows protein extracted from whole *B. napus* buds at various developmental stages, separated on a SDS polyacrylamide gel and immunoblotted with the A6 (a) and PR (b) antibodies. Both antibodies cross reacted with a protein of approximately 60 kD which showed temporal regulation. The quantity of immunoreactive protein reached a peak in 1.5–2.5 mm buds (lane 2), the developmental stage at which microspore release occurs in *B. napus*. The fact that both the antibodies cross reacted with a 60 kD protein throughout anther development suggests that the A6 protein retains most or all of its C-terminal tail. The increase in size from the predicted 53 kD to approximately 60 kD may be the result of glycosylation as there are eight possible glycosylation sites within the A6 sequence. The apparent increase in size brings the A6 protein into line with the size of the lily exoglucanase (62 kD). To confirm that the immunoreactive bands seen in Figure 7 correspond to the A6 protein, extracts were separated on a wide range (3–10.5 pl) isoelectric focusing (IEF) gel and immunoblotted. Figure 8 shows that both antibodies cross reacted with several developmentally regulated bands in the IEF gel. However, in 1.5–3.5 mm buds, two bands with pls of 8.5–9.0 are more strongly

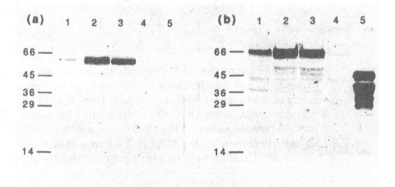

Fig. 7. The A6 and PR antibodies cross react with a temporally regulated 60 kD protein in *B. napus* buds. Wild-type *B. napus* protein extracts were separated on duplicate SDS polyacrylamide gels and probed with anti-A6 antibody (a) and anti-PR-β-1,3-glucanase antibody (b). Lane 1: <1.5 mm buds; lane 2: 1.5–2.5 mm buds; lane 3: 2.5–3.5 mm buds; lane 4: 3.5–4.5 mm buds; lane 5: wild-type leaf protein. The position of molecular size markers is indicated on the left of each blot.

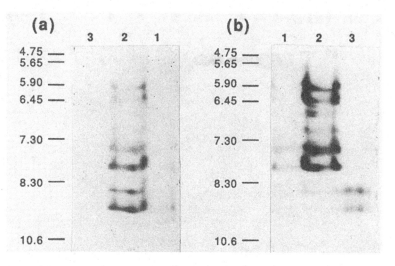

Fig. 8. A6 antibody preferentially cross reacts with two basic proteins in *B. napus* buds. Wild-type *B. napus* protein extracts were separated on a 3–10.5 pl range IEF gel and probed with anti-A6 antibody (a) and anti-PR-β-1,3-glucanase antibody (b). Lane 1: <1.5 mm buds; 1.5–3.5 mm buds; >3.5 mm buds. The position of pl markers is indicated on the left of each blot.

highlighted by the A6 antibody. These values are consistent with the predicted pl of the A6 protein. Again, the blots demonstrate the temporal nature of A6 protein expression, with maximum levels of A6 protein present at the time of microspore release. The plethora of developmentally regulated bands suggests the presence of an A6-related protein family in *B. napus*. This supports the result of the Southern blot (Fig. 5) where a complex hybridisation pattern suggested the presence of an A6 multigene family.

To investigate the distribution of A6-like proteins further, *B. napus* buds were dissected and the separate floral organs probed with the A6 antibody. Figure 9 shows an immunoblot of protein samples taken from anthers, carpels and sepals plus petals separated on an SDS polyacrylamide gel. Again, a temporal 60 kD band was detected in the anther samples. The A6 antibody also cross reacted with a temporal 60 kD band in sepals plus petals and, to a lesser extent, in carpels. It is interesting to note that the developmental pattern is similar and that the immunoreactive protein in all three cases is 60 kD in size. However, it is unlikely that the immunoreactive protein present in carpels and sepals plus petals

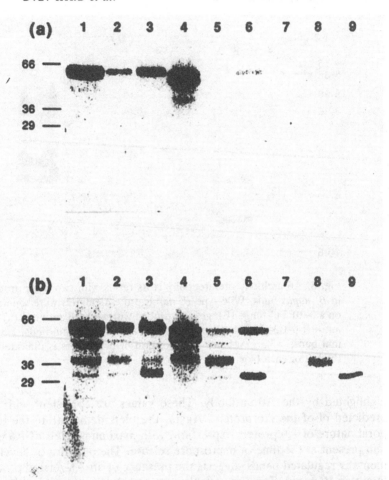

Fig. 9. A6-related proteins are expressed in several floral organs in *B. napus*. Wild-type *B. napus* protein extracts were separated on duplicate SDS polyacrylamide gels and probed with anti-A6 antibody (a) and anti-PR-β-1,3-glucanase antibody (b). Lane 1: anthers from 2.0 mm buds; lane 2: carpels from 2.0 mm buds; lane 3: sepals and petals from 2.0 mm buds; lane 4: anthers from 3.0 mm buds; lane 5: carpels from 3.0 mm buds; lane 6: sepals and petals from 3.0 mm buds; lane 7: anthers from 5.0 mm buds; lane 8: carpels from 5.0 mm buds; lane 9: sepals and petals from 5.0 mm buds. The position of pl markers is indicated on the left of each blot.

samples is A6 protein. The A6-barnase tobacco transformants demonstrate that A6 promoter expression is apparently restricted to the tapetum. If the amount of cross reacting protein detected in the carpels and sepals plus petals of *B. napus* reflected A6 promoter activity, this would almost certainly result in ablation of these tissues. As this was not the case, the immunoreactive proteins in these tissues are unlikely to be the A6 protein, although they may represent members of the A6-related family. The results in Fig. 9 suggest that members of the A6 family are expressed throughout floral development. Labrador and Nevins, (1989) proposed a role for exoglucanases in the loosening of cell walls to facilitate cell expansion during growth. During floral development the whole bud is undergoing rapid expansion, and members of the A6 family may be involved in this process.

Expression of the *A. thaliana* A6 gene under the A9 tapetum-specific promoter in tobacco

Having demonstrated the developmental pattern of A6 expression throughout floral development in *B. napus,* the next aim was to assay the A6 protein directly for glucanase activity. In addition to confirming the sequence data which suggests that A6 is a glucanase, assays could also allow exo- and endoglucanase activity to be distinguished. To this end, the entire coding region of the *A. thaliana* G62 genomic clone was linked to the A9 tapetum-specific promoter (Paul *et al.*, 1992) and transferred into tobacco. The A9 promoter is active prior to A6 in anther development (Paul *et al.*, 1992). The resulting A9–A6 transformants expressed the A6 protein and were phenotypically normal (fortuitously the A6 antibody does not cross react with any tobacco proteins, thus providing an immunologically clean background in which to study A6 protein expression in tobacco). Figure 10 shows locular fluid and anther wall protein extracts isolated from three representative A9–A6 transformants and a wild-type tobacco. Isolation of locular fluid and anther wall proteins was performed using a method devised by R. Scott (Leicester). The protein samples, isolated from buds at the time of microspore release, were separated on an SDS polyacrylamide gel and immunoblotted with the A6 and PR antibodies. A 60 kD band is present in the locular fluid, and to a lesser extent in anther wall protein, of the A9–A6 plants but not the wild-type plant. Firstly, this demonstrated that the transformants expressed the A6 protein, and secondly that the protein was secreted into the locule of the anther. This supports the theory that A6 is part of the callase enzyme complex since callase must be released into the locule in order to act on the callose wall of the tetrads.

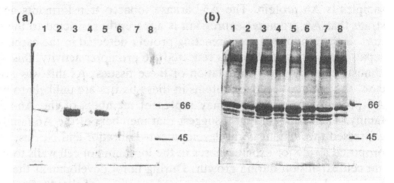

Fig. 10. A6 antibody cross reacts with transgenic A6 protein in locular fluid and anther walls of the A9–A6 transformants. Locular fluid and anther wall proteins extracted from three representative A9–A6 transformants and a wild-type tobacco were separated on an SDS polyacrylamide gel and probed with anti-A6 antibody (a) and the anti-PR antibody (b). Lane 1: locular fluid extract from A9–A6 transformant a; lane 2: anther wall extract from A9–A6 transformant a; Lane 3: locular fluid extract from A9–A6 transformant b; lane 4: anther wall extract from A9–A6 transformant b; lane 5: locular fluid extract from A9–A6 transformant c; lane 6: anther wall extract from A9–A6 transformant c; lane 7: locular fluid from wild-type tobacco; lane 8: anther wall extract from wild-type tobacco. The position of molecular size markers is indicated on the right of each blot.

Assays for glucanase activity were performed on these plants. The exoglucanase assay used was devised by D. Worrall (Leicester) based on a method by Sock, Rohringer & Kang (1990). The protein samples were first separated on a wide range IEF gel after which the gel was covered with a 1.5 mm agarose gel overlay. The overlay contained the substrate, laminarin, and the intermediary chemicals required for the assay. The gel and its overlay were incubated at 30 °C overnight to develop. The glucose produced as a result of enzyme activity produces a dark blue zone on a clear background. Figure 11 shows duplicate wide range IEF polyacrylamide gels of A9–A6 transformed tobacco, wild-type tobacco, wild-type brassica protein samples and laminarinase, a positive control for the enzyme assay. One gel was immunoblotted with the A6 antibody (Fig. 11b). The range of immunoreactive bands in the *B. napus* samples is similar to that observed previously (see Fig. 8). The A6 antibody also highlights a band in the A9–A6 transformant which migrates to the predicted pI of the A6 protein. Figure 11a shows that both *B. napus* samples contain a wide range of exoglucanase activities, some of which can be

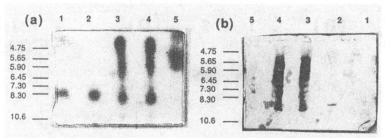

Fig. 11. Exoglucanase activity is present in *B. napus* and tobacco samples. Several bands of activity can be correlated to proteins detected by anti-A6 antibody. Proteins were extracted at the time of microspore release and separated on a 3–10.5 pI range IEF gel. (a) Exoglucanase gel overlay assay. Dark regions represent glucose produced as a result of exoglucanase activity. (b) Immunoblot of the gel probed with anti-A6 antibody. Lane 1: wild-type tobacco; lane 2: a representative A9–A6 transformant; lane 3: *B. napus* anthers; lane 4: *B. napus* buds; lane 5, laminarinase. The position of pI markers is indicated on the left of each blot.

correlated to bands highlighted by the A6 antibody. The tobacco samples show a single area of activity which corresponds to the position of the A6 protein in the A9–A6 transformant. It could therefore be argued that exoglucanase activity associated with the A6 protein was masked by the endogenous tobacco glucanase activity. However, when the samples are separated on a narrow range (8.5–10 pI) IEF polyacrylamide gel, it was clear that this was not the case. Figure 12 shows wild-type and A9–A6 transformed tobacco samples separated on a narrow range IEF gel. In this case the immunoreactive A6 band did not correspond to the position of the exoglucanase activity seen in tobacco. This suggests that no enzyme activity is associated with the transgenic A6 protein. This may be because the assay itself has not been optimised, since at present the assay is not consistent and the degree of staining varies greatly between experiments. Alternatively, the substrate, laminarin, may be unsuitably large and too little free glucose is produced by the A6 protein activity.

However, subsequent assays using much smaller substrates (laminaribiose a β-1,3 linked glucan dimer and cellobiose a β-1,4 linked glucan dimer) have given identical results to those seen in Figure 11 (data not shown). This, in addition to the fact that there are active exoglucanases within the samples isolated, suggests that the lack of A6 activity is due to the characteristics of A6 rather than to the assay techniques used. There are several possible explanations for these observations. Firstly, the A6 protein may be extremely prone to denaturation and is therefore rendered

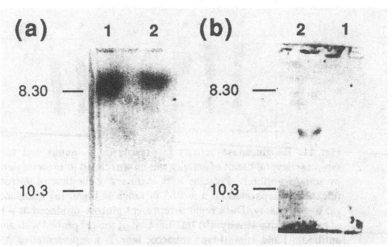

Fig. 12. Exoglucanase activity in the A9–A6 transformants does not correspond to the A6 protein. Proteins were extracted at the time of microspore release and separated on a 8.5–10 pl range IEF gel. (a). Exoglucanase gel overlay assay. Dark regions represent glucose produced as a result of exoglucanase activity. (b) Immunoblot of the gel probed with anti-A6 antibody. Lane 1: wild-type tobacco; lane 2, a representative A9–A6 transformant. The position of pl markers is indicated on the left of each blot.

inactive by the extraction and assay conditions. Secondly, A6 activity may require particular reaction conditions, for example, salt concentration or pH, which are not created in the assays described. Finally, the cloning procedures used to construct the A9–A6 transgene may have introduced sequence mutations which prevent enzyme activity. However, the transgenic protein was of the predicted size and pI. Endoglucanase assays (Pan, Ye & Kue, 1989) were also performed on these samples and again no activity was associated with the transgenic A6 protein (data not shown).

Other work on anther-specific β-1,3-glucanases in tobacco and *B. napus* has shown that contrary to work performed in lily (Stieglitz, 1977), only very low levels of enzyme activity were associated with locular fluid and anther extracts (Scott, personal communication). This further suggests that the β-1,3-glucanases of callase are extremely sensitive to denaturation and may explain the failure to detect an enzyme activity associated with the A6 protein in the A9–A6 transformed plants.

Summary

In summary, the *B. napus* A6 cDNA and its corresponding *A. thaliana* genomic clone show sequence similarity to β-1,3-glucanases and residues essential for catalytic activity are conserved in the A6 sequences. However, the presence of a unique and extensive C-terminal tail and dendrogram analyses suggest that the A6-related proteins may form a distinct class of plant glucanase. A6 promoter fusions to the *gus* and *barnase* reporter genes demonstrate that the A6 gene is tapetally expressed and is active in the anther at the time of microspore release. The temporal pattern of GUS activity seen in both the tobacco transformants and *B. napus* transformant is similar to that of the callase β-1,3-glucanases described by Stieglitz (1977). Immunoblots of *B. napus* protein samples probed with the A6 antibody demonstrate that the A6 protein is approximately 60 kD in size. Furthermore, immunoblots of the A9–A6 tobacco transformants show that the A6 transgenic protein is secreted into the anther locule at the time of microspore release. Taken together, these data strongly suggest that A6 is a component of callase enzyme complex, although to date the enzymatic nature of the protein remains elusive.

References

Bevan, M.W. (1984). Binary *Agrobacterium* vectors for plant transformation. *Nucleic Acids Research,* **12**, 8711–21.

Boller, T. & Vögeli, U. (1984). Vacuolar isolation of ethylene-induced chitinase in bean leaves. *Plant Physiology,* **74**, 442–4.

Chambers, R.S., Broughton, M.J., Cannon, R.D., Carne, A., Emerson, G.W. & Sullivan, P.A. (1993). An exo-1,3-β-glucanase of *Candida alicans:* purification of the enzyme and molecular cloning of the gene. *Journal of General Microbiology,* **139**, 325–34.

Chapman, G.P. (1987). The Tapetum. *International Review of Cytology,* **107**, 111–25.

Cline, K. & Albersheim, P. (1981). Host–pathogen interactions. XVI. Purification and characterization of a β-glucosyl hydrolase/transferase present in the walls of soyabean cells. *Plant Physiology,* **68**, 207–20.

Dong, X., Mindrinos, M., Davis, K.R. & Ausubel, F.M. (1991). Induction of *Arabidopsis* defense genes by virulent and avirulent *Pseudomonas syringae* strains and by a cloned avirulence gene. *Plant Cell,* **3**, 61–72.

Edington, B.V., Lamb, C.J. & Dixon, R.A. (1991). cDNA cloning and characterisation of a putative 1,3-β-D-glucanase transcript induced by a fungal elicitor in bean cell suspension cultures. *Plant Molecular Biology,* **16**, 81–94.

Frankel, R., Izhar, S. & Nitsan, J. (1969). Timing of callase activity and cytoplasmic male sterility in *Petunia. Biochemical Genetics,* **3**, 451–5.

Høj, P.B., Hartman, D.J., Morrice, N.A., Doan, D.N. P. & Fincher, G. (1989a). Purification of (1,3)-β-glucan endohydrolase isoenzyme II from germinated barley and determination of its primary structure from a cDNA clone. *Plant Molecular Biology,* **13**, 31–42.

Høj, P.B., Rodriguez, E.B., Stick, R.V.& Stone, B.A. (1989b). Differences in active site structure in a family of β-glucan endohydrolases deduced from the kinetics of inactivation by epoxyalkyl β-oligoglucosides. *Journal of Biological Chemistry*, **264**, 4939–47.

Izhar, S. & Frankel, R. (1971). Mechanism of male sterility in *Petunia*: the relationship between pH, callase activity in the anthers, and the breakdown of the microsporogenesis. *Theoretical and Applied Genetics*, **41**, 104–8.

Klebl, F. & Tanner, W. (1989). Molecular cloning of a cell wall exo-β-1,3-glucanase from *Saccharomyces cerevisiae*. *Journal of Bacteriology*, **171**, 6259–64.

Kurosaki, F., Tokitoh, Y. & Nishi, A. (1991). Purification and characterization of wall-bound β-1,3-glucanases in cultured carrot cells. *Plant Science*, **77**, 21–8.

Labrador, E. & Nevins, D.J. (1989). An exo-β-D-glucanase derived from *Zea* coleoptile walls with a capacity to elicit cell elongation. *Physiologia plantarum*, **77**, 479–86.

Liénart, Y., Comtat, J. & Barnoud, F. (1986). A wall-bound exo-1,3-β-D-glucanase from *Acacia* cultured cells. *Biochimica et Biophysica Acta*, **883**, 353–60.

Linthorst, H.J.M., Melchers, L.S., Mayer, A., van Roekel, J.S.C., Cornelissen, B.J.C. & Bol, J.F. (1990). Analysis of gene families encoding acidic and basic β-1,3-glucanases of tobacco. *Proceedings of the National Academy of Sciences, USA*, **87**, 8756–60.

de Loose, M., Alliotte, T., Gheysen, G., Genetello, C., Soetaert, P., Van Montagu, M. & Inze, D. (1988). Primary structure of a hormonally regulated β-glucanase of *Nicotiana plumbaginifolia*. *Gene*, **70**, 13–23.

Lotan, T., Ori, N. & Fluhr, R. (1989). Pathogenesis-related proteins are developmentally regulated in tobacco flowers. *Plant Cell*, **1**, 881–7.

Macgregor, E.A. & Ballance, G.M. (1991). Possible secondary structure in plant and yeast β-glucanase. *Biochemical Journal*, **274**, 41–3.

Meins, F. Jr., Neuhaus, J.-M., Sperisin, C. & Ryals, J. (1991). The primary structure of plant pathogenesis-related glucanohydrolases and their genes. In *Genes Involved in Plant Defense*, ed. Boller, T. and Meins, F.Jr. pp. 245–282. New York: Springer.

Moore, A.E. & Stone, B.A. (1972). A β-1,3-glucan hydrolase from *Nicotiana glutinosa*. II. Specificity, action patterns and inhibitor studies. *Biochimica et Biophysica Acta*, **258**, 248–64.

Mrsa, V., Klebl, F. & Tanner, W. (1993). Purification and characterisation of the *Saccharomyces cerevisiae* BGL2 gene product, a cell wall endo-β-1,3-glucanase. *Journal of Bacteriology*, **175**, 2102–6.

Muthukumar, G., Suhng, S.-H., Magee, P.T., Jewell, R.D. & Pimerano, D.A. (1993). The *Saccharomyces cerevisiae* SPR1 gene encodes a sporulation-specific exo-1,3-β-glucanase which contributes to ascospore thermoresistance. *Journal of Bacteriology,* **175**, 386–94.

Pan, S.-Q., Ye, X.-S. & Kué, J. (1989). Direct detection of β-1,3-glucanase isozymes on polyacrylamide electrophoresis and isoelectric focusing gels. *Analytical Biochemistry,* **182**, 136–40.

Parent, J.G. & Asselin, A. (1984). Detection of pathogenesis-related proteins (PR or b) and of other proteins in the intercellular fluid of hypersensitive plants infected with tobacco mosaic virus. *Canadian Journal of Botany,* **62**, 564–659.

Paul, W., Hodge, R., Smartt, S., Draper, J. & Scott, R. (1992). The isolation and characterisation of the tapetum-specific *Arabidopsis thaliana* A9 gene. *Plant Molecular Biology,* **19**, 611–22.

Payne, G., Ward, E., Gaffney, T., Ahl Goy, P., Moyer, M. Harper, A., Meins, F. Jr & Ryals, J. (1990). Evidence for a third structural class of β-1,3-glucanase in tobacco. *Plant Molecular Biology,* **15**, 797–808.

Roberts M. R., Foster, G.D., Blundell, R.P., Robinson, S.W., Kumar, A., Draper, J. & Scott, R. (1993). Gametophytic and sporophytic expression of an anther-specific *Arabidopsis thaliana* gene. *Plant Journal,* **3**, 111–21.

Scott, R., Dagless, E., Hodge, R., Paul, W., Soufleri, I. & Draper, J. (1991). Patterns of gene expression in developing anthers of *Brassica napus. Plant Molecular Biology,* **17**, 195–207.

Slakeski, N., Baulcombe, D.C., Devos, K.M., Ahluwalia, B., Doan, D.N.P. & Fincher, G.B. (1990). Stucture and tissue-specific regulation of genes encoding barley 1,3;1,4-β-glucan endohydrolases. *Molecular and General Genetics,* **224**, 437–49.

Sock, J., Rohringer, R. & Kang, Z. (1990). Extracellular β-1,3-glucanases in stem rust-affected and abiotically stressed wheat leaves. *Plant Physiology,* 94, 1376–89.

Stieglitz, H. (1977). Role of β-1,3-glucanase in postmeiotic microspore release. *Developmental Biology,* **57**, 87–97.

Stieglitz, H. & Stern, H. (1973). Regulation of β-1,3-glucanase activity in developing anthers of *Lilium. Developmental Biology,* **34**, 169–73.

van Loon, L.C. & van Kammen, A. (1970). Polyacrylamide disc electrophoresis of the soluble leaf proteins from *Nicotiana tabacum* var. 'Samsun' and 'Samsun NN'. Changes in protein constitution after infection with tobacco mosaic virus. *Virology,* **40**, 199–211.

Vazquez de Aldana, C.R., Correa, J., Segundo, P.S., Bueno, A., Nebreda, A.R., Mendez, E. & del Rey, F. (1991). Nucleotide sequence of the exo-1,3-β-glucanase-encoding gene, EXG1, of the yeast *Saccharomyces cerevisiae. Gene,* **97**, 173–82.

von Heijne, G. (1983). Patterns of amino-acids near signal-sequence cleavage sites. *European Journal of Biochemistry,* **1** 33, 17–21.

Warmke, H.E. & Overman, M.A. (1972). Cytoplasmic male sterility in sorghum. I. Callose behaviour in fertile and sterile anthers. *Journal of Heredity,* **63**, 103–8.

Worrall, D., Hird, D.L., Hodge, R., Paul, W., Draper, J. & Scott, R.J. (1992). Premature dissolution of the microsporocyte causes male sterility in transgenic tobacco. *Plant Cell*, **4**, 759–71.

EMBL Data Library accession numbers X69887 (A6), X69888 (A11), X69889 (A20), X69890 (A28) X70409 (G62).

V. FERRANT, J. VAN WENT and M. KREIS

Ovule cDNA clones of *Petunia hybrida* encoding proteins homologous to MAP and shaggy/zeste-white 3 protein kinases

Summary

The genetic regulation of the development of the female gametophyte and early embryo of flowering plants is an area of considerable scientific interest. Cloning genes that are determinant in megagametogenesis and embryogenesis will be a major challenge for the future.

Using an enzymatic maceration technique, intact and viable embryo sacs have been isolated from mature ovules of *Petunia hybrida*. Total RNA was prepared and used as a template for the synthesis of cDNA. Due to the small amounts of material obtained, the cDNA was amplified, using the polymerase chain reaction (PCR), prior to cloning into λZAPII.

Using degenerate oligonucleotides, targeting conserved motifs in protein kinases expressed both maternally and zygotically in other organisms, ovule and embryo sac cDNA from *P. hybrida* was subjected to PCR amplification. cDNA fragments of about 160 bp were amplified and cloned. Two classes of 'mini-clones' have been characterised and shown to encode, respectively peptide sequences similar to the subdomains VI, VII and VIII of the protein kinases shaggy/zeste-white 3 and MAP/ERK (mitogen activated protein kinase or extracellular signal-regulated kinase). Probes derived from the above 'mini-clones' were used to screen cDNA libraries and several clones have been isolated and are being characterised.

In other systems, MAP and shaggy/zeste-white 3 kinases have been associated respectively with mitotic stimulation and establishment of embryo segment polarity. By analogy it might be possible that the homologous plant kinases are required for developmental processes during megagametogenesis and embryogenesis.

Society for Experimental Biology Seminar Series 55: *Molecular and Cellular Aspects of Plant Reproduction*, ed. R.J. Scott & A.D. Stead.
© Cambridge University Press, 1994, pp. 159—172.

Introduction

During angiosperm development, a spore-producing generation altern-
ates with a gamete-producing generation. The two generations are
respectively referred to as the sporophyte (diploid) and the gametophyte
(haploid). The pollen grain constitutes the male gametophyte and the
embryo sac the female gametophyte.

The sequential stages of embryo sac development are schematically
shown in Fig. 1. A megaspore mother cell ($2n$) produced within the ovule
undergoes meiosis to form 4 haploid cells. In polygonum types, three of
these cells abort and the remaining one forms the megaspore (or unicellu-
lar embryo sac). The haploid nucleus divides three times to form at
least eight nuclei, four at each end of the embryo sac. Subsequently, the
8-nucleate embryo sac becomes cellular. Three haploid cells are formed
near the micropyle, one of which is the female gamete while the other
two develop into the synergids. Three haploid cells, called the antipodals
are formed at the chalazal side of the embryo sac. The two remaining
nuclei migrate to the centre of the embryo sac and are known as the
polar nuclei of the central cell.

Fertilisation occurs when one of the two sperm cells fuses with the
female gamete to form a zygote, and the other male gamete fuses with
the central cell to form the primary endosperm cell. The zygote divides
and differentiates to form the embryo. Thus far the molecular mechanisms involved in these processes are not understood.

The female gametophyte and the young embryo are deeply embedded
in sporophytic tissue and, as a consequence, molecular and cell biology
studies are very difficult to perform. This is one of the reasons why the
biology of the female gametophyte, the formation of the zygote and its

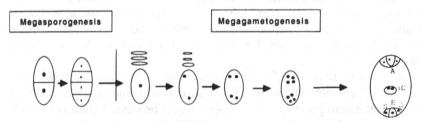

Fig. 1. Schematic representation of the development of the embryo sac
of flowering plants. Central cell (CC), egg cell (E), synergids (S) and
antipodal (A). For the megagametogenesis the megaspore, stage 2
nuclei, stage 4 nuclei, stage 8 nuclei undifferentiated and stage 8 nuclei
differentiated are shown.

early development have not been as well investigated as that of the male gametophyte. Probably no other organs of the flowering plant have been less studied than the ovule and the embryo sac.

Recently, work aimed at identifying and studying genes that control major events during the early stages of plant embryogenesis and gametogenesis has been initiated. Molecular, genetical and biochemical approaches are used to identify and study genes activated at critical stages of megagametogenesis and early embryogenesis (see these proceedings and Wagner *et al.*, 1989; Ferrant *et al.*, unpublished observations; Mayer *et al.*, 1991; Jürgens *et al.*, 1991; de Jong, Schmidt & de Vries, 1993).

In order to study the molecular biology of the developing embryo sac, it is necessary to isolate sufficient material free of surrounding sporophytic tissue.

Isolation and characterisation of embryo sacs from *Petunia*

The isolation of intact and living embryo sacs has been a difficult and time-consuming task. Various methods have been tested including squashing, manual dissectioning and enzymatic digestion (for a review see Theunis, Pierson & Cresti, 1991).

Bradley (1948) developed the very first method for the isolation of embryo sacs. Fixed ovules of *Nicotiana* and *Petunia* were hydrolysed with HCl prior to dissection. This method was, however, only suitable for structural studies. In the 1970s, Tyrnov, Enaleeva & Knokhlov (1975) were the first to work on the enzymatic isolation of embryo sacs of angiosperms using pectinase. This opened the way for the *in vitro* culture of isolated embryo sacs, isolation of female gametophytic cells and *in vitro* manipulation of fertilisation. Since then, different isolation procedures, including enzymatic digestion, squashing or micro-dissection have been developed for over 48 species (Theunis *et al.*, 1991).

The micro-dissection technique consists of dissecting the ovule under a microscope using a micro-knife, a needle or a microcapillary. Although a simple technique, it is time consuming and does not allow the isolation of large amounts of viable embryo sacs.

Currently, the most suitable method for the isolation of viable and polarised embryo sacs is enzymatic digestion of whole ovules. This consists of macerating the ovule in a mixture of different enzymes (e.g. pectinase, cellulase or crude mixtures of enzymes such as driselase, macerozyme and snailase) supplemented with an osmoticum (e.g. sorbitol, sucrose or mannitol). During the first hours of digestion, the embryo sac is protected by the presence of a layer consisting of cuticle-like and cell wall material, but a prolonged incubation of the ovules will liberate individual gametophytic cells.

Using an enzymatic maceration technique, viable embryo sacs have been prepared from *Petunia* (Van Went & Kwee, 1991), *Nicotiana* (Huang *et al.*, 1990) and *Zea* (Wagner *et al.*, 1989). Staining with fluorescein diacetate (FDA) (Heslop-Harrison & Heslop-Harrison, 1970) showed plasma membrane integrity and individual cell viability.

Ferrant *et al.* (unpublished observations) have slightly modified and adapted the enzymatic maceration technique (see Fig. 2), developed by Van Went & Kwee (1990) for the isolation of large amounts of viable embryo sacs for molecular studies. Using flowers of *P. hybrida* line dBBF1 as a starting material, a 60% yield has been obtained. The isolated embryo sacs (Fig. 3) are intact and the various cell types show the same characteristics as *in situ* embryo sacs (see Van Went & Kwee, 1990). This sets the stage for a molecular analysis of *P. hybrida* gametogenesis.

In contrast to the male gametophyte, there are at present no published data available on the biochemical characterisation of female gametophytes of higher plants. Only cytological observations are available, which show that the embryo sac undergoes a series of differentiation events during its development, and a number of gene products should necessarily be required to programme these fundamental processes. Although probes for these specific genes will be indispensable tools for the investigation of embryo sac development at the molecular level, no such gene sequences are as yet available. In other systems, however, it is known that gene products which accumulate during gametogenesis and are non-randomly distributed, contribute to the differentiation of cell lineages in the early embryo.

Fundamental regulatory pathways in mammals, *Drosophila* and yeast, such as the determination of cell identity during embryo development and the cell responses to external stimuli, are regulated by reversible protein phosphorylation (Neiman *et al.*, 1993; Pfeifer & Bejsovec, 1992; Meek & Street, 1992). Nearly 200 protein kinases involved in a wide range of regulatory processes have been described (Hunter, 1991; Hanks, 1991). Fundamental pathways seem to have been evolutionarily conserved (e.g. protein kinases controlling mitogen stimulation) and have been found in organisms ranging from yeast to man (Nurse, 1990; Neiman *et al.*, 1993). Only recently have plant genes involved in protein phosphorylation been isolated and characterised (see Kreis *et al.*, 1994). In plants, protein phosphorylation may regulate pathways common to eukaryotes in addition to unique processes.

ISOLATION OF EMBRYO SACS OF PETUNIA

Petunia hybrida L.

(dBBF1, BBF1, H1F4)

DISSECTION OF 6 OVARIES FROM MATURE FLOWERS

DIGESTION OF 1000 OVULES IN 1 ml MACERATION MEDIUM CONTAINING 3% DRISELASE, 8% SORBITOL 0.1% MES BUFFER, pH5.5, FOR 2h at 30°C WITH SHAKING (150 rpm)

CONTINUE DIGESTION OF OVULES FOR 1h WITHOUT SHAKING AT ROOM TEMPERATURE

REPLACE TWICE MACERATION MEDIUM WITH STABILIZING SOLUTION CONSISTING OF 10% SORBITOL, pH7.5

LIBERATION OF THE EMBRYO SACS BY GENTLE AGITATION OF THE SUSPENSION

COLLECT EMBRYO SACS WITH A MICROPIPETTE UNDER A DISSECTING MICROSCOPE USING TRANS-ILLUMINATION

STORE IN LIQUID NITROGEN

Fig. 2. Flow chart for the enzymatic isolation of *Petunia* embryo sacs.

Construction of *Petunia* ovule and embryo sac specific cDNA libraries

Total RNA was extracted from ovules and embryo sacs to be used as a template to synthesise cDNA. Only small amounts were obtained, and

**OVULE AFTER 2h MACERATION IN 3% DRISELASE
8% SORBITOL, 0.1% MES BUFFER, pH5.5.
THE EGG (E) AND CENTRAL (CC) CELL ARE INDICATED**

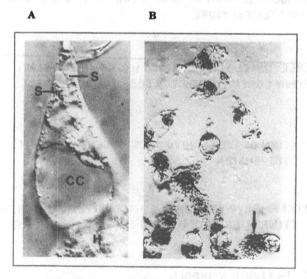

**LIBERATED EMBRYO SACS AFTER 3 h DIGESTION
A. EMBRYO SAC WITH ATTACHED HYPOSTASE (H)
B. POPULATION OF LIBERATED EMBRYO SACS**

Fig. 3. Partially digested ovule after 2 h of incubation, and embryo sac almost completely free of sporophytic tissue. The hypostase is still attached. The arrow shows the cuticle-like material (see also van Went & Kwee, 1991).

these could not be cloned efficiently. Therefore the cDNA was amplified (Domec *et al.*, 1990; Belyavsky, Vinogradova & Rajewsky, 1989) using the polymerase chain reaction (PCR) prior to cloning into the λZAP vector as described by Ferrant *et al.* (unpublished observations). Randomly selected clones showed inserts ranging from 400 to 2000 bp.

Isolation and characterisation of cDNAs coding for MAP and shaggy/zeste-white 3 related protein kinases in plants

Petunia cDNA sequences encoding proteins homologous to the protein kinases shaggy/zeste-white 3 (the gene is known as *shaggy* or *zeste-white* 3)(*sgg/zw3*) (Bourouis *et al.*, 1990; Siegfried *et al.*, 1990) and MAP/ERK (mitogen activated protein kinase or extracellular signal-regulated kinase) (Boulton *et al.*, 1991; Posada *et al.*, 1991; Courchesne, Kunisawa & Thorner, 1989) were isolated using a 'homology cloning' approach (see Lawton *et al.*, 1989; Lawton, 1990; Bianchi *et al.*, 1994). A mixture of degenerate oligonucleotides corresponding to sgg/zw3 and MAP/ERK catalytic subdomains VI and VIII (Hanks, Quinn & Hunter, 1988; Hanks, 1991) were used as primers for PCR amplification of cDNA sequences from ovule and embryo sac-specific libraries of *P. hybrida* (Ferrant *et al.*, unpublished observations; Kreis *et al.*, 1994). Fragments of 160 bp were amplified, cloned into a plasmid vector and characterised. Nucleotide sequence analysis of these 'mini-clones' showed that the inserts contained sequences encoding the amino acids targeted by the oligonucleotide primers. The encoded peptides contained residues and motifs characteristic of protein kinases, including the conserved subdomain VII (DFG). Two classes of clones have been studied. A detailed analysis of the predicted protein sequences of the first class revealed extensive sequence identity with the protein kinase sgg/zw3 from *Drosophila*, the second to ERK/MAP protein kinases. Related sequences have also been identified in *Medicago sativa* (Pay *et al.*, 1993; Jonak *et al.*, 1993; Duerr *et al.*, 1993) and *Arabidopsis thaliana* (Bianchi *et al.*, 1994). These results suggested that genes encoding homologues of sgg/zw3 and ERK/MAP kinases are also present in plants and expressed in the ovule.

Subsequently, probes derived from each of the two classes of mini-clones were used to screen a *P. hybrida* ovule cDNA library. One MAP/ERK and 5 sgg/zw3 hybridising clones were purified and characterised. The cDNA inserts were excised *in vivo* and sequenced using ExoIII deleted clones and oligonucleotides. The sgg/zw3 and ERK/MAP homologous genes from Petunia were, respectively, termed *PSK* (Petunia *shaggy* related *k*inase) and *PMEK* (Petunia *MAP/ERK* related *k*inase).

Comparisons of the deduced sequence of the catalytic domain of *Petunia* PSK-6 kinase to other kinases showed, as expected, highest identities to the members of the GSK-subfamily of protein kinases (Hanks, 1991). An alignment of the catalytic domains (as defined by Hanks *et al.*, 1988) of the Petunia PSK-6, Arabidopsis ASK-γ (Bianchi *et al.*, 1994), alfalfa MsK-1 (Pay *et al.*, 1993), Drosophila sgg/zw3 (Siegfried *et al.*, 1990) and GSK-3 (glycogen synthase kinase3) (Woodgett, 1990) presented in Fig. 4 shows that the plant sequences are about 70% identical to sgg/zw3 from *Drosophila* and GSK-3 from rat. Recent results have shown that *GSK-3* and *sgg/zw3* are functionally homologous *in vivo*, since expression of GSK-3 in sgg/zw3 mutant embryo restores these embryos to normal activity. In *Drosophila*, seven different transcripts are produced at specific stages of development, two being expressed maternally and zygotically (Bourouis *et al.*, 1990; Siegfried *et al.*, 1990; Ruel *et al.*, 1993).

The *Petunia* PMEK-1 clone was sequenced and the derived amino acid sequence of the catalytic domain compared to other protein kinases. High identities were revealed with members of the ERK/MAP kinase family of serine/threonine protein kinases (Fig. 5), namely alfalfa MsERK1 and MsK7 (Duerr *et al.*, 1993; Jonak *et al.*, 1993), rat ERK1 and ERK2 (Boulton *et al.*, 1991), and yeast KSS1 (Courchesne *et al.*, 1989).

The amino acid sequence of the catalytic domain of PMEK1 contains the consensus amino acids and motifs characteristic of the subdomains of the MAP serine/threonine kinases. The PMEK-1 subdomain VIII contains a short sequence (FMTEYVVTR) similar (but not identical) to the sequence FL*TE*Y*V*ATR present in the pp42/MAP kinase which is also conserved in ERK1, KSS1, MsERK1 and MsK7. The underlined tyrosyl and threonyl residues require phosphorylation for enzymatic activity of the MAP kinase (Payne *et al.*, 1991). PMEK-1 is not a member of the cdc2 subgroup because its sequence lacks the motif PSTAIR in subdomain III, characteristic of all p34cdc2 kinases.

Conclusions

In yeast, mammals and *Drosophila*, reversible phosphorylation is known to control a wide range of cellular processes and recent observations suggest that protein kinase cascades are probably involved in gametogenesis and embryogenesis. Protein phosphorylation might be one of the very early steps in the signal transduction pathways turned on at fertilisation when the sperm binds to its receptor (Ruderman *et al.*, 1991). It has now been established that, in clam, the protein that becomes phosphorylated (on tyrosine) within minutes of fertilisation is a member of the MAP/ERK family.

```
          I                    II              III            IV
1. YMAERVVGTGSFGVVFQAKCLETGESVAIKKVLQDRRYKNRELQIMRTLDHPNVVKLRHC
2. *********H*********************T*********************T**L*******S*K**
3. *********H*********************T**********K**********T**L*******S*K**
4. *TDTK*I*N*******Y***LCDS**L*********K*F**********K***C*I*R**YF
5. *TDTK*I*N**********LCD***L************F*******T*K*E*C*I***LYF


          V                                      VIa
1. FYSTTEK-NEVYLNLVLEYVSETVYRVSRHYSRMNQHMPIIYVQLYTYQICRALNYMHSV
2. *F*****-D*L**********P***H**IK**NKL**R**LV**K******F*S*S*I*RC
3. *F*****-D*L**********P***S**I***NK******M*****S******A*I*NS
4. ***SG**KD********D**P******A*****AK*TL*V******M**LF*S*A*I*-S
5  ***SG**RD**F*******IP****K*A*Q*AKTK*TI**NFIR**M**LF*S*A*I*-S


        VIb               VII                  VIII
1. LHVCHRDIKPQNLLVNPHTHQLKLCDFGSAKMLVPGEPNISYICSRYYRAPELIFGATEY
2. IG*********************V**********V**K*********************
3. IG*********************I**********K*********************
4. FGI***********LD*D*AV**********Q**R****V*****************D*
5. *GI***********LD*E*AV**********Q*LH****V****************IN*


          IX                              X
1. TTAIDMWSAGCVMAELLLGQPLFPGESGVDQLVEIIKILGTPTREEIRCMNPNYTEFKFP
2. *****V******L********************V********K***********
3. *****I******LG*********************V********K***********
4. *SS*********L********I***D***********V*******Q**E**********
5. **K**V******L********I***D********V**V*******Q**E**********


                    XI
1. EIKAHPWHKIFQKKMPPEAVDLVSRLLQYSPTLRCTALEACAHPFF
2. Q**********H*R***************N***A**DSLV****
3. Q**********H*R***************N**S*******V***Y
4. Q******T*V*RPRT****IA*C****E*T*A*L*P******S**
5. Q**S***Q*V*RIRT*T**IN***L**E*T*SA*I*P*K*******
```

Fig. 4. Alignment of the catalytic domain of *Petunia* (1) PSK-6, *Arabidopsis* (2) ASK-γ (Bianchi *et al.*, 1994), *Alfalfa* (3) MsK-3 (Pay *et al.*, 1993), rat GSK-3β (4)(Woodgett, 1990) and *Drosophila* (5) zw3-A protein kinase sequences (Siegfried *et al.*, 1990). Residues conserved between PSK-6 and the other sequences are marked (*). Gaps (−) are introduced to optimise sequence similarity. The subdomains are indicated by roman numbers (see Hanks *et al.*, 1988).

```
        I                    II              III
1.  YKYVPIKPIGRGAYGIVCSSVNRETNEKVAIKKINNAFENRIDALRTLRELKLLRHLRHE
2.  KYKP**M***K********AH*S****H**V***A***D*K***K***********MD**
3.  RYKTQLQY**E****M*S*AYDHVRKTR*****ASP**HQTYCQ*****IQI*LGF***
4.  RYKTNLSY**E****M***AYDNLNKVR*****ASP**HQTYCQ*******I*LRF***
5.  RYKTNLSY**E****M***AHCNINKVR*****ASP**HQTYCQ*****I****MD**
6.  QYKLVDL***E****T***AIHKPSGI*******VLT*DPQ*SVT**I**I****YFHE*


        IV                   V
1.  NVIALKDVMMPIQRRSFKDVYLVYELMDTDLHQIIKSS----QTLSNDH-CQYFLFQLLR
2.  **V*IR*IVP*P**EV*N***IA************R*N----*A**EE*-*****Y****
3.  ***GIR*ILRAPTLEAMR***I*QD**E***YKLL**Q----*A******IV****Y*I**
4.  *I*GIN*IIRAPTIEQM****I*QD**ET**YKLL*TQH----A*****I*****Y****
5.  *I*GIN*IIRAPTIEQM******QD**E***YKLL*TQH----A*****I*****Y*I**
6.  *I*SIL*KVR*VSIDKLNA****E***E***QKV*NNQNSGFS***D**-ARLFMY****


        VI  a/b              VII
1.  GLKYLHSANILHRDLKPGNLLINANCDLKICDFGLAR-------TSSGKD--QFMTEYVV
2.  ****I****V*******S***L**************------V**ET*----*******
3.  ****I****V*******S****TT*************------IADPDH*HTG*L****A
4.  ****I****V*******S***L*TT************------V----H*HTG*****A
5.  ****I****V*******S***L*TT************------VADPDH*HTG*L****A
6.  ****I***QVI***I**S***L*S*C***V*******CLASSSDSR---SLVG*****A


    VIII              IX                    X
1.  TRWYRAPELLLCCDNYGTSIDVWSVGCIFADVLGRKPVFPGTECLNQLKLIINILGSQRE
2.  **********NSSD*TAA**********MELMD***L***RDHVH**R*LMELI*TPS*
3.  ********IM*NSDY*TK***I*****L*EM*SNR*I***KHY*D**NH*LG***TPSQ
4.  ********IM*NSKG*TK***I*****L*EM*SNR*I***KHY*D**NH*LG***PSQ
5.  ********IM*NSKG*TK********L*EM*SNR*I***KHY*D**NH*LG***PSQ
6.  ********IM*TFQE*T*AM*I**C***L*EMVSG**L***RDYHH**W**LEV**TPSF


                                        XI
1.  EDIEFIDNPKARKYIKS---LPYSPGTPFSRLYPNAHPLAIDLLQRMLVFDPSKRISVMI
2.  D*LG-FL*EN*KR**RQ---**PYRRQS*QEKF*HV**E****VEK**T***R***TVED
3.  **LNC*I*M***N*LQ*----**SKTKVAWAK*F*KSDSK*L***D****T*N*N***T*EE
4.  **LNC*I*L***N*LL*----**HKNKV*WN**F***DSK*LD**DK**T*N*H***E*EQ
5.  **LNC*I*L***N*L*----**HKNKV*WN**F***D*K*LD**DK**T*N*H***E*EA
6.  **FNQ*KSKR*KE**ANPMRLPPL*WETVWSKFTDLN*DM*D**DK**Q*N*D****AAE
```

Fig. 5. Alignment of the catalytic subdomains I to XI of *Petunia* (1) PMEK-1, Alfalfa (2) Msk7 (Jonak *et al.*, 1993), Rat ERK1 (3) (Boulton *et al.*, 1991), rat ERK2 (4)(Boulton *et al.*, 1991), Xenopus p42 (5) (Posada *et al.*, 1991) and yeast KSS1 MAP kinase sequences (Courchesne *et al.*, 1990). Residues conserved between PMEK-1 and other sequences are marked (*). Gaps (-) are introduced to optimise sequence similarity. The subdomains are indicated by roman numbers (see Hanks *et al.*, 1988).

Recent genetic and molecular analyses have shown that the protein kinase shaggy/zeste white 3 (sgg/zw3) from *Drosophila* is required for the formation of the anterior–posterior compartment boundaries of the embryo. The members of the GSK-3 subfamily of protein kinases have been implicated in the modulation of a wide range of key regulatory elements all of which are probably components of signalling pathways in the transduction of hormonal or growth factor stimuli (Siegfried, Chou, & Perrimon, 1992; Ruel *et al.*, 1993).

The high protein sequence conservation between the kinases from plants and other organisms presumably reflects a conservation of function. A challenging and interesting task will be to uncover their precise function during plant gametogenesis and embryogenesis. The manipulation of the kinase activity in transgenic plants should help us to elucidate their function.

Acknowledgements

VF thanks ICI Seed-UK for their generous support. This work was partly supported by a grant from the Ministère de la Recherche et de la Technologie and Ministère des Affaires Etrangères. The project was initiated with the help of an EEC grant (BRIDGE N°.PL 890201). We are grateful to Dr Raquin and for the help and advice with the growth of the *Petunia* plants.

References

Belyavsky, A., Vinogradova, T. & Rajewsky, A. (1989). PCR-based cDNA library construction: general cDNA libraries at the level of a few cells. *Nucleic Acids Research,* **17**, 2919–32.

Bianchi, M., Guivarc'h, D., Thomas, M., Woodgett, J.R. & Kreis, M. (1994). *Arabidopsis* homologs of the shaggy and GSK-3 protein kinases: molecular cloning and functional expression in *E. coli. Molecular and General Genetics* **242**, 337–45.

Boulton, T.G., Nye, S.H., Robbins, D.J., Ip, N.Y., Radziejewska, E., Morgenbesser, S.D., DePinho, R.A. Panayotatos, N., Cobb, M.H. & Yancopoulous, G.D. (1991). ERKs: A family of protein-serine/threonine kinases that are activated and tyrosine phosphorylated in response to insulin and NGF. *Cell,* **65**, 663–75.

Bourouis, M., Moore, P., Ruel, L., Grau, Y., Heitzler, P. & Simpson, P. (1990). An early embryonic product of the gene shaggy encodes a serine/threonine protein kinase related to the CDC28/cdc2 subfamily. *EMBO Journal,* **9**, 2877–84.

Bradley, M.V. (1948). An aceto carmine squash technique for mature embryo sacs. *Stain Technology*, **23**, 29–40.

Courchesne, W., Kunisawa, R. & Thorner, J. (1989). A putative protein kinase overcomes pheromone-induced arrest of cell cycling in *S. cerevisiae*. *Cell*, **58**, 1107–19.

de Jong, A.J., Schmidt, E.D.L. & de Vries, S.C. (1993). Early events in higher-plant embryogenesis. *Plant Molecular Biology*, **22**, 367–77.

Domec, C., Garbay, B., Fournier, M. & Bonnet, J. (1990). cDNA library construction from small amounts of unfractionated RNA: association of cDNA synthesis with polymerase chain reaction amplification. *Analytical Biochemistry*, **188**, 422–6.

Duerr, B., Gawienowski, M., Ropp, T. & Jacobs, T. (1993). MsERK1: A mitogen-activated protein kinase from a flowering plant. *The Plant Cell*, **5**, 87–96

Hanks, S.K. (1991). Eukaryotic protein kinases. *Current Opinion in Structural Biology*, **1**, 369–83.

Hanks, S.K., Quinn, A.M. & Hunter, T. (1988). The protein kinase family: conserved features and deduced phylogeny of the catalytic domains. *Science*, **241**, 42–52.

Heslop-Harrison, J. & Heslop-Harrison, Y. (1970). Evaluation of pollen viability by enzymatically-induced fluorescence: intracellular hydrolysis of fluorescine diacetate. *Stain Technology*, **45**, 115–20.

Huang, B.Q., Russell, S.D., Strout, G.W. & Mao, L.J. (1990). Organization of isolated embryo sacs and eggs of *Plumbago zeylanica* (Plumbaginacea) before and after fertilization. *American Journal of Botany*, **77**, 1401–10.

Hunter, T. (1991). Protein kinase classification. *Methods in Enzymology*, **200**, 3–37.

Jonak, C., Pay, A., Boegre, L., Hirt, H. & Heberle-Bors, E. (1993). The plant homologue MAP kinase is expressed in a cell cycle dependent and organ specific manner. *The Plant Journal*, **3**, 611–717.

Jürgens, G., Mayer, U., Tores Ruiz, R.A., Berleth, T. & Miséra, S. (1991). Genetic analysis of pattern formation in the *Arabidopsis* embryo. *Development Supplement*, **1**, 27–38.

Kreis, M., Bianchi, M.W., Ferrant, V., Le Guen, L., Thomas, T., Halford, N.G., Barker, J.H.A., Hannappel, U., Vicente-Carbajosa, J. & Shewry, P.R (1993). Plant genes encoding homologues of the SNF1 and shaggy protein kinases. *Plant Molecular Biology, NATO/ ASI Meeting*, Mallorca.

Lawton, M.A. (1990). Plant protein kinases: a molecular genetic approach. *Current Topics in Plant Biochemistry and Physiology*, **9**, 373–82.

Lawton, M.A., Yamamoto, R.T., Hanks, S.K. & Lamb, C. (1989). Molecular cloning of plant transcripts encoding protein kinase homologs. *Proceedings of the National Academy of Sciences, USA*, **86**, 3140–4.

Mayer, U., Tores Ruiz, R.A., Berleth, T., Miséra, S. & Jürgens, G. (1991). Mutations affecting body organization in the *Arabidopsis* embryo. *Nature*, **353**, 402–7.

Meek, D.W. & Street, A.J. (1992). Nuclear protein phosphorylation and growth control. *Journal of Biochemistry,* **287**, 1–15.

Neiman, A.M., Stevenson, B.J., XU, H.-P., Sprague, F.F., Hereskowitz, I., Wigler, M. & Marcus, S. (1993). Functional homology of protein kinases required for sexual differentiation in *Schizosaccharomyces pombe* and *Saccharomyces cerevisiae* suggests a conserved signal transduction module in eukaryotic organisms. *Molecular Biology of the Cell,* **4**, 107–20.

Nurse, P. (1990). Universal control mechanism regulating onset of M-phase. *Nature,* **344**, 503–7.

Pay, A., Jonak.C., Bögre, L., Meskiene, I., Mairinger, T., Szalay, A., Heberle-Bors, E. & Hirt, H. (1993). The Msk family of alfalfa protein kinase genes encodes homologues of *shaggy/glycogen synthase kinase-3* and shows differential expression patterns in plant organs and development. *The Plant Journal,* **3**, 847–56.

Payne, D., Rossomando, A., Martino, P., Erickson, A., Her, J-H., Shabanowitz, J., Hunt, D., Weber, M. & Sturgill, T. (1991). Identification of the regulatory phosphorylation sites in PP42/mitogen-activated protein kinase (MAP kinase). *EMBO Journal,* **10**, 885–92.

Peifer, M. & Bejsovec, A. (1992). Knowing your neighbors: cell interactions determine intrasegmental patterning in *Drosophila. Trends in Genetics,* **8**, 243–9.

Posada, J., Sanghera, J., Pelech, S., Aebersold, R. & Cooper, J. (1991). Tyrosine phosphorylation and activation of homologous protein kinases during oocyte maturation and mitogenic activation of fibroblasts. *Molecular Cell Biology,* **11**, 2517–28.

Ruderman, J., Luca, F., Shibuya, E., Gavin, K., Boulton, T. & Cobb, M. (1991). Control of the cell cycle in early embryos. *Cold Spring Harbor Symposia on Quantitative Biology,* Vol. LVI, 495–502.

Ruel, L., Pantesco, V., Lutz, Y., Simpson P. & Bourouis, M. (1993). Functional significance of a family of protein kinases encoded at the shaggy locus in *Drosophila. EMBO Journal,* **12**, 1657–69.

Siegfried, E., Perkins, L.A., Capaci, T.M. & Perrimon, N. (1990). Putative protein kinase product of the *Drosophila* segment-polarity gene *zeste-white* 3. *Nature,* **345**, 825–9.

Siegfried, E., Chou, T.B. & Perrimon, N. (1992). *wingless* Signaling acts through *zeste-white 3,* the *Drosophila* homolog of glycogen synthase kinase-3, to regulate engrailed and establish cell fate. *Cell,* **71**, 1167–79.

Theunis, C.H., Pierson, E.S. & Cresti, M. (1991). Isolation of male and female gametes in higher plants. *Sexual Plant Reproduction,* **14**, 145–54.

Tyrnov, V.S., Enaleeva, N.K.H. & Knokhlov, S.S. (1975). A study of *in vitro* isolated embryo sacs in Angiosperms. In *Theses of Reports XII International Botanical Congress Nauika, Leningrad,* p. 266.

Van Went, J. & Kwee, H.-S. (1990). Enzymatic isolation of living embryo sacs of *Petunia*. *Sexual Plant Reproduction,* **3**, 257–62.

Wagner, V.T., Song Yuan, C., Matthys-Rochon, E. & Dumas, C. (1989). Observations on the isolated embryo sac of *Zea mays* L. *Plant Science,* **59**, 127–32.

Woodgett, J.R. (1990). Molecular cloning and expression of glycogen synthase kinase-3/Factor A. *EMBO Journal,* **9**, 2431–8.

F.C.H. FRANKLIN, K.K. ATWAL, J.P. RIDE and V.E. FRANKLIN-TONG

Towards the elucidation of the mechanisms of pollen tube inhibition during the self-incompatibility response in *Papaver rhoeas*

Introduction

Self-incompatibility (SI) is the single most important outbreeding device found in Angiosperms. It is a mechanism under genetic control which regulates the acceptance or rejection of pollen during fertilisation. In many species SI is controlled by a single, multi-allelic (S-) gene (Darlington & Mather, 1949; de Nettancourt, 1977). Self-fertilisation is prevented when pollen carrying an S-allele which is genetically identical to that carried by the stigma on which it lands is discriminated from pollen carrying an S-allele which is not carried by the stigma; the former is inhibited, whereas the latter grows normally to achieve fertilisation. As such, self-incompatibility systems play a fundamental and crucial role in the processes involved in Angiosperm sexual reproduction.

SI systems may be conveniently divided into two groups: one is under sporophytic control; the other, which appears to be much more widespread, is under gametophytic control. Over the last ten years there has been considerable accumulation of new information, revealing much about the molecular and cellular processes involved in the two major SI systems. Most of the work on SI has concentrated on *Brassica* (which is under sporophytic control) and *Nicotiana,* together with other Solanaceous species (which are under gametophytic control). We are primarily interested in the SI system of *Papaver rhoeas,* which is determined by a single, multi-allelic gametophytically controlled S-gene (Lawrence, Afzal & Kenrick, 1978). In recent years we have begun to dissect the molecular events involved in SI in *Papaver rhoeas.*

Our development of a system which permits the SI reaction of *Papaver rhoeas* to be reproduced *in vitro* (Franklin-Tong, Lawrence & Franklin, 1988) has been invaluable in enabling us to demonstrate biological

Society for Experimental Biology Seminar Series 55: *Molecular and Cellular Aspects of Plant Reproduction*, ed. R.J. Scott & A.D. Stead.

activity of putative S-gene products. In earlier studies, by analysing extracts made from hundreds of pooled stigmas, we determined that the active products of the S-alleles were small glycoproteins with a variable pI (Franklin-Tong, Lawrence & Franklin, 1989). However, the yield of these proteins was extremely low, which limited further progress in this direction. Recent modifications to the original extraction procedures, primarily by using only the stigmatic papillae has helped to resolve this. N-terminal sequence from the purified protein has enabled us to clone and identify the stigmatic S-gene (Foote *et al.*, 1994). In this chapter we will confine ourselves to describing some of the recent advances made with respect to determining some of the key elements involved in the pollen–stigma interaction during the SI response.

The *Papaver rhoeas* S-protein is not a ribonuclease

Before we had cloned the S-gene from *Papaver*, it was found that the S-glycoproteins of *N. alata*, *Lycopersicon peruvianum*, *Petunia hybrida*, and *Solanum chacoense* and other self-incompatible Solanaceous species, were ribonucleases (RNases) (McClure *et al.*, 1989; Broothaerts & Vendrig, 1990; Xu *et al.*, 1990). It was proposed that these gametophytic SI systems might operate by the S-specific stylar RNases functioning cytotoxically. Uptake into the pollen tube cytoplasm and degradation of pollen rRNA was demonstrated (McClure *et al.*, 1990; Gray *et al.*, 1991); this interference with protein synthesis was proposed to cause the inhibition of incompatible pollen tube growth.

Given that *P. rhoeas* has, genetically at least, the same SI system as *N. alata*, it was initially expected that stigmatic S-proteins of *P. rhoeas* would also be RNases. However, several differences were found (Franklin-Tong *et al.*, 1991). First, RNase activity levels in fractions from pistil tissue of *P. rhoeas* stigmas were several hundred-fold lower than those found in *N. alata* and even lower than those found in styles of self-compatible *N. tabacum*. Second, RNase activity did not appear to be correlated with acquisition of self-incompatibility since the RNase activity levels exhibited by immature stigmas (which are not self-incompatible) were indistinguishable from those of the mature (self-incompatible) stigmas. Third, the specific pollen inhibitory activity of the stigmatic S-protein from *P. rhoeas* was not correlated with RNase activity in *P. rhoeas*. Finally, there was no evidence that RNases inhibited pollen tube growth in *P. rhoeas*. Thus, no detectable ribonuclease activity that correlated with the presence of the functional S-protein in *P. rhoeas* was found, indicating that the S-gene in this species does not encode a ribonuclease (Franklin-Tong *et al.*, 1991).

These data from *Papaver* suggested that a different SI mechanism is operating in this species, which, taken to its logical conclusion, suggests that the S-genes from these two species are different. The cloning and sequencing of the S-gene from *P. rhoeas* has confirmed this idea (Foote *et al.*, 1994). This being the case, we needed to construct alternative hypotheses in order to determine which mechanisms might be involved in the SI response in *P. rhoeas* and thus give us an insight into how pollen is inhibited in this system. We have taken several lines of enquiry, which are outlined below.

Phytoalexins do not appear to be involved in the SI response in *P. rhoeas*

Although the SI response has often been compared to the host–pathogen response (Bushnell, 1979; Lewis, 1980; Hodgkin, Lyon & Dickinson, 1988), there have been no thorough studies investigating this possibility to date. We therefore decided to see whether a type of hypersensitive response similar to that found in host–pathogen responses might provide a model for the mechanism of pollen inhibition. Phytoalexins are low molecular weight compounds induced in an incompatible reaction between a resistant plant in response to an avirulent pathogen. Their production may also be induced by abiotic elicitors. Phytoalexins are antimicrobial and limit the advance of many potential pathogens, but their production is also usually associated with the rapid hypersensitive death of the plant cell in which they accumulate. We proposed that accumulation of similar molecules might be elicited in pollen as a result of an incompatible reaction, which would inhibit tube growth and possibly cause cell death. The attraction of this model was that it would account for the highly specific half-compatible reactions, since each pollen tube would be affected individually.

A series of experiments was carried out to test this hypothesis, using the phytoalexin sanguinarine, which occurs naturally in the Papaveraceae, and also phytoalexin-like molecules extracted from tissue obtained from *P. rhoeas* using an ethyl acetate extraction procedure (Atwal, 1993). The effects of these extracts were tested on pollen tube growth *in vitro* and also on fungal germ-tube growth (using *Botrytis cinerea* spores germinated *in vitro*). This latter test was to check if the low molecular weight molecules tested were phytoalexin-like in their action.

We found that both *Papaver* pollen tube growth and *Botrytis* hyphal growth was highly susceptible to inhibition by sanguinarine (Table 1a), indicating that phytoalexins might be capable of playing a role in pollen tube inhibition (Atwal, 1993). Using an extraction procedure suited to

Table 1. a. *Effect of the phytoalexin sanguinarine, which occurs naturally in the Papaveraceae, on both pollen tube growth and tube growth of spores from the fungus* Botrytis cinerea. *Although there is a differential in their sensitivities, both are susceptible to inhibition by the phytoalexin.* b. *and* c. *Ethylacetate extracts were tested on pollen tube grown* in vitro *and growth of germ tubes from spores from the fungus* Botrytis cinerea.

	Papaver rhoeas pollen tube length ($n=12$) ($\pm$s.e.m.)	*Botrytis cinerea* germ tube length ($n=12$) ($\pm$s.e.m.)
a. Sanguinarine (μg ml^{-1})		
0.0	897 ± 25	268 ± 17
3.9	766 ± 25	254 ± 17
16.0	541 ± 26	90 ± 13
62.5	169 ± 42	40 ± 5
b. Extracts (mg tissue per assay)		
dry pollen: 1	181 ± 26	234 ± 1.4
dry pollen: 2	69 ± 4.2	245 ± 6.4
unpollinated stigma: 1.25	623 ± 135	155 ± 3.5
unpollinated stigma: 2.5	570 ± 38	166 ± 11
control	1653 ± 272	130 ± 8.5
c. Extracts (mg tissue per assay)		
Compatible pollination: 0.75	469 ± 150	200 ± 50
Compatible pollination: 1.5	311 ± 148	185 ± 86
Incompatible pollination: 0.75	578 ± 137	174 ± 91
Incompatible pollination: 1.5	359 ± 198	169 ± 99
control	1759 ± 84	137 ± 67

a.: Extracts were made from pollen and unpollinated, mature stigmas, c.: Extracts were made from mature stigmas pollinated with compatible and incompatible pollen and left for 24 h. Similar effects were obtained with extracts made from pollinated stigmas left for 3 h. Results presented are each the mean of the effect of two separate extracts.

this type of molecule, we then determined whether inhibitors could be extracted from incompatible and compatible pollinations, using dry pollen and unpollinated stigmas as controls (Table 1b and 1c). The effect of the plant tissue extracts on *Botrytis* was always stimulatory and no significant differences were observed between pollinations. However, inhibitors of pollen tube growth were present in both unpollinated stigmas and dry pollen of *P. rhoeas*, with extracts from pollen being far more active (Table 1b). When extracts of compatible and incompatible pollinations were studied, they both contained compounds which were highly inhibitory to pollen tube growth (Table 1c). However, there was no indication that any new inhibitors were produced specifically as a result of an incompatible response, there being no clear difference between the levels of inhibition induced by the two extracts. While these results were disappointing, in that they did not produce any good evidence that this type of mechanism might be operating in the self-incompatibility response, they do provide sufficient information to effectively rule out this particular model for the operation of pollen inhibition during the SI response in this species.

De novo transcription occurs in incompatible pollen

Germination and early growth of pollen from *P. rhoeas* does not require transcription; thus, no new RNA species are synthesised during the first few hours of growth (Franklin-Tong *et al.*, 1990; Franklin-Tong & Franklin, 1992). This is quite a common phenomenon (Mascarenhas, 1975). With regard to the pollen–stigma interaction, *de novo* transcription of pollen genes occurs as a result of an incompatible reaction (Franklin-Tong *et al.*, 1990). There are several pieces of evidence to support this. Firstly, when pollen from *P. rhoeas* was challenged with the stigmatic S-protein in the presence of actinomycin D the effect of SI appears to be alleviated, though not completely (Franklin-Tong *et al.*, 1990; Franklin *et al.*, 1992). Since transcription is required in incompatible pollen for the SI response to operate, this implies that pollen gene expression is induced as a consequence of the response. Secondly, we have identified a number of new RNA transcripts specifically produced in incompatible pollen as a consequence of the SI response. *in vitro* translations from RNA extracted from incompatible pollen revealed a cluster of about 16 novel proteins of molecular weight approximately 21–23 kD, with variable, but essentially neutral pIs, which was absent in dry, germinated and compatible pollen samples (Franklin-Tong *et al.*, 1990). There also appeared to be 'down-regulation' of some transcripts in an incompatible response. Thirdly, a number of pollen 'response' genes, some of which are up-regulated and some which are 'down-regulated' during the SI response have been cloned (Franklin-Tong & Franklin, 1992; Franklin-Tong, Lawrence & Franklin, 1992). Our studies of the pattern of differential expression of some of these genes have shown that they are clearly associated with the SI response. The 'down-regulated' genes are expressed at high levels in mature dry, ungerminated and compatible pollen, but at low levels in incompatible pollen. The up-regulated genes are expressed at high levels in incompatible pollen and at low levels, or not at all in compatible pollen (see Fig. 1). Expression of these genes appears to be

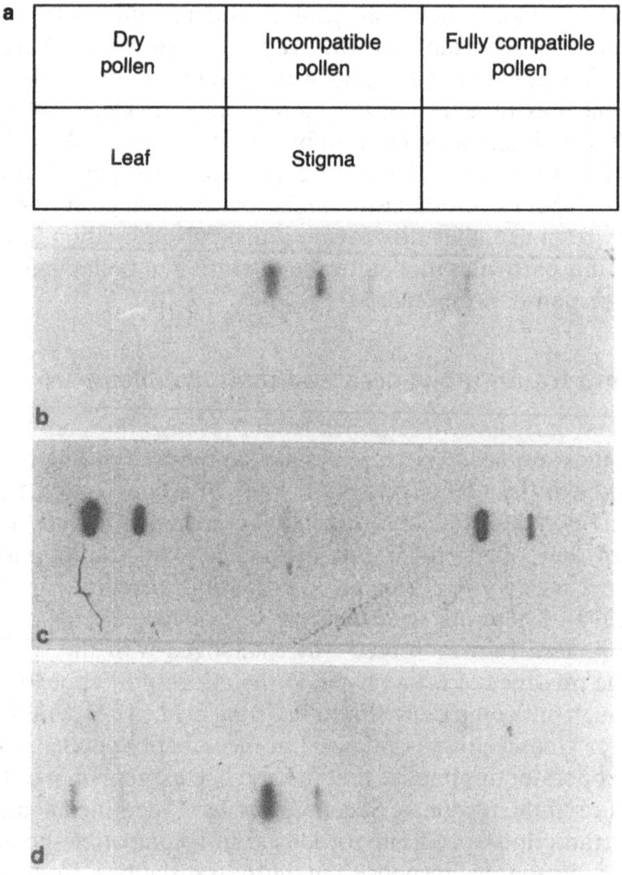

Fig. 1. Slot blots showing the pattern of expression of pollen 'response' genes. a. Diagram of the slot blot, showing the position and type of samples; b. Blot probed with up-regulated clone (pPUR 208); c. Blot probed with down-regulated clone (pPDR 12); d. Blot probed with a stigma-abundant gene (SA-6). RNA isolated from incompatible and compatible pollen, dry pollen, stigma and leaf tissue was amplified by PCR and slot blotted on to nylon filters. The filters were probed with selected pollen 'response' genes. A previously characterised stigma-abundant gene was used as a control. The results show that pPUR 208 (pollen up-regulated) hybridises only with the sample obtained from incompatible pollen. pPDR 12 (pollen down-regulated) hybridises strongly with samples from both compatible and dry pollen. Their expression appears to be pollen-specific, since neither hybridise to leaf or stigma samples.

pollen specific. The implication of these results is that the products of these transcripts play a role in the inhibition following an incompatible reaction in this species. However, at this stage a degree of caution in the interpretation of these data is in order since, at present, we cannot be certain that these events are the primary cause of the arrest of pollen tube growth. Although the transcription inhibitor study suggests that these genes are important in the final termination of growth, it is obviously conceivable that an event temporally downstream of their expression is the initial cause of the arrest of pollen tube growth and that it is this event that induces their expression. If so, the expression of these genes may not be specific to the SI response, but have a more general association with various types of stress responses as a result of pollen tube inhibition. We are currently characterising these genes further, in order to understand more fully how SI operates in this species.

Signal transduction pathways involved in the SI response

During the last few years, compelling evidence has been gathered that suggests that the signal transduction pathways found in animal systems also play an important role in plant cells. Although it has been commonly assumed that self-incompatibility must involve cell–cell recognition and signalling, since there must be some way in which the interaction between stigma and pollen is mediated, very few studies have addressed the question as to how this is achieved. Our evidence that genes are switched on or off within the pollen grain or pollen tube as the result of an external signal, provides important evidence for the involvement of a 'second messenger' signalling system in the SI response. It is only quite recently that the various signal transduction components involved in SI have begun to be analysed and information is rather sparse. However, recently, Walker and Zhang (1990) found that the stigmatic S-glycoprotein in *Brassica oleracea* had homology to a receptor kinase from maize. Stein *et al.* (1991) went on to clone an S-receptor kinase (SRK) gene from *B. oleracea* which resides at the S-locus. The presence of a receptor kinase in the stigma provided the first firm evidence implicating the involvement of a signal transduction pathway in pollination events. We therefore, have recently begun to investigate which signal transduction pathways and their components are involved in the SI response in *P. rhoeas*.

Protein phosphorylation events take place in pollen

One major mechanism for signal transduction in animal cells, which appears likely to be used by plant cells, is the 'phosphatidylinositol response' (Michell, 1975). This pathway is of particular interest to us, since it leads to activation of protein kinase C, which causes gene expression via protein phosphorylation. Since we appear to have gene expression switched on in pollen as a result of the SI response, this seemed an appropriate pathway to examine initially. Results indicate a possible role for protein phosphorylation (and therefore, indirectly, for inositol lipids) in the SI reaction in *P. rhoeas*. Preliminary experiments revealed that pollen, labelled with ^{32}P-orthophosphate, exhibited rapid and transient phosphorylation of several pollen proteins when pollen is challenged with stigma extracts. Distinct differences were detectable between a challenge with an incompatible and a compatible stigma extract (Franklin-Tong *et al.*, 1992). Some proteins are newly phosphorylated within 1 minute of the SI interaction, and there is evidence that some of these are de-phosphorylated within 6 minutes (see Fig. 2). There are also some phosphoproteins which appear in the compatible pollen by 6 minutes after challenge, which are not present in incompatible pollen. These results suggest a role for a signal transduction pathway in the SI system, involving activation of both protein kinases and phosphatases. Calcium-

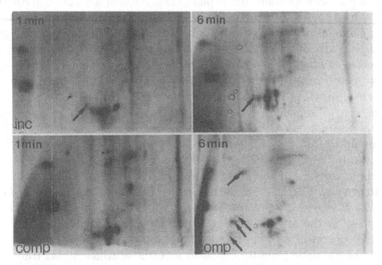

Fig. 2. Protein phosphorylation patterns in pollen of *P. rhoeas*. Close-up of autoradiographs of 2D gels of pollen proteins which have been labelled with ^{32}P-orthophosphate and challenged with incompatible and compatible stigma extracts. Samples were taken at time intervals after this. Phosphoproteins illustrated here come from: incompatible reactions (inc) and compatible reactions (comp), 1 min and 6 min after challenge.

dependent protein kinases have been identified in germinated pollen of *Nicotiana alata* (Polya *et al.*, 1986) and we have recently cloned a number of putative kinase genes from pollen of *P. rhoeas* (F.C.H. Franklin, unpublished observations) This suggests that a Ca^{2+}-mediated signal transduction pathway may operate during pollination and/or SI responses. This led to initiation of studies investigating the possible involvement of Ca^{2+} as a second messenger in the SI response (see below). We are currently analysing the phosphorylation of pollen proteins in more detail.

Involvement of a Ca^{2+}-mediated signal transduction pathway in SI

It is well established that cytosolic free Ca^{2+} ($[Ca^{2+}]i$) is involved as a second messenger in many signal transduction processes in plants (Hepler & Wayne, 1985; Trewavas & Gilroy, 1991). A number of studies has already indicated that Ca^{2+} plays an important role in pollen germination and tube growth (Brewbaker & Kwack, 1963; Mascarenhas and Lafountain, 1972; Jaffe, Weisenseel & Jaffe, 1975; Picton & Steer, 1983; Heslop-Harrison & Heslop-Harrison, 1992). More recently, a gradient of $[Ca^{2+}]i$ has been observed and measured in the tips of growing, but not non-growing pollen tubes (Obermeyer & Weisenseel, 1991; Rathore, Cork & Robinson, 1991; Miller *et al.*, 1992), suggesting that a $[Ca^{2+}]i$ gradient is an important feature of growing pollen tubes. This may indicate that Ca^{2+} signalling occurs in normally growing tubes.

We have recently begun to investigate the involvement of $[Ca^{2+}]i$ acting as a second messenger during the SI response in pollen tubes of *P. rhoeas*. In order to conclusively demonstrate signal-response coupling via Ca^{2+}, what is ideally required is the measurement of changes in $[Ca^{2+}]i$ in living cells experiencing the stimulus and exhibiting the biological response. This is possible through the use of Ca^{2+}-binding fluorescent dyes, which have high affinity and specificity for free Ca^{2+} and which undergo a marked change in fluorescence upon binding to it. We have used this technique to image $[Ca^{2+}]i$ in living, growing pollen tubes of *P. rhoeas* by microinjecting the single wavelength dye Calcium Green-1 (Franklin-Tong *et al.*, 1993*a*). Laser scanning confocal microscopy (LSCM) was used in order to visualise the changes in fluorescence, which indicate $[Ca^{2+}]i$ changes; this not only allows monitoring over time, but also enables good spatial resolution of the changes to be made.

Pollen tubes were monitored before, during and after the addition of the stigmatic S-proteins. As can be seen in Fig. 3a, 3b and the first part of Fig. 3c, no rise in $[Ca^{2+}]i$ was detectable in normally growing pollen tubes or in pollen tubes after addition of either compatible or heat-denatured incompatible stigmatic S-proteins and growth of these pollen tubes continued at normal rates. In contrast, addition of incompatible stigmatic S-glycoproteins induced transient increases in the levels of $[Ca^{2+}]i$ in pollen tubes (second half of Fig. 3c) and pollen tube growth was specifically inhibited in incompatible reactions following the elevation of $[Ca^{2+}]i$. These results all suggested that a Ca^{2+}-mediated signal transduction pathway was operating in the SI response (Franklin-Tong *et al.*, 1993*a*).

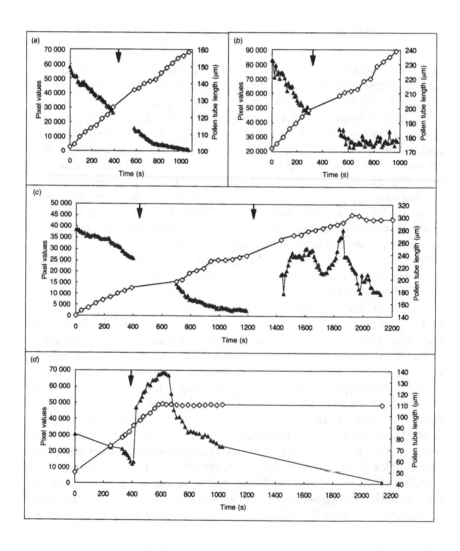

In order to obtain further evidence that $[Ca^{2+}]i$ might be acting as a second messenger in pollen tubes, we tried to obtain a link between the elevation of $[Ca^{2+}]i$ in pollen tubes and the biological response of pollen tube inhibition. This was achieved by microinjecting caged Ca^{2+} together with Calcium-Green-1, followed by photolabilisation of the cage to release free Ca^{2+} into the pollen tube, and monitoring events (Fig. 3d). This resulted in the inhibition of pollen tube growth. Since the SI response could be mimicked by the artificial raising of Ca^{2+} this provided an important link between a transient rise in $[Ca^{2+}]_i$ and the biological response. Together, these results demonstrate, for the first time, that the SI response in *P. rhoeas* is mediated by cytosolic calcium acting as a 'second messenger' (Franklin-Tong *et al.*, 1993a).

As well as being able to follow compatible reactions with incompatible reactions (Fig. 3c), sequential responses could also be induced in a single

---

Fig. 3. Changes in $[Ca^{2+}]i$ and growth rates of pollen tubes microinjected with Calcium Green-1 and treated in different ways. The filled triangles represent the total fluorescence from the first 100 μm of each pollen tube. This was calculated by adding together the total pixel values from this region of a pollen tube in successive digitised images of a time course. The open squares represent the relative length of the pollen tube at each corresponding time point. It should be noted that the x-axes of 3(c) and 3(d) are slightly different to those of 3(a) and 3(b). The arrows indicate the point(s) at which each pollen tube was subjected to treatment. (From Franklin-Tong *et al.*, 1993a). (a) Addition of compatible stigmatic S-glycoproteins. No significant increase in $[Ca^{2+}]i$ was observed and pollen tube growth continues normally after treatment. (b) Addition of heat-denatured incompatible stigmatic S-glycoproteins. Fluorescence decreases over time. both before and after treatment; thus no increase in $[Ca^{2+}]i$ is observed. Pollen tube growth is not significantly affected by the addition of heat-denatured stigma extract. (c) Addition of compatible stigmatic S-glycoproteins (at first arrow), followed by addition of incompatible stigmatic S-glycoproteins (at second arrow). No increase in $[Ca^{2+}]i$ and no significant inhibition of pollen tube growth is observed after treatment with compatible S-glycoproteins. After addition of incompatible stigmatic S-glycoproteins a transient rise in $[Ca^{2+}]i$, consisting of two major peaks is observed and these $[Ca^{2+}]i$ transients are followed by the cessation of pollen tube growth at around 2000s, after which point it is clear that growth has completely ceased, since the final four points for pollen tube length are identical. (d) Photoactivation of caged Ca^{2+} with 25s UV irradiation (indicated by arrow). Pollen tube growth ceases virtually immediately after the initial elevation in $[Ca^{2+}]_i$ is observed.

pollen tube by repeated addition of sub-optimal concentrations of stigmatic S-proteins, provided pollen inhibition was not complete. This usually resulted in a smaller $[Ca^{2+}]i$ transient, followed by the slowing down of growth, after which growth subsequently recovered to near normal rates before addition of another dose of S-protein. This suggests that there is a threshold level which the $[Ca^{2+}]i$ must reach for a full SI response to be achieved and that pollen tubes remain responsive to further $[Ca^{2+}]i$ transients. This implies that the inhibition is not a simple 'all or nothing' phenomenon. Indeed, this observation could explain why some incompatible pollen tubes achieve longer tube lengths than others. In a number of cases, transient rise in $[Ca^{2+}]i$ elicited by incompatible S-proteins was followed later by a second rise in $[Ca^{2+}]i$, which may be interpreted as a $[Ca^{2+}]i$-induced $[Ca^{2+}]i$ release. This phenomenon is often detected in animal cells (Fabiato, 1983), but only a few cases have been observed in plant cells (Fricker *et al.*, 1991).

It has been possible to determine the cellular distribution of the $[Ca^{2+}]i$ transients observed in the incompatible responses due to the use of LSCM which enables increased spatial resolution. This approach has revealed the $[Ca^{2+}]i$ transient is quite localised, in a region corresponding almost exactly with the position, shape and size of the nuclear complex. Since it is well documented that the pollen nuclei are often closely associated with endoplasmic reticulum and other organelles (Cresti *et al.*, 1990; Lancelle & Hepler, 1992), we propose that these organelles are good candidates for the source of the Ca^{2+}.

Involvement of IP_3 in pollen tube growth

We have already alluded to the possible involvement of the 'phosphatidylinositol response' in the SI response. Now, since the $[Ca^{2+}]i$ transient exhibited by the incompatible pollen tubes appears to be localised around the nucleus, the evidence suggests that the $[Ca^{2+}]i$ released in the pollen tube comes from intracellular stores. This suggests a role for the phosphoinositide pathway in the SI response, since it is generally believed that intracellular Ca^{2+} release as the result of a stimulus is mediated through the release of IP_3 (Berridge & Irvine, 1989; Einspahr & Thompson, 1990). It will be remembered that our pollen protein phophorylation studies also indicated that this pathway may be implicated in the SI response. We have begun to investigate whether it is possible to obtain more direct evidence that the phosphatidylinositol response is involved in the inhibition of pollen tube growth.

Preliminary experiments, involving the labelling of growing pollen of *P. rhoeas* with ^{3}H-myoinositol, indicated that at least some of the components of the phosphoinositide pathway are present in pollen of *P. rhoeas* (B.K. Drøbak and V.E. Franklin-Tong, unpublished observations). We have recently obtained additional evidence that IP_3 is likely to act as a second messenger in growing pollen tubes and may well turn out to be involved in the SI response too, although we have no conclusive evidence for this to date. We have imaged $[Ca^{2+}]i$ in living, growing pollen tubes of *P. rhoeas* microinjected with Calcium-Green-1 and caged IP_3, in conjunction with laser scanning confocal microscopy (LSCM), in a manner similar to that described above. Release of IP_3 from the cage by photolabilisation, using UV light, resulted in a transient rise in $[Ca^{2+}]i$. This release of $[Ca^{2+}]i$ was followed by a slowing down of cytoplasmic streaming, slowed growth rates, distorted pollen tube tip growth patterns and finally cessation of pollen tube growth (Franklin-Tong *et al.*, in preparation). This provides evidence that there are IP_3 receptors in the pollen tube tip region that are sensitive to the release of IP_3. In addition, this demonstrates that, in these pollen tubes, release of IP_3 results in the release of Ca^{2+}, presumably from intracellular stores. Finally, as a result of this trigger, pollen tube growth is affected quite markedly, resulting in total inhibition. Thus, there is good evidence that there is a phosphoinositide pathway operating in pollen tubes of *P. rhoeas* and that it is involved in the control of pollen tube growth. Whether IP_3 is actually involved in the SI response remains to be established at this stage. Nevertheless, this represents real progress in our understanding of the signal transductions pathways involved in the regulation of pollen tube growth.

A model for the SI response in *P. rhoeas*

We have already provided evidence that the SI system in *P. rhoeas* is different to that in *Nicotiana* and self-incompatible Solanaceous species. We, therefore, needed to put forward an alternative mechanism for the operation of SI in this species. Since investigations into the analogy between host–pathogen responses and SI reaction appeared to lack evidence, we pursued other lines of investigation. From the various pieces of evidence collected to date, we have outlined a model for the SI system in *P. rhoeas* (Fig. 4). We propose that the stigmatic S-proteins act as signal molecules, which interact with the pollen receptor (as yet unidentified). This step would govern the S-specificity of the response, since presumably in a compatible 'reaction' the S-protein will not have the correct specificity to bind the pollen receptor, while in incompatible reactions, the stigmatic molecule would bind the pollen receptor. Thus, we postulate that only in an incompatible pollination does the stigmatic

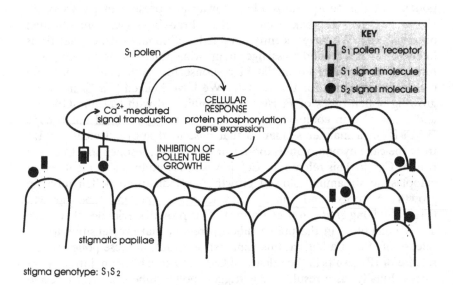

Fig. 4. Model for the SI response in pollen of *Papaver rhoeas*, based on the evidence collected to date. Evidence suggests that the stigmatic S-protein, secreted on the stigma surface acts as a signal molecule. In this example model, we show an S_1 pollen grain landing on an S_1S_2 stigma. It is proposed that the pollen carries S-specific receptors, which, in incompatible S-allele combinations allows recognition of the stigmatic S-protein. In this case, S_1 receptors are carried by the pollen, which recognise and bind the S_1 signal molecules secreted by the stigma. This binding triggers an intracellular signal transduction pathway within the pollen grain/tube. Ca^{2+} is known to act as a second messenger, resulting in the inhibition of pollen tube growth. The phosphoinositide pathway may also be involved though evidence for this is, as yet, limited. Pollen proteins are phosphorylated, suggesting the involvement of protein kinases, and expression of pollen 'response' genes occurs as a consequence of the receptor binding; however, whether this is directed by the initial signal is not known yet. In a compatible 'reaction', no Ca^{2+} signal is generated and pollen response genes are not expressed. The most simple explanation is that no receptor binding takes place in this situation and since no 'recognition' takes place, the train of signal transduction events is not triggered. We have illustrated this with the S_1 pollen receptor faced with an S_2 stigma signal molecule, which, since it does not have the correct specificity, cannot bind. (From Franklin-Tong & Franklin, 1993*b*.)

signal molecule bind to the pollen receptor. When a signal molecule binds the receptor, it triggers an intracellular signalling pathway.

The exact pathways, and the elements involved, have not yet all been identified, but evidence collected to date indicates that at least one of them is mediated by Ca^{2+}, which acts as a second messenger. A transient in cytosolic $[Ca^{2+}]i$ is generated in the pollen grain or tube, leading either directly or indirectly to the inhibition of pollen tube growth. The phosphoinositide pathway is strongly implicated as a signal transduction pathway leading to pollen tube inhibition, and possibly with the SI response if the $[Ca^{2+}]i$ signal is truly generated intracellularly. We also have evidence of the involvement of protein kinases and phosphorylation of phosphoproteins early in the response.

We know that expression of genes in the pollen results as a consequence of the SI response and that inhibition of transcription of pollen during the SI reaction results in the partial alleviation of the SI response. Thus, at least some of the pollen 'response' genes should be expected to be triggered directly as a consequence of the $[Ca^{2+}]i$ signal. Others may be triggered as a result of the inhibition of pollen tube growth. Characterisation of the various pollen response genes, which is currently being undertaken, should give us a more detailed insight into some of the other processes involved in SI reaction in this species. At present, we are making more detailed temporal studies in an attempt to try and clarify the sequence of some of the events involved in the pollen response, together with more of the components involved.

Acknowledgements

Much of this work was funded by AFRC. We also wish to acknowledge support from the Gatsby Charitable Foundation. Part of this work (Ca^{2+}-imaging) was carried out at the University of Edinburgh and we would like to thank Tony Trewavas and Nick Read for their help and advice regarding this work. We could not have carried out these experiments without the dedicated work of our gardening staff.

References

Atwal, K.K. (1993). Studies on the mechanism of pollen inhibition during the self-incompatibility response of *Papaver rhoeas*. PhD Thesis, University of Birmingham.

Berridge, M.J. & Irvine, R.F. (1989). Inositol phosphates and cell signalling. *Nature,* **341**, 197–205.

Brewbaker, J.L. & Kwack, B.H. (1963). The essential role of calcium ions in pollen germination and pollen tube growth. *American Journal of Botany,* **50**, 859–63.

188   F.C.H. FRANKLIN *et al.*

Broothaerts, W.J. & Vendrig, J.C. (1990). Self-incompatibility proteins from *Petunia hybrida* have ribonuclease activity. *Physiologia Plantarum,* **79**, A43.

Bushnell, W.R. (1979). The nature of basic compatibility: comparisons between pistil–pollen and host–parasite interaction. In *Recognition and Specificity in Plant Host–Parasite Interactions,* pp. 211–227.

Cresti, M., Milanesi, C., Salvatici, P. & van Aelst, A.C. (1990). Ultrastructural observations of *Papaver rhoeas* mature pollen grains. *Botanica Acta,* **102**, 349–54.

Darlington, C.D. & Mather, K. (1949). *The Elements of Genetics.* London: Allen and Unwin.

Einspahr, K.J. & Thompson, G.A. Jr. (1990). Transmembrane signalling via phosphatidylinositol 4,5-bis phosphate hydrolysis in plants. *Plant Physiology,* **93**, 361–6.

Fabiato, A. (1983). Calcium-induced calcium release from the cardiac sarcoplasmic reticulum. *American Journal of Physiology,* **245**, C1–C14.

Foote, H.C.C., Ride, J.P., Franklin-Tong, V.E., Walker, E.A., Lawrence, M.J. & Franklin, F.C.H. (1994). Cloning and expression of a distinctive class of self-incompatibility (S-) gene from *Papaver rhoeas* L. *Proceedings of the National Academy of Sciences, USA,* **91**, 2265–9.

Franklin, F.C.H., Hackett, R.M. & Franklin-Tong, V.E. (1992). The molecular biology of self-incompatible responses. In *Perspectives in Plant Cell Recognition,* ed. Callow, J.A. & Green, J.R., pp. 79–103. Cambridge: Cambridge University Press..

Franklin-Tong, V.E., Lawrence, M.J. & Franklin, F.C.H. (1988). An *in vitro* bioassay for the stigmatic product of the self-incompatibility gene in *Papaver rhoeas* L. *New Phytologist,* **110**, 109–18.

Franklin-Tong, V.E., Lawrence, M.J. & Franklin, F.C.H. (1989). Characterization of a stigmatic component from *Papaver rhoeas* L. which exhibits the specific activity of a self-incompatibility (S-) gene product. *New Phytologist,* **112**, 307–15.

Franklin-Tong, V.E., Lawrence, M.J. & Franklin, F.C.H. (1990). Self-incompatibility in *Papaver rhoeas* L.: inhibition of incompatible pollen is dependent on pollen gene expression. *New Phytologist,* **116**, 319–24.

Franklin-Tong, V.E., Lawrence, M.J. & Franklin, F.C.H. (1991). Self-incompatibility in *Papaver rhoeas* L.: there is no evidence for the involvement of stigmatic ribonuclease activity. *Plant, Cell and Environment,* **14**, 423–9.

Franklin-Tong, V.E. & Franklin, F.C.H. (1992). Gametophytic self-incompatibility in *Papaver rhoeas* L.: *Sexual Plant Reproduction,* **5**, 1–7.

Franklin-Tong, V.E., Lawrence, M.J. & Franklin, F.C.H. (1992). Recognition, signals and pollen responses in the incompatibility reaction in *Papaver rhoeas* L. In *Angiosperm Pollen and Ovules: Basic and Applied Aspects*, ed. Mulcahy, D.J., Bergamini-Mulcahy, G. & Ottaviano, E. pp. 84–93. New York: Springer-Verlag.

Franklin-Tong, V.E., Ride, J.P., Read, N.D., Trewavas, A.J. & Franklin, F.C.H. (1993*a*). The self-incompatibility reaction in *Papaver rhoeas* is mediated by cytosolic calcium. *The Plant Journal,* 163–77.

Franklin-Tong, V.E. & Franklin, F.C.H. (1993*b*). Gametophytic self-incompatibility systems: contrasting mechanisms for *Nicotiana* and *Papaver. Trends in Cell Biology*, 340–5.

Fricker, M.D., Gilroy, S., Read, N.D. & Trewavas, A.J. (1991). Visualization and measurement of the calcium message in guard cells. In *Molecular Biology of Plant Development*, ed, G.I. Jenkins & W. Schuch, pp. 177-190. Cambridge: The Company of Biologists.

Gray, J.E., McClure, B.A., Bonig, I., Anderson, M.A. & Clarke, A.E. (1991). Action of the style product of the self-incompatibility gene of *Nicotiana alata* (S-RNase) on *in vitro* grown pollen tubes. *The Plant Cell,* 3, 271–83.

Hepler, P.K. & Wayne, R.O. (1985). Calcium and plant development. *Annual Review of Plant Physiology,* 36, 397–439.

Heslop-Harrison, J. & Heslop-Harrison, Y. (1992). Germination of monocolpate pollen: effects of inhibitory factors and the Ca^{2+} channel blocker, nifedipine. *Annals of Botany,* 69, 395–403.

Hodgkin, T., Lyon, G.D. & Dickinson, H.G. (1988). Recognition in flowering plants: a comparison of the *Brassica* self-incompatibility system and plant–pathogen interactions. *New Phytologist,* 110, 557–69.

Jaffe, L.A., Weisenseel, M.H. & Jaffe, L.F. (1975). Calcium accumulation within the growing tips of pollen tubes. *Journal of Cell Biology,* 67, 488–92.

Lancelle, S.A. & Hepler, P.K. (1992). Ultrastructure of freeze-substituted pollen tubes of *Lilium longiflorum. Protoplasma,* 167, 215–30.

Lawrence, M.J., Afzal, M. & Kenrick, J. (1978). The genetical control of self-incompatibility in *Papaver rhoeas* L. *Heredity,* 40, 239–85.

Lewis, D.H. (1980). Are there inter-relations between the metabolic role of boron, synthesis of phenolic phytoalexins and the germination of pollen? *New Phytologist,* 84, 261–70.

Mascarenhas, J.P. (1975) The biology of the Angiosperm pollen grain. *The Botanical Review,* 41, 259–319.

Mascarenhas, J.P. & Lafountain, J. (1972). Protoplasmic streaming, cytochalasin B and growth of the pollen tube. *Tissue and Cell,* 4, 11–14.

McClure, B.A., Haring, V., Ebert, P.R., Anderson, M.A., Simpson, R.J., Sakiyama, F. & Clarke, A.E. (1989). Style self-incompatibility gene products of *Nicotiana alata* are ribonucleases. *Nature,* 342, 955–7.

McClure, B.A., Gray, J.E., Anderson, M.A. & Clarke, A.E. (1990). Self-incompatibility in *Nicotiana alata* involves degradation of pollen rRNA. *Nature,* **347,** 757–60.

Michell, R.H. (1975). Inositol phospholipids and cell surface receptor function. *Biochimica et Biophysica Acta,* **415,** 81–147.

Miller, D.D., Callaham, D.A., Gross, D.J. & Hepler, P.K. (1992). Free Ca^{2+} gradient in growing pollen tubes of *Lilium*. *Journal of Cell Science,* **101,** 7–12.

Nettancourt, D. de. (1977). *Incompatibility in Angiosperms.* Berlin: Springer Verlag.

Obermeyer, G. & Weisenseel, M.H. (1991). Calcium channel blocker and calmodulin antagonists affect the gradient of free calcium ions in lily pollen tubes. *European Journal of Cell Biology,* **56,** 319–27.

Picton, J.M. & Steer, M.W. (1983). Evidence for the role of calcium ions in tip extension in pollen lubes. *Protoplasma,* **115,** 11–17.

Polya, G.M., Rae, A.L., Harris, P.J. & Clarke, A.E. (1986). Ca^{2+}-dependent protein phosphorylation in germinated pollen of *Nicotiana alata,* an ornamental tobacco. *Physiologia Plantarum,* **67,** 151–7.

Rathore, K.S., Cork, R.J. & Robinson, K.R. (1991). A cytoplasmic gradient of Ca^{2+} is correlated with the growth of lily pollen tubes. *Developmental Biology,* **148,** 612–9.

Stein, J.C., Howlett, B., Boyes, D.C., Nasrallah, M.E. & Nasrallah, J.B. (1991). Molecular cloning of a putative receptor protein kinase gene encoded at the self-incompatibility locus of *Brassica oleracea. Proceedings of the National Academy of Sciences, USA,* **88,** 8816–20.

Trewavas, A.J. & Gilroy, S. (1991). Signal transduction in plant cells. *Trends in Genetics,* **7,** 356–61.

Walker, J.C. & Zhang, R. (1990). Relationship of a putative receptor from maize to the S-locus glycoprotein of *Brassica. Nature,* **345,** 743–6.

Xu, B., Mu, J., Nevins, D., Grun, P. & Kao, T.-H. (1990). Cloning and sequencing of cDNAs encoding two self-incompatibility associated proteins in *Solanum chacoense. Molecular and General Genetics,* **224,** 341–6.

J. HESLOP-HARRISON
and Y. HESLOP-HARRISON

Intracellular movement and pollen physiology: progress and prospects

Introduction

It is a truism of biology that most of the key processes typifying the living state in eukaryotic cells involve intracellular movement. Growth, morphogenesis, division, secretion, interaction with other cells, all require continuous physical redistribution of cell components, whether of organelles, membranes, vesicles, nuclei or chromosomes: in this respect the motility systems of the cell may be said to lie near to the heart of eukaryotic physiology. Our purpose in this outline review is to summarise some features of the life of the male gametophyte generation of the flowering plants as seen from this point of view.

Overt vectorial movement in the pollen tube itself is expressed in three main ways: in the circulation along its length of organelles and other cytoplasmic inclusions; in the net movement away from the parent pollen grain of the vegetative nucleus, the generative cell and the sperms formed from it; and in the transport and release through the plasmalemma of secretory vesicles concerned with the insertion of precursor materials into the growing wall and the release of enzymes into the environment. All of these processes are conducted within the confines of a single tip-growing, partly heterotrophic, cylindrical cell.

It is well established for both plant and animal cells that intracellular motility depends on interaction with cytoskeletal elements. Two systems are positively known from the pollen tube: one microtubule based, and the other actin-fibril based. While the role of the former in the movement of cytoplasmic components still remains a matter for speculation, that of the latter, the actin cytoskeletal system, has been explored in some detail in recent years. Indeed, it now seems likely that all of the principal movements within the pollen tube are dependent on the interaction between a plant equivalent of myosin, the motility ATPase first known

Society for Experimental Biology Seminar Series 55: *Molecular and Cellular Aspects of Plant Reproduction*, ed. R.J. Scott & A.D. Stead.
© Cambridge University Press, 1994, pp. 191–201.

from muscle, and actin. It is with this aspect that we are concerned in the present article.

The actomyosin system

Actin was first described from the pollen tube by Condeelis (1974), who showed that in the growing tube actin microfilaments are present in bundles forming extended fibrils. Yen *et al.* (1986) reported the extraction of a putative pollen myosin, the second expected component of an actomyosin system, and myosin-like molecules have been localised to the surfaces of pollen-tube organelles, vegetative nuclei and generative cells by immunofluorescence (Heslop-Harrison & Heslop-Harrison, 1989*b*; Tang, Hepler & Scordilis, 1989). The general architecture of the actin-fibril system has been explored by electron microscopy (e.g. Franke *et al.*, 1972), and latterly in three-dimensional detail by optical microscopy using fluorochrome-labelled phalloidin as a specific actin probe (e.g., Perdue & Parthasarathy, 1985; Pierson, Derksen & Traas, 1986). The fibrils extend in longitudinal arrays throughout the older part of the tube in the core cytoplasm. At the periphery, finer fibrils and individual microfilaments lie in close association with the plasmalemma, often linked by protein bridges with similarly oriented microtubules (Lancelle, Cresti & Hepler, 1987). This is, in fact, the main site of the microtubule cytoskeleton, although individual microtubules or microtubule groups occur occasionally in the central zone (Raudaskoski *et al.*, 1987).

Organelle, nucleus and cell movement in the pollen tube

The circulation of organelles in the tube can be correlated directly with the disposition of the actin fibril system, notably in the older parts of the growing tube where the fibrils often lie apart. In such favourable circumstances the organelles can be seen in living tubes to move along individual fibrils. These can be shown to correspond to the actin fibrils revealed by phalloidin labelling (Heslop-Harrison & Heslop-Harrison, 1989*c*), while immunofluorescence shows that the associated organelles bear a myosin-like molecule. The traffic along individual fibrils is unidirectional, to be expected from the observations of Condeelis (1974) who showed by heavy-meromyosin labelling that the microfilaments composing a single bundle are uniformly polarised.

The circulation in the growing tube conforms in the main to a standard pattern, with the principal acropetal and basipetal fluxes extending from near the apex all the way into the grain early during growth, or – later – back to the first callose plug, the blockage that defines the proximal end of the vegetative-cell protoplast as it migrates into new stretches of the tube. Maintaining such a circulation obviously requires that a balance should exist in the number of forwardly and backwardly polarised fibrils.

This last consideration finally provides the basis for interpreting the much-debated movement of the vegetative nucleus, generative cell and the sperms it produces. It is often loosely suggested that these larger bodies travel towards the apex of the tube, but this is a misapprehension. What they actually do after germination is to move away from the parent pollen grain, tracking the apex as it grows forward, but maintaining on average the same distance from it until the tube grows into the micropyle of the ovule. The essence of the problem is how this roughly constant position in the advancing protoplast is maintained, given that in the extending tube organelles and other inclusions of the contiguous cytoplasm are in active two-way movement. A dominant theme in the literature, even up to the present, is that the larger inclusions must be capable of independent movement, 'swimming', so to speak, through the bidirectional currents, either individually, or linked to one another. It is of historic interest that Russian workers, including notably Navashin (review, 1969), contested this view on hydrodynamic grounds. Navashin proposed that the apparent movement is in fact passive, resulting from a preferential affinity of the bodies concerned with one of the opposed currents in the cytoplasm of the growing tube. This interpretation conveys elements of the truth, but is still unsatisfactory. However, an adequate explanation is now available for the observed behaviour of nuclei and cells in the tube on the basis of the interaction of their surface membranes, known to carry myosin or a plant equivalent, directly with linear fibrils of the actin cytoskeleton (Heslop-Harrison & Heslop-Harrison, 1989*b*). The vegetative nucleus and the generative cell or sperms are held in a more or less constant location within the active stretch of the protoplast, and the most likely explanation is that their positioning in the midst of the cytoplasm depends on a dynamic balance between acropetal and basipetal forces applied to their surfaces at points or zones of contact with actin fibrils of different polarities. The vegetative nucleus continuously changes shape in the tube, and is often drawn out in both forward and backward directions. This directly illustrates that opposed forces are applied to its surface; and that these result from interaction with the actin system is dramatically illustrated when the nucleus contracts into a condensed form after experimental disruption of the actin system by cytochalasin, an actin inhibitor (Heslop-Harrison & Heslop-Harrison, 1989*a*).

Thus it may be said that interactions between the myosin-bearing inclusions and the actin cytoskeleton govern most of the movements within the vegetative cell protoplast. However, there remains the question of how the advance of the protoplast itself is managed during growth of the tube. Extension may proceed at a rate of several microns per second, and continue for ten or more centimeters in the case of plants like maize, with the protoplast maintaining its position in the leading stretch. When a pollen tube protoplast is released enzymically into a medium of low tonicity it rapidly expands isodiametrically, indicating that the content is under hydrostatic pressure, and that the originally cylindrical form must be preserved by the constraint imposed by the strength of the older wall, with its outer cellulosic and inner callosic layers. But this alone does not explain how the migration into the new parts of the tube is driven. Behind the apex itself, discussed in the next section, the protoplast moves in relation to the older wall, 'crawling' along it, so to speak. This means that the key interactions must be between the polysaccharide fabric of the wall and the plasmalemma apposed to it, with its contiguous microtubules and microfilaments. This is an aspect now requiring urgent attention if a full account of how the pollen tube grows is to be pieced together.

The mechanism of tip growth and the establishment of zonation

Pollen tubes grow by extension at the tip, where wall material is added by localised transfer into the wall of precursor polysaccharides conveyed in vesicles mainly derived from dictyosome activity (Rosen, Gawlick & Siegesmund, 1964; Van der Woude, Morré & Bracker, 1971). The functioning of this system has provided some of the most intriguing problems of pollen biology, involving, as it does, the establishment and maintenance of the polarity that determines the cylindrical form of the cell. It is now apparent that the intracellular motility system plays a major part in this.

Contrast-enhanced video recording shows that, at the proximal, or basal, end of the active protoplast, whether in the grain or at a callose plug, the basipetally oriented streams of moving organelles are simply inverted, becoming thereby acropetally oriented. This is, of course, the prerequisite for circulatory flux. The situation at the extending apex is different. Immediately proximal to the growth cone proper where the principal insertion of the wall-precursor particles proceeds, a regular form of movement is maintained, first described by Iwanami (1956): here the acropetal flow is confined to the periphery of the tube, and the basipetal to the core. Larger organelles such as plastids and mitochondria pass forward in the peripheral cylinder, inverting behind the apex to return along the core, a pattern of movement often referred to as being comparable to an inverse fountain.

Whereas at the proximal end of the protoplast, the actin fibrils loop around where the streams of moving organelles are reversed, at the apex the fibrils associated with both the acropetal and basipetal flow pathways pass into a zone where actin exists as a dense mass of shorter fibrils (Heslop-Harrison & Heslop-Harrison, 1991). While it is difficult to discern preferred orientations in this mass, 'streak' micrographs from contrast-enhanced video records of the movement of the secretory vesicles in this region show a projection of the inverse fountain beyond the zone where the larger organelles change direction, suggesting that there is an overall, if obscure, pattern in the disposition of the short fibrils well into the apex. The interpretation we prefer is that the wall-precursor vesicles pass along the flanks of the tube into the growth cone from the sites of synthesis or storage, with individuals discharging their contents into the nascent wall when random Brownian movement brings them to the surface. Those prevented from reaching the wall because of space competition are then jostled into the basipetal stream in the centre of the tube, eventually transferring to an acropetal pathway to be recycled.

The nature of the flow pathways in the apical region of the tube together with the change in structure of the actin cytoskeleton provide the explanation for the organelle zonation frequently described from electron micrographs, which in fact merely provide a snapshot of a dynamic process (Heslop-Harrison & Heslop-Harrison, 1990).

While most of the foregoing features of the tip-growth system are reasonably well established, it would be a mistake to suppose that we are anywhere near understanding the totality of events. For example, one intriguing problem concerns the propagation of the two polarities of the actin fibrils behind the apex. The fact that the tube is constantly growing means that the fibrils of each polarity must extend in harmony, the F-actin of the microfilaments incorporating G-actin in the appropriate sense at matching rates. Elements of the actin cytoskeleton remain in parts of the tube cut off by plug formation, indicating that there is no recycling. Might it soon be possible to visualise the two polarities in the sites where incorporation proceeds?

Initiation and maintenance of polarity

How, then, is tip growth established and the cylindrical growth pattern initiated? In the main, angiosperm pollens fall into two major classes, with a few intermediates. In families usually regarded as phylogenetically primitive, mature, dormant pollen is bicellular, the vegetative cell containing its own nucleus and the generative cell. The haploid nucleus of the latter is already at the 2C DNA level, often with the chromosomes resolved and held in a state of suspended pre-metaphase until division takes place after germination. The situation in certain advanced families is different. Here, the generative cell divides before pollen maturity to form the two sperms, each with a nucleus at the 1C DNA level. It has long been known that this difference in the timing of development is associated with many other physiological differences: notably, tricellular pollens tend to be short-lived, to have a higher respiratory rate before germination associated with more fully developed mitochondria, and to germinate more rapidly on the stigma (for a review, see Heslop-Harrison, 1987). Strikingly, the difference is also reflected in the degree of activity in the ungerminated pollen. Bicellular pollens have a distinct dormancy period during which intracellular movement is suspended, while in tricellular pollens organelle movement continues after dispersal, often very actively, as in the grasses (Heslop-Harrison & Heslop-Harrison, 1992a).

The two pollen types provide an interesting contrast in the way the polarity, so much a characteristic of the pollen tube, is established initially. The tricellular grasses have pollen grains with a single germination aperture, and during later development in the anther, after the formation of the sperms, the actin cytoskeleton of the vegetative cell develops as a system of extended fibrils focused upon this aperture. These fibrils demarcate the main pathways along which the organelles, including the numerous amyloplasts which form the principal nutritional reserve, will move during the emergence of the tube. Liliaceous pollen also has a single apertural zone, but here actin exists during the dormancy phase in a 'storage' form, in quasi-crystalline spiculate bodies, or apposed to the surfaces of the vegetative nucleus and the generative cell. Polarity does not develop in the actin cytoskeleton until after re-hydration and restoration of intracellular movement, when once again actin fibrils become focused on the potential germination sites (Heslop-Harrison & Heslop-Harrison, 1992a, 1992b).

In both situations, polarity is established in relation to structural features of the pollen wall, and it is interesting to reflect that these features can in turn be traced back to events during meiosis when an originally isodiametric cell produces four spores each with inherent polarity (Heslop-Harrison, 1971).

How, then, do the structural characteristics of the pollen wall determine re-orientations in the vegetative cell, including notably the polarisation of the actin cytoskeleton? It is significant that the inner cellulosic layer of the wall is modified at the sites of potential tube emergence, with the microfibrils disposed in such a way as to weaken the texture. This ensures that it is in these areas the wall will give first as the vegetative cell hydrates and expands. A necessary effect of this is that the plasmalemma of the vegetative cell is allowed to bulge at the aperture sites, not only facilitating water uptake, but probably also enabling localised ingress of exogenous solutes by opening mechanosensitive ion channels. Each of these circumstances could result in the establishment of new polarities in the vegetative cell.

During germination the emergent tube tip, at first hemispherical, is devoid of larger organelles, but contains a dense population of wall-precursor-vesicles, seemingly in random movement. The transition to cylindrical growth marks the initiation of the tube and is accompanied by the establishment of the regular pattern of intracellular movement, indicating a probable causal link. It is reasonable in such circumstances to suspect that this involves some specific factor, and latterly the calcium ion has been singled out as a likely candidate.

Calcium and polarity

The possible role of the calcium ion in the growth of the pollen tube has been discussed in several publications (for a review, see Steer & Steer, 1989). The presence of a Ca^{2+} gradient declining along the tube from a peak near the tip has been demonstrated using a variety of techniques. Recently fluorescence ratio-imaging with Ca^{2+}-specific probes has been used to demonstrate the presence of such gradients in living tubes, leaving little doubt of their reality. A popular view is that the calcium ion may be concerned centrally with governing exocytosis at the tube tip and regulating the disposition of elements of the actin cytoskeleton. Evidence has been presented indicating that calmodulin co-localises approximately with calcium in the tube (Tirlapur & Cresti, 1992), suggesting this may be part of a system modulating Ca^{2+} function.

Most investigations of the Ca^{2+}-gradient have been carried out with tubes growing in a fluid medium. This has led to the proposition that the gradient results from a locally restricted influx of the ion from an exogenous source, and that this circumstance is the major factor in establishing and maintaining growth polarity. Supporting evidence for this proposition comes from observation of the effects of Ca^{2+}-blockers, which effectively destroy polarity in germinating grains and growing tubes bathed in liquid medium (Reiss & Herth, 1985).

However, many pollens will germinate and produce normal tubes in saturated air without access to fluid water or solutes. If a Ca^{2+}-gradient is maintained in such extending tubes in the absence of external calcium, it follows that there must be a mechanism that permits the continuous re-cycling of the resource held by the parent grain. How this might operate is a clearly central question, but one neglected hitherto. There seem to be two principal possibilities. Calcium is held in pollen-grain walls, and it is conceivable that a fraction of this external calcium might be mobilised and diffuse in some soluble form to the apex during early tube growth. We prefer an alternative interpretation, namely that the secretory vesicles concerned with wall formation at the tip may themselves be important carriers. Various measurements of the calcium gradients in pollen tubes indicate a high level of calcium where these are concentrated, and high levels can be detected in isolated vesicles (Heslop-Harrison et al., 1985). Perhaps the vesicles release calcium as they discharge into the wall, providing a pool which, in the absence of a bathing medium, is ultimately retrieved and passed back into the tube. Solving this problem does not seem to be entirely beyond the competence of modern technology.

The inter-relationships are, however, extremely complex. As we have already noted, it is by no means clear how the polarities and sites of the actin fibrils in the tip zone are actually determined, yet the spatial disposition of the actin cytoskeleton certainly establishes the characteristic circulation pattern at the tip and throughout the tube. But this, by governing the movement of membranes, organelles and secretory vesicles will necessarily affect the calcium flux. A puzzling circle of interaction indeed.

The microtubule system: an enigma

Microtubules, while present in the growing pollen tube, have little to do with the main types of intracellular movement described above. Franke et al. (1972) showed that colchicine at concentrations eliminating all microtubules from the vegetative cell had little effect on organelle movement, and only at concentrations well above those necessary to disrupt the microtubule cytoskeleton is there any appreciable effect on extension growth or cell shape (Heslop-Harrison & Heslop-Harrison, 1988). Yet the relative positions of the larger inclusions of the vegetative cell – the nucleus, the generative cell and the sperms formed from it – are radically changed when the microtubule cytoskeleton is removed. Recently we have found that anti-microtubule agents change the behaviour of pollen tubes during recovery from short pulses of cytochalasin treatment, indicating an interaction during restoration of the normal cylindrical growth pattern. It is tempting to relate this to the normal events during germination. Organised microtubules are absent from the vegetative cell in ungerminated pollen (although present in the generative cell), but appear around the periphery of the emerging tube at time of transition to cylindrical growth, suggesting that they may have some function in this process. Paradoxically, however, normal germination and tube initiation proceed in the presence of anti-microtubule agents. Still another aspect of pollen biology ripe for fresh investigation!

References

Condeelis, J.S. (1974). The identification of F-actin in pollen tubes and protoplasts of *Amaryllis belladonna*. Experimental Cell Research, **88**, 435–9.

Franke, W.W., Herth, W., Van Der Woude, W.J. & Morré, D.J. (1972). Tubular and filamentous structures in pollen tubes: possible involvement as guide elements in protoplasmic streaming and vectorial migration of vesicles. *Planta,* **105**, 317–41.

Heslop-Harrison, J. (1971). Wall pattern formation in angiosperm microsporogenesis. *Symposium of the Society Experimental Biology,* **25**, 277–300.

Heslop-Harrison, J. (1987). Pollen germination and pollen-tube growth. *International Revue of Cytology,* **107**, 1–78.

Heslop-Harrison, J. & Heslop-Harrison, Y. (1988). Sites of origin of the peripheral microtubule system of the vegetative cell of the angiosperm pollen tube. *Annals of Botany* **62**, 455–61.

Heslop-Harrison, J. & Heslop-Harrison, Y. (1989*a*). Conformation and movement of the vegetative nucleus of the angiosperm pollen tube: association with the actin cytoskeleton. *Journal of Cell Science,* **93**, 299–308.

Heslop-Harrison, J. & Heslop-Harrison, Y. (1989*b*). Myosin associated with the surfaces of organelles, vegetative nuclei and generative cells in angiosperm pollen grains and tubes. *Journal of Cell Science,* **94**, 319–25.

Heslop-Harrison, J. & Heslop-Harrison, Y. (1989*c*). Actomyosin and movement in the angiosperm pollen tube: an interpretation of some recent results. *Sexual Plant Reproduction,* **2**, 199–208.

Heslop-Harrison, J. & Heslop-Harrison, Y. (1990). Dynamic aspects of apical zonation in the angiosperm pollen tube. *Sexual Plant Reproduction,* **3**, 187–94.

Heslop-Harrison, J. & Heslop-Harrison, Y. (1991). The actin cytoskeleton of unfixed pollen tubes following microwave accelerated DMSO-permeabilisation and TRITC-phalloidin staining. *Sexual Plant Reproduction,* **4**, 6–11.

Heslop-Harrison, J. & Heslop-Harrison, Y. (1992*a*). Intracellular motility, the actin cytoskeleton and germinability in the pollen of wheat (*Triticum aestivum* L.). *Sexual Plant Reproduction,* **5**, 247–55.

Heslop-Harrison, J. & Heslop-Harrison, Y. (1992*b*). Cyclical transformations of the actin cytoskeleton of hyacinth pollen subjected to recurrent vapour-phase hydration and dehydration. *Biology of the Cell*, **75**, 245–52.

Heslop-Harrison, J., Heslop-Harrison, Y., Cresti, M., Tiezzi, A. & Moscatelli, A. (1988). Cytoskeletal elements, cell shaping and movement in the angiosperm pollen tube. *Journal of Cell Science*, **91**, 49–60.

Heslop-Harrison, J.S., Heslop-Harrison, J., Heslop-Harrison, Y. & Reger, B.J. (1985). The distribution of calcium in the grass pollen tube. *Proceedings of the Royal Society of London B*, **225**, 315–27.

Heslop-Harrison, Y. & Heslop-Harrison, J. (1992). Germination of monocolpate angiosperm pollen: evolution of the actin cytoskeleton and wall during hydration, activation and tube emergence. *Annals of Botany*, **69**, 385–94.

Iwanami, Y. (1956). Protoplasmic movement in pollen grains and tubes. *Phytomorphology*, **6**, 288–96.

Lancelle, S.S., Cresti, M. & Hepler, P.K. (1987). Ultrastructure of the cytoskeleton in freeze-substituted pollen tubes of *Nicotiana alata*. *Protoplasma*, **140**, 141–50.

Navashin, M.S. (1969). On the nature of the movement of the generative cell in the pollen tube and the problem of the localisation of cell elements. *Revue de Cytologie et de Biologie Végétales*, **32**, 141–148.

Perdue, T.D. & Parthasarathy, M.V. (1985). *In situ* localisation of F-actin in pollen tubes. *European Journal of Cell Biology*, **39**, 13–20.

Pierson, E.S., Derksen, J. & Traas, J.A. (1986). Organisation of microfilaments and microtubules in pollen tubes grown *in vitro* or *in vivo* in various angiosperms. *European Journal of Cell Biology*, **41**, 14–18.

Raudaskoski, M., Astrom, M., Perrtilla, K., Virtanen, I. & Louhetainen, J. (1987). Role of the microtubule cytoskeleton in pollen tubes: an immunochemical and ultrastructural approach. *Biology of the Cell*, **61**, 177–88.

Reiss, H.D. & Herth, W. (1985). Nifedipine–sensitive calcium channels are involved in polar growth of lily pollen. *Journal of Cell Science*, **76**, 247–54.

Rosen, W.G., Gawlick, W.V. & Siegesmund, K.A. (1964). Fine structure and cytochemistry of *Lilium* pollen tubes. *American Journal of Botany*, **51**, 61–71.

Steer, M.W. & Steer, J.M. (1989). Pollen tube tip growth. *New Phytologist*, **111**, 323–58.

Tang, X., Hepler, P.K. & Scordilis, S.P. (1989). Immunochemical and immunocytochemical identification of a myosin heavy chain polypeptide in *Nicotiana* pollen tubes. *Journal of Cell Science*, **92**, 569–74.

Tirlapur, U.K & Cresti, M. (1992). Computer-assisted video image analysis of spatial variations in membrane-associated Ca^{2+} and calmodulin during pollen hydration, germination and tip growth in *Nicotiana tabacum* L. *Annals of Botany,* **69**, 503–8.

Van der Woude, W.J., Morré, D.J. & Bracker, C.E. (1971). Isolation and characterisation of secretory vesicles in germinated pollen of *Lilium longiflorum. Journal of Cell Science,* **8**, 331–51.

Yen, Y.J., Wang, X.Z., Tong, Z.Y., Ma, Y.Z. & Liu, G.Q. (1986). Actin and myosin in pollens and their role in the growth of pollen tubes. *Kexue Tongbao,* **31**, 267–72.

J.R. GREEN, C.J. STAFFORD, P.J. WRIGHT
and J.A. CALLOW

Organisation and functions of cell surface molecules on gametes of the brown alga *Fucus*

Introduction

Fertilisation in the brown alga *Fucus* involves species-specific interactions between biflagellate sperm and spherical eggs (Bolwell *et al.*, 1977; Evans, Callow & Callow, 1982; Callow, Callow & Evans, 1985; Callow, Stafford & Green, 1992). We are interested in two related aspects of *Fucus* gamete cell surfaces: 1) How are the cell surface molecules organised? and 2) What is the molecular basis of recognition and the associated cell responses that occur within a few seconds or minutes of gamete fusion? Such studies in higher plants are difficult because the gametes are embedded within tissues, and plasma membrane based receptors have limited accessibility because of intervening cell walls. In addition, it is still relatively difficult to obtain gametes in sufficient numbers from higher plants compared with *Fucus* from which naked gametes are released in large enough quantities to allow detailed biochemical studies (Bolwell, Callow & Evans, 1980; Stafford, Callow & Green, 1992*a*). Thus the *Fucus* system has much to offer, and hopefully the findings will be relevant to gamete interactions in higher plants. This review will focus on how we have used a combination of biochemical and immunological approaches to study: 1) the organisation of the *Fucus* egg cell surface and 2) the role of sperm proteins in egg binding and the triggering of cell wall release.

Fertilisation in *Fucus*

The general properties of the *Fucus* system have been reviewed extensively (Evans *et al.*, 1982; Callow *et al.*, 1985; Green *et al.*, 1990). Fertilisation is a species-specific interaction between biflagellate, sperm cells (approx. 5 μm long; the shorter, anterior flagellum bears hair-like mastigonemes), and brown, spherical, non-motile, eggs,

Society for Experimental Biology Seminar Series 55: *Molecular and Cellular Aspects of Plant Reproduction*, ed. R.J. Scott & A.D. Stead.
© Cambridge University Press, 1994, pp. 203–214.

(approx. 60–80 µm in diameter). The eggs differ from those of animals in not having the equivalent of a vitelline layer, jelly coat or zona pellucida outside the plasma membrane, and in addition they are not surrounded by a cell wall.

Fertilisation in *Fucus* involves several stages. After release from the conceptacles, sperm are attracted to the eggs by a chemoattractant, secreted by the latter. This is a hydrocarbon pheromone (octatriene) and has been named fucoserraten (Maier & Muller, 1986). The attraction, which is not species specific, causes the sperm to swarm around the eggs. The sperm then appear to probe the surface of the eggs with their anterior flagella before fusion (Friedmann, 1961). Fertilisation is species-specific and it is likely that the recognition and/or binding between sperm and eggs is responsible for this specificity. This is followed by plasmogamy (fusion of eggs and sperm) and blocks to polyspermy. The earliest perceived response of fucoid eggs to fertilising sperm is a membrane depolarisation from approximately $-60\,mV$ to $-25\,mV$ (Brawley, 1991). This activation or fertilisation potential serves as a fast block to polyspermy and it appears that there is a transient elevation of cytoplasmic Ca^{2+}. The plasma membrane of the unfertilised egg contains Ca^{2+} channels which are opened by depolarisation, and at least part of the elevated Ca^{2+} appears to result from influx (C. Brownlee, pers. comm.). After the egg is fertilised, polysaccharides stored in cortical vesicles just below the plasma membrane are released providing a further block to polyspermy. A wall rich in alginic acid is initially released after 1–2 min fertilisation, and after 1–4 h additional polysaccharides are secreted, including fucoidan and cellulose, which cover the fertilised egg in a cell wall. A rhizoid begins to emerge from the polarised zygote 12–14 h after fertilisation (Evans *et al.*, 1982; Quatrano, 1982).

An assay involving the use of Calcofluor White to visualise wall polysaccharides released by fertilised eggs has been used to quantify fertilisation (Bolwell *et al.*, 1979). Experiments in which this fertilisation assay has been perturbed with lectins, proteases, polysaccharides, glycosidases and fractions prepared from *Fucus* sperm and eggs have led to a model for recognition which involves the binding of a sperm protein to a mannose- and fucose-containing glycoconjugate on the egg surface; the latter could be a glycoprotein, proteoglycan or (less likely) a polysaccharide (Bolwell *et al.*, 1979, 1980; Catt *et al.*, 1983; Green *et al.*, 1990; Callow *et al.*, 1992).

Organisation of the *Fucus* egg cell surface

Spatial organisation

We have used monoclonal antibodies (MAbs) and lectins to study the organisation and functions of cell surface proteins and glycoproteins of

Fucus sperm and eggs. The data for sperm have been reviewed previously and will only be considered where relevant (Jones, Callow & Green, 1988; 1990; Green *et al.*, 1990; Callow *et al.*, 1992). Of 12 MAbs (FS1-FS12) that were raised to *Fucus* sperm, three (FS2, FS4 and FS5) have been shown to recognise *Fucus* egg plasma membrane glycoproteins (Stafford, Green & Callow, 1992*b*; Table 1). FS2 and FS5 compete for binding to the same set of glycoproteins on the egg, which is different to a set recognised by FS4. FS2/5 and FS4 recognise common carbohydrate epitopes of different antigens present on eggs and sperm (Table 1). Immunofluorescence in combination with the high resolution imaging obtained with confocal laser scanning microscopy (CLSM) has allowed detailed analysis of the binding of these antibodies to the egg surface (Stafford *et al.*, 1992*b*). In particular it has been possible to observe the organisation and spatial distribution of surface components on the egg, by assembling composite images comprising many individual sections. Observations have shown that the three cross-reacting MAbs produce heterologous binding patterns on *Fucus* eggs: FS2 and FS5 bind to small patches (spatial domains) on the egg surface, whereas FS4 binds to larger areas. EM-immunogold labelling of egg sections confirmed that FS2 bound to patches in the form of small protuberances. The domains observed on the egg are not induced by patching or capping since they are found on both fresh and fixed eggs and Fab (fragment antigen binding) fragments prepared from FS5 (an IgM antibody) gave similar binding patterns to the intact IgM (Stafford, Callow & Green, 1993).

Double labelling experiments with fluorescein isothiocyanate- and gold-conjugated probes have shown that the regions labelled by the MAbs FS2/5 and FS4 are mainly exclusive with some areas of overlap whereas other areas are unlabelled by any of the antibodies. Thus regions can be described as being FS2/5$^+$FS4$^-$, FS2/5$^-$FS4$^+$, FS2/5$^+$FS4$^+$ and FS2/5$^-$FS4$^-$ (Fig 1). It is interesting that the antibodies FS2/5 and FS4 also show different binding patterns on sperm and vegetative tissue (Table 1).

The lectins concanavalin A (ConA) and fucose binding protein (FBP from *Lotus tetragonolobus)* have also been used to label the egg cell surface and they both bind to small discrete domains (Catt *et al.*, 1983; Stafford *et al.*, 1992*b*). Lectin-blotting experiments, and Western blots with the MAbs, have shown that ConA recognises a subset of the glyco-proteins recognised by FS2/5; FBP binds to a 62 kDa glycoprotein that is also recognised by ConA and FS2/5, but not FS4 (Stafford *et al.*, 1992*b*). The localisation and blotting experiments show that the FBP domains lie within the ConA domains which are located within the FS2/5$^+$ regions on the egg surface (Fig. 1). Overall the results show that the

Table 1. *Binding characteristics of MAbs FS2, FS4 and FS5 on* Fucus serratus *gametes and vegetative tissue*

MAbs	Sperm		Eggs		Vegetative Tissue	
	Localisation of binding	Molecular nature of antigens recognised	Localisation of binding	Molecular nature of antigens recognised	Localisation of binding	Molecular nature of antigens recognised
FS2, FS5	Anterior flagellum (mastigonemes)	Set of glycoproteins 90–250 kD	Small domains on cell surface	Set of glycoproteins* 47–185 kD	Epidermal/cortical layer, mucilage on thallus surface; medullary cell walls***	Fucose-containing polysaccharides
FS4	Body, both flagella	205 kD glycoprotein	Large domains on cell surface	Set of glycoproteins** 43–170 kD	Mucilage plug in ostiole	Fucose-containing polysaccharides

Data for sperm taken from Jones et al. (1988, 1990), data for eggs taken from Stafford et al. (1992b), data for vegetative tissue taken from Green et al. (1993).
* Different from glycoproteins recognised on sperm. ** Different from glycoproteins recognised by other MAbs on eggs. *** Labelling after EGTA treatment.

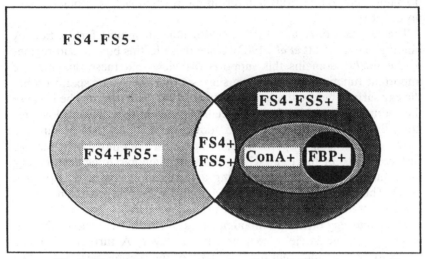

Fig. 1. Interpretive diagram showing the spatial domains on the *Fucus* egg cell surface identified by MAbs and lectin probes.

surface plasma membrane of the *Fucus* egg is organised in a complex fashion which is highly unusual for a morphologically undifferentiated cell.

Functional significance of the spatial domains

The functional significance of the organisation of the egg cell surface into spatial domains has been assessed by studying the effects of the MAb and lectin probes on the fertilisation bioassay. FS2 and FS5 inhibited fertilisation after pre-incubation with eggs, whereas FS4 had no effect (Stafford *et al.*, 1993). Although these antibodies also bind to sperm, control experiments showed that their effects were due to binding to the egg. Since FS5 is an IgM (M_r approx. 900 kDa) it could be argued that the inhibitory effect on eggs was due to steric hindrance near putative receptors for sperm. However, purified IgM and Fab (fragment antigen binding) fragments prepared from FS5 IgM were also able to block fertilisation, showing that cross-linking and spatial interference are probably not involved, and suggest that more direct effects are occurring. Since FS4 presumably completely masks access to the set of glycoproteins recognised by this antibody, it is unlikely that the FS4$^+$ domains have an important role in sperm recognition. Thus there is a clear functional difference between the spatial domains on the egg surface: the FS2/5$^+$

domains seem to be involved in fertilisation, whereas FS4$^+$ domains have no direct role.

The lectins ConA and FBP are also able to inhibit fertilisation by binding to eggs (Catt *et al.*, 1983); since these lectins bind to sub-regions of the FS2/5$^+$ domains this supports the view that these domains are important functionally. Evidence suggests that the *Fucus* sperm probes the egg surface before binding (Friedman, 1961) and this may be because the sperm are searching for the specific ligands in the relevant domains. Once they have reached the FS2/5$^+$ domains they may then be directed to the correct ligand. Sperm of *Pelvetia canaliculata* move over the egg surface for 10s or more, attach to one spot, then gyrate against the egg for a further 10-20s before evoking a fertilisation potential (Brawley, 1991). This behaviour supports the view that sperm receptors on the egg plasma membrane are present in restricted domains.

The *Fucus* egg differs from animal eggs in that it lacks the equivalent of a zona pellucida, jelly coat or vitelline layer. A further difference seems to be the functional heterogeneity of the plasma membrane. The FS2/5$^+$ domains on the *Fucus* egg could be important in recognition and binding of sperm, sperm-egg fusion, events leading to blocks in polyspermy, egg activation or a combination of some or all of these. Since FBP recognises one major component of 62kDa (also recognised by ConA and FS2/5) and inhibits fertilisation, attention is now being focused on the role of this glycoprotein in the fertilisation process.

Temporal expression of egg cell surface glycoproteins

The development of spatial domains on the *Fucus* egg plasma membrane has been studied recently using a combination of micromanipulation and CLSM (Stafford, Callow & Green, unpublished observations). The cytology of oogonial maturation in *Fucus* has previously been described in detail (McCully, 1968). Gametes develop within cup-shaped conceptacles which develop in the receptacles at the tips of the branches of mature plants. Female gametes are produced in large sacs called oogonia which arise from the conceptacle wall; the oogonial mother cell usually develops as an outgrowth of an epithelial cell. After two meiotic divisions oogonia reach the 4-nucleate stage and ensuing mitosis gives rise to the 8-nucleate stage. This is followed by cleavage into eight haploid eggs which subsequently become separated from each other by the formation of a thin membrane around each cell. The 4-nucleate and 8-nucleate/cell stages can be released from the fertile conceptacles at the tips of the fronds, and immature oogonia can be dissected from the 8-celled structures (Stafford, Callow & Green, unpublished observations).

Oogonia at different stages of development have been immunolabelled with FS2/5 and FS4. The results showed that the domains recognised by FS2/5 and FS4 arise at different times during egg maturation. Thus, FS2/5 labels small patches on the plasma membrane at the 4-nucleate stage, whereas FS4 does not show any labelling. All the MAbs label the 8 nucleate/celled stage and the labelling with FS2/5 resembles that seen on mature eggs. However, FS4 labels domains which are much smaller than those which occur on mature eggs. This suggests that the glycoproteins recognised by FS2/5 are synthesised and exported very early in egg development. Glycoproteins labelled by FS4 are not incorporated into the plasma membrane until later in development and are built up into larger and larger domains. These results are analagous to studies of eggs and maturing ovaries of *Strongylocentrotus purpuratus*, where specific antigens important for fertilisation are synthesised early on in oogenesis and rapidly moved to the egg cell surface (Ruiz-Bravo, Janak & Lennarz, 1989).

Formation and maintainance of egg cell surface domains

Overall, the results on the *Fucus* egg show that FBP, ConA and FS2/5 appear to recognise progressively larger families of glycoproteins held within small discrete domains on the egg cell surface and these are spatially, developmentally, functionally and in molecular characteristics largely distinct from a family of glycoproteins recognised by FS4.

Questions remain as to the cellular mechanisms involved in assembling and maintaining the spatial domains on the egg cell surface. Organisation of surface domains is not unusual for polarised or morphologically highly differentiated cells, for example epithelial cells or sperm cells (Primakoff & Myles, 1983; Simons & Fuller, 1985). In these instances specific proteins are maintained within the domains by barriers to free diffusion (Gumbiner & Louvard, 1985; Cowan, Myles & Koppel, 1987). This is unlikely to operate in spherical, apolar cells like an egg and it is more probable that the cytoskeleton has a part to play. Thus, possible attachment (directly or indirectly) of *Fucus* egg cell surface molecules to microtubules or actin filaments is under investigation.

FS2/5 and FS4 bind to distinct carbohydrate epitopes on different sets of glycoproteins (Table 1). These antibodies also bind to carbohydrate epitopes of fucose-containing polysaccharides (including fucoidan and ascophyllan) which occur in *Fucus* vegetative tissue (Green *et al.*, 1993). Localisation experiments have shown that FS2/5 bind to the mucilage on the outside of the thallus and also to the walls of cells in the central medulla, whereas FS4 only labels the mucilage on the outside of the ostiole (through which gametes are released from the conceptacle). Thus glycoproteins or polysaccharides containing the carbohydrate epitopes defined by FS2/5 and FS4 are targeted to distinct spatial locations in sperm, eggs and vegetative tissue. It is possible that the different carbohydrate epitopes identified by these MAbs may be involved in targeting these molecules to particular regions, or in maintaining them in those regions once assembled.

Identification of sperm components involved in egg binding and the triggering of cell wall release

An *in vitro* binding assay between sperm and egg fractions

Attempts to characterise the *Fucus* egg and sperm receptors involved in the species-specific binding and/or fusion between gametes have generally involved testing for inhibitory effects of certain surface probes, for example MAbs or lectins, in the fertilisation bioassay (Catt *et al.*, 1983; Stafford *et al.*, 1993). This assay presumably integrates a number of stages in the egg–sperm interaction at which discrete recognition events could be involved, and a more refined approach has been attempted in which the various stages may be separately analysed, and the molecules involved characterised (Wright, *et al.*, 1994).

An *in vitro* assay has been developed which monitors sperm/egg recognition and binding. This assay utilises a KCl-soluble extract of *Fucus* sperm (Bolwell *et al.*, 1980), that contains cell surface proteins/glycoproteins, and an enriched plasma membrane (PM) fraction from eggs (Stafford *et al.*, 1992a). Both these fractions inhibit fertilisation in a species-preferential manner when added to the previously described bioassay, that is *F. serratus* egg or sperm components are more effective at inhibiting homologous fertilisation than heterologous fertilisation. Binding of the egg PM vesicles to the sperm KCl extract has been quantified as follows (Wright *et al.*, 1994): The KCl-extracted proteins are bound to wells of a microtitre plate, and then incubated with biotinylated egg PM vesicles. Binding is assessed after the addition of streptavidin conjugated to alkaline phosphatase followed by the appropriate substrate, and the OD is read at 410 nm. Using this binding assay the following findings have emerged:

1. Binding of the biotinylated egg PM vesicles to the sperm KCl extract is saturable.
2. There is little or no binding of egg PM vesicles to *Fucus* vegetative material: thus binding is gamete-specific.

3. *F. serratus* egg vesicles bind to both *F. serratus* and *F. vesiculosus* sperm KCl extracts: thus binding is not species-specific.
4. Binding is blocked after pre-treatment of the sperm KCl extract with proteases, but not periodate, suggesting that the binding principle on sperm is a protein.
5. Binding is blocked after pre-treatment of the KCl extract with certain sulphated polysaccharides including native fucoidan and dextran sulphate, but is not affected by de-sulphated fucoidan, chondroitin sulphate or other polysaccharides.

The latter results suggest that the protein in the KCl extract responsible for binding to eggs recognises specific spatial arrangements of sulphate groups on carbohydrates. The importance of sulphation is emphasised since these reagents also inhibit the fertilisation bioassay. Previous studies have suggested that fucose and/or mannose residues on the egg components are responsible for sperm binding and it is therefore likely that it is these carbohydrate residues which are sulphated (Callow *et al.*, 1985; Green *et al.*, 1990). These results with *Fucus* are analagous to findings on *bindin,* a fucose-binding protein from sea urchin, and its mammalian analogue proacrosin (De Angelis & Glabe, 1987, 1988; Jones, 1990). For both of these molecules, sulphation of the egg ligand is a critical feature recognised by basic amino acids within the complementary binding molecules.

Though the sperm and egg fractions inhibit fertilisation in a species-preferential manner, the binding between these fractions in the *in vitro* assay is not species-specific. Other results on the effects of a variety of reagents on the fertilisation assay have shown that the MAbs FS2/5, the lectins ConA and FBP, the polysaccharides fucoidan and mannan, and glycosidases inhibit fertilisation in both *F. serratus* and *F. vesiculosus;* thus their effects are not species-specific (Bolwell *et al.*, 1979; Callow *et al.*, 1985; Stafford *et al.*, 1993). It is possible that they produce non-specific effects by interfering with conserved parts of receptor molecules rather than regions involved in species specific recognition. In addition, gamete adhesion might have a low level of specificity which may be a function of other, as yet undetected receptor molecules.

Identification of a sperm protein involved in egg binding

The binding assay described above has been used to identify the sperm protein involved in egg binding (Wright, Callow & Green, unpublished observations). High performance gel filtration (HPGF), in high ionic strength conditions, separated the sperm KCl extract into several major fractions. These fractions were tested in the binding assay and a protein (apparent M_r 60 kD) has been identified as being involved in binding to the egg PM vesicles. This protein ran on denaturing SDS gels as a doublet with apparent molecular weight of 27 kD. This suggests that either the native form of the protein is a dimer or the molecular weight on HPGF is an artefact caused by high ionic strength buffer promoting hydrophobic interactions. When KCl-soluble proteins were separated by SDS-polyacrylamide gel electrophoresis, blotted on to nitrocellulose and incubated with biotinylated egg PM vesicles, these bound to a band at 27 kD, confirming the role of this protein.

Addition of the *Fucus* sperm KCl extract to eggs, in the absence of sperm, induced the release of polysaccharides on to the surface as judged by Calcofluor White labelling (Wright, Callow & Green, unpublished observations). This labelling is patchy, in contrast to the uniform release of polysaccharides observed when sperm are added to eggs (Evans *et al.*, 1982; Callow *et al.*, 1985). The effect is rapid, with most eggs being triggered to release polysaccharides within minutes of the addition of the sperm extract. Evidence suggests that the sperm component responsible for this triggering effect is the 27 kD protein previously identified as being involved in sperm–egg binding. This protein therefore has a dual function since it is involved in sperm–egg recognition/adhesion and induces a response in the egg to binding. Future studies will include: 1) the determination of the sequence of the 27 kD sperm protein and comparisons between *F. serratus* and *F. vesiculosus*; 2) determining whether the sperm protein binds to particular domains on the egg surface and identification of the egg components involved and 3) determining the signals that occur in the egg after the sperm 27 kD protein has bound and how these compare with the effects of intact sperm during fertilisation.

Acknowledgement

This work has been supported by grants and studentships from AFRC and SERC.

References

Bolwell, G.P., Callow, J.A., Callow, M.E. & Evans, L.V. (1977). Cross-fertilisation in fucoid seaweeds. *Nature,* **268**, 626–7.
Bolwell, G.P., Callow, J.A., Callow, M.E. & Evans, L.V. (1979). Fertilisation in brown algae. II. Evidence for lectin sensitive complementary receptors involved in gamete recognition in *Fucus serratus*. *Journal of Cell Science,* **36**, 19–30.

Bolwell, G.P., Callow, J.A. & Evans, L.V. (1980). Fertilisation in brown algae. III. Preliminary characterisation of putative gamete receptors from eggs and sperm of *Fucus serratus. Journal of Cell Science,* **43**, 209–24.

Brawley, S.H. (1991). The fast block against polyspermy in fucoid algae is an electrical block. *Developmental Biology,* **144**, 94–106.

Callow, J.A., Callow, M.E. & Evans, L.V. (1985). Fertilisation in *Fucus.* In *Biology of Fertilisation, vol. 2, Biology of the Sperm,* ed. C.B. Metz & A. Monroy, pp. 389–407. Academic Press.

Callow, J.A., Stafford, C.J. & Green, J.R. (1992). Gamete recognition and fertilisation in the fucoid algae. In *Perspectives in Plant Recognition. SEB Seminar Series* **48**, ed. J.A. Callow & J.R. Green, pp. 19–33. Cambridge: Cambridge University Press.

Catt, J.W., Vithanage, H.I.M.V., Callow, J.A., Callow, M.E. & Evans, L.V. (1983). Fertilisation in brown algae. V. Further investigations of lectins as surface probes. *Experimental Cell Research,* **147**, 127–33.

Cowan, A.E., Myles, D.G. & Koppel, D.E. (1987). Lateral diffusion of the PH-20 protein on guinea pig sperm: evidence that barriers to diffusion maintain plasma membrane domains in mammalian sperm. *Journal of Cell Biology,* **104**, 917–23.

De Angelis, P.L. & Glabe, C.G. (1987). Polysaccharide structural features that are critical for the binding of sulphated fucans to bindin, the adhesive protein from sea urchin sperm. *Journal of Biological Chemistry,* **262**, 13946–52.

De Angelis, P.L. & Glabe, C.G. (1988). Role of basic amino acids in the interaction of binding with sulphated fucans. *Biochemistry,* **27**, 8189–94.

Evans, L.V., Callow, J.A. & Callow, M.E. (1982). The biology and biochemistry of reproduction and early development in *Fucus.* In *Progress in Phycological Research vol. 1.,* ed. F.E. Round & D.J.Chapman, pp. 67–110. Amsterdam: Elsevier.

Friedmann, I. (1961). Cinemicrography of spermatozoids and fertilisation in *Fucales. Bulletin of the Research Council Israel,* **10D**, 73–83.

Green, J.R., Jones J.L., Stafford, C.J. & Callow J.A. (1990). Fertilisation in *Fucus:* exploring the gamete cell surfaces with monoclonal antibodies. In *NATO ASI Series Vol. H45, Mechanism of Fertilisation: Plants to Humans,* ed. B. Dale, pp. 189–202. Springer-Verlag: Berlin Heidelberg.

Green, J.R., Stafford, C.J., Jones, J.L., Wright, P.J. & Callow, J.A. (1993). Binding of monoclonal antibodies to vegetative tissue and fucose-containing polysaccharides of *Fucus serratus* L. *New Phytologist,* **124**, 397–408.

Gumbiner, B. & Louvard, D. (1985). Localised barriers in the plasma membrane: a common way to form domains. *Trends in Biochemical Sciences,* **10**, 435–8.

Jones, R. (1990). Identification and functions of mammalian sperm-egg recognition molecules during fertilisation. *Journal of Reproduction and Fertility Supplement*, **42**, 89–105.

Jones, J.L., Callow, J.A. & Green, J.R. (1988). Monoclonal antibodies to sperm surface antigens of the brown alga *Fucus serratus* exhibit region-, gamete-, species and genus preferential binding. *Planta*, **176**, 298–306.

Jones, J.L., Callow, J.A. & Green, J.R. (1990). The molecular nature of *Fucus serratus* sperm surface antigens recognised by monoclonal antibodies FS1 to FS12. *Planta*, **182**, 64–71.

McCully, M.E. (1968). Histological studies on the genus *Fucus*. II. Histology of the reproductive tissues. *Protoplasma*, **66**, 205–30.

Maier, I. & Muller, D.G. (1986). Sexual pheromones in algae. *Biological Bulletin*, **170**, 145–75.

Primakoff, P. & Myles, D.G. (1983). A map of the guinea pig sperm surface constructed with monoclonal antibodies. *Developmental Biology*, **98**, 417–28.

Quatrano, R.S. (1982). Cell wall formation in *Fucus* zygotes: a model system to study the assembly and localisation of wall polymers. In *Cellulose and Other Natural Polymer Systems*, ed. R.M. Brown Jr., pp. 45–59. New York: Plenum Press.

Ruiz-Bravo, N., Janak, D.J. & Lennarz, W.J. (1989). Immunolocalisation of the sea urchin sperm receptor in eggs and maturing ovaries. *Biology of Reproduction*, **41**, 323–34.

Simons, K. & Fuller, S.D. (1985). Cell surface polarity in epithelia. *Annual Review of Cell Biology*, **1**, 243–88.

Stafford, C.J., Callow, J.A. & Green, J.R. (1992*a*). Isolation and characterisation of plasma membranes from *Fucus serratus* eggs. *British Phycological Journal*, **27**, 429–34.

Stafford, C.J., Green, J.R.& Callow, J.A. (1992*b*). Organisation of glycoproteins into plasma membrane domains on *Fucus serratus* eggs. *Journal of Cell Science*, **101**, 437–48.

Stafford, C.J., Callow, J.A. & Green, J.R. (1993). Inhibition of fertilisation in *Fucus* (Phaeophyceae) by a monoclonal antibody that binds to domains on the egg cell surface. *Journal of Phycology*, **29**, 325–30.

Wright, P.J., Green, J.R. & Callow, J.A. (1994). An *in vitro* binding assay to identify fertilisation specific receptors from sperm of the brown alga *Fucus serratus*. *Journal of Phycology*, in press.

A.D. STEAD and W.G. VAN DOORN

Strategies of flower senescence – a review

Introduction

Senescence is a process usually considered to encompass those events which lead irreversibly to death. However, whilst the term 'flower senescence' is frequently encountered, it rarely relates to the flower as a whole but more particularly to those parts of the flower which are regarded as attractive, that is the perianth parts. In fact, as some parts of the flower senesce, other floral organs are still developing. For example, in species exhibiting protandry, the anthers may well senesce as the stigma expands and becomes receptive to pollination. A further contrast is to be found when petal and leaf senescence are compared; the degeneration of the chloroplast would normally be regarded as senescence in leaves, however, in petals such a process occurs early on and is considered part of the process of petal development.

The process of flower senescence differs greatly from species to species; in some the petals wilt and may eventually abscise, in others abscission of the perianth occurs whilst fully turgid; in yet others a change in colour of all, or part, of the perianth portends the ageing process. From this it is clear that both the physiology and structure of the petals undergo changes during senescence. This review endeavours to describe some of the physiological and structural changes that occur in these varied strategies and the factors, particularly pollination, which may initiate the senescence of the perianth parts.

Longevity of flowers

The longevity of the flowers of many species has been reported by several authors. The time to wilting in the monocots (Table 1) varies from less than 60 minutes after anthesis in the orchid *Desmotrichum appendicula-*

Society for Experimental Biology Seminar Series 55: *Molecular and Cellular Aspects of Plant Reproduction*, ed. R.J. Scott & A.D. Stead.
© Cambridge University Press, 1994, pp. 215–237.

Table 1. *Time to wilting in monocotyledonous flowers showing perianth wilting as the first sign of senescence*

Desmotrichum appendiculatum	< 1 h	(Orchidaceae)
Tradescantia virginica	6–7 h	(Commelinaceae)
Hemerocallis fulva	12–14 h	(Liliaceae)
Asphodelus luteus	12–24 h	(Liliaceae)
Guzmannia tricolor	24–30 h	(Bromeliaceae)
Iris pallida / I. ensata	24–36 h	(Iridaceae)
Iris gueldenstaedtia	3 d	(Iridaceae)
Billburgia thyrsoidae	3 d	(Bromeliaceae)
Fritillaria meleagris	5 d	(Liliaceae)
Gladiolus (hybrid)	5–6 d	(Iridaceae)
Hemerocallis flava	6 d	(Liliaceae)
Lilium alba	6 d	(Liliaceae)
Narcissus poeticus	8 d	(Amaryllidaceae)
Clivia nobilis	10–12 d	(Amaryllidaceae)
Cattleya labiata	30 d	(Orchidaceae)
Cypripedium insigne	40 d	(Orchidaceae)
Phalaenopsis grandiflora	50 d	(Orchidaceae)
Oncidium cruentum	60 d	(Orchidaceae)
Odontoglossum rossii	80 d	(Orchidaceae)

Data from Kerner von Marilaun (1891), Kränzlin (1910), Wacker (1911), and unpublished observations. The observations were all made in the botanical garden or in greenhouses, and are therefore only approximate.

tum (Kränzlin, 1910) to about 80 days in *Odontoglossum rossii*, also an orchid (Kerner von Marilaun, 1891). The time to wilting is remarkably different even within some genera: examples are *Hemerocallis fulva*, which wilts about half a day after anthesis, and *Hemerocallis flava*, in which wilting only occurs after about 6 d (Kerner von Marilaun, 1891). It must be pointed out, however, that many of these observations relate to plants growing in glasshouses or in botanical gardens, and it is therefore quite possible that the plants were growing in situations where pollination was unlikely to occur; the recorded flower longevities are thus likely to be the maximum possible. Time to wilting in the dicots (Table 2) was found to vary from about 4–5 h in *Portulaca grandiflora* (Wacker, 1911) to about 20 days in *Gerbera* and *Dendranthema*. The variability in floral longevity between the monocots and dicots is therefore very similar with the exception that some members of the Orchidaceae have very long-lived flowers. Of the shorter-lived flowers the period of flower opening may be either by day or night depending upon the activity of the

Table 2. *Time to wilting in dicotyledonous flowers showing perianth wilting as the first sign of senescence*

Portulaca grandiflora	4–5 h	(Portulacaceae)
Ipomoea purpurea	5–6 h	(Convolvulaceae)
Calandrinia grandiflora	7–8 h	(Portulacaceae)
Hibiscus trionum	9–12 h	(Malvaceae)
Oenothera lamarckiana	11–12 h	(Onagraceae)
Raphanus maritimus	2–3 d	(Cruciferae)
Clerodendron thomsonii	3 d	(Verbenaceae)
Cardamine pratensis	3–4 d	(Cruciferae)
Anoda hastata	3–4 d	(Malvaceae)
Cobaea scandens	4–5 d	(Polemoniaceae)
Caltha palustris	4–5 d	(Ranunculaceae)
Nicotiana purpurea	4–5 d	(Solanaceae)
Althoea rosea	5–6 d	(Malvaceae)
Viola odorata	9–11 d	(Violaceae)
Gerbera jamesonii	16–24 d	(Asteraceae)
Dendranthema grandiflora	17–22 d	(Asteraceae)

Data from Kerner von Marilaun (1891), Wacker (1911), Fitting (1921), and unpublished observations. The observations were all made in the botanical garden or in greenhouses, and are therefore only approximate.

pollinator, thus several moth-pollinated flowers (e.g. *Oenothera*) open only at night (Table 3).

In species in which abscission terminates the functional life of the flower there is also a great deal of variability in floral longevity (Table 4), although there are no exceptionally long-lived flowers (cf. wilting in some

Table 3. *The longevity of some ephemeral flowers, in hours, either opening by day or during the night*

Duration of opening (h)			
Flowers open by day		Flowers open by night	
Hibiscus trionum	3	*Cereus nycticalus*	5
Oxalis stricata	8	*Mirabilis longiflora*	7
Tradescantia virginica	12	*Cereus grandiflorus*	7
Hemerocallis fulva	14	*Oenothera biennis*	12

Data from Molisch (1928).

Table 4. *Time to abscission in flowers showing petal abscission as the first sign of senescence*

Monocotyledonae		
Heliconia psittacorum	12 h	(Heliconiaceae)
Dicotyledonae		
Linum perenne	3 h	(Linaceae)
Verbena officinalis	5–6 h	(Verbenaceae)
Helianthemum vulgare	5–6 h	(Cistaceae)
Erodium cicutarium	8 h	(Geraniaceae)
Cistus creticus / C. ladaniferus	12 h	(Cistaceae)
Aphelandra aurantiaca	1–2 d	(Acanthaceae)
Borago officinalis	1–2 d	(Boraginaceae)
Impatiens noli tangere	2–3 d	(Balsaminaceae)
Scrophularia nodosa	3–4 d	(Scrophulariaceae)
Cuphea vicossissima	3–5 d	(Lythraceae)
Digitalis purpurea	6–7 d	(Scrophulariaceae)
Pelargonium zonale	7–8 d	(Geraniaceae)

Data from Kerner von Marilaun (1891), Wacker (1911), Fitting (1921), Pfeiffer (1928), and Criley & Broschat (1992). Observations were made in the botanical garden or in greenhouses, and are therefore only approximate.

Orchidaceae). Some flowers, such as *Linum perenne*, show abscission of the petals within 3 h of anthesis whilst the longest time to real abscission without prior wilting is apparently about one week, as in *Pelargonium zonale*.

Although there is little or no relationship between flower longevity and the taxonomic classification of species (Primack, 1985), the pattern of flower senescence is somewhat more consistent within taxonomic groups (Woltering & van Doorn, 1988). Thus the flowers of species from the Caryophyllaceae generally wilt, whilst those of the Scrophulariaceae usually abscise. However, there are examples where the petals of some species wilt whilst those of a closely related species may abscise whilst still turgid. For example, within the Verbenaceae the petals of *Cleronendron thomsonii* are reported to wilt 3–4 days after anthesis whilst those of *Verbena officinalis* abscise 5–6 h after anthesis (Wacker, 1911).

Fitting (1921) remarked that the various times to senescence may have evolved in relation with the probability of pollination. If the probability of pollination is high then the time to senescence can be short, but when the probability is low, for example in the often very specialised orchid flowers, the flowers have to remain to ensure pollination. However, Fitting (1921) added that this explanation may not always hold, as his observations in the botanical garden of Buitenzorg (now Bogor) in Indonesia showed that the time to wilting in *Dendrobium crumenatum* was not as long as in several other orchids, but the chance of pollination was apparently similar both in *D. crumenatum* and the other orchids, in that particular location. None the less, the exceedingly long period of flowering in some orchids may have an evolutionary relationship with their degree of specialisation in attracting one animal species. It might be expected that flower longevity would be correlated to habitat as pollinator frequency might vary, particularly with altitude, however Primack (1985) was unable to demonstrate any such relationship.

Other factors may also be of importance in the time to wilting or abscission. One such factor is the rapid growth and development of many bulbous flowers, which often have a short life-cycle, and perhaps in consequence the flowers too are short-lived. It should be noted that several species from the Liliaceae and the Iridaceae, to mention only two families, show a very short time to wilting.

Stiles (1975, 1978) and Dobkin (1987) mentioned that practically all the hummingbird-pollinated flowers of the wet lowland tropics they observed lasted for only half a day. The genera studied included *Aechmea, Alloplectus, Aphelandra, Calathea, Cephaelis, Columnea, Costus, Guzmannia, Heliconia, Passiflora*, and *Pentagonia*. Over a period of several months the inflorescence of *Heliconia*, for example, contains numerous buds but only one open flower each day. Two types of flower senescence are found within the genus: in those species in which water remains standing in the coloured bracts, the flowers open in the early morning and wither in the afternoon, thereby slowly bending into the water, and in the species in which no water is standing in the bracts, the flower opens in the early morning and abscises in the late afternoon (Dobkin, 1984, 1987; Criley & Broschat, 1992). The flowers of *H. psittacorum, H. trinidatis* (and *Costus cylindricus*) abscise within 0.5 day of anthesis, whether they are pollinated or not, and, remarkably, within a population of these species the abscission occurs almost synchronously, i.e. within about 15 min each afternoon, starting between about 15:30 and 17:00 (Dobkin, 1987).

Stiles (1975, 1978) hypothesised that the ephemeral character of these hummingbird-pollinated flowers might be the result of selection against destruction by various insects and birds. The flowers are very rich in nectar, hence might be easy prey for non-pollinators. Abscission or withering of the perianth in the water prevents attack and apparently protects the ovary. Dobkin (1987) speculated that the floral 'half-dayness' of these flowers might relate to prevention of self-fertilisation, for example by the mite population found in the floral bracts. We might add that the presentation of one flower per individual on each day, given a high availability of the pollinators, is a mechanism to ensure high genetical diversity in the offspring.

The ability to influence and control flower longevity has considerable commercial potential; cut flowers with a longer vase life will have a greater commercial value, whilst the length of the effective pollination period will influence fruit and seed set. In the natural habitat, the duration of flower opening will determine the probability of successful pollination and therefore influence seed set and the fecundity of any particular genotype. Variations in flower longevity within a population will therefore have an effect on the survival, and possibly the evolution, of genotypes within that population. Furthermore, the type of senescence process may well influence the frequency of self-pollination, for example, in *Mimulus guttatus* during the shedding of the turgid corolla, the anthers come into contact with the stigma and if pollen remains in the anthers, self-pollination will occur (Dole, 1990). A similar claim was made by Charles Darwin (1891) with respect to *Digitalis purpurea*, although in glasshouse experiments the frequency of self-pollination was found to be very low (Stead & Moore, 1979); indeed other work by Dudash and Ritland (1991) suggests that in *Mimulus* too self-pollination in the field by corolla shedding is rare. Similarly, in *Papaver rheoas*, flower longevity is greatest when seed set is lowest (Stead, unpublished observations). Thus during winter months, even when grown in environmental chambers, pollinator activity is low and few seeds are set and petals only abscise 4–5 days after flower opening. In summer, when numerous small insects pollinate the flowers, seed set is high and petals abscise within 1–2 days of flower opening.

Physiological and structural aspects of petal senescence

Ephemeral flowers

Amongst the species having very short-lived flowers there are examples where the flowers wilt or abscise. For example, in *Linum perenne* the petals abscise within a day of flower opening (Wacker, 1911) whilst those of *Ipomoea purpurea* (Matile & Winkenbach, 1971) and *Hemerocallis fulva* (Lukaszewski & Reid, 1989) wilt within one day.

a) Flowers showing petal wilting

The physiological mechanisms which control these processes are diverse; *Hemerocallis* flowers, for example, are ethylene-insensitive (Lay-Yee, Stead & Reid, 1992) but those of *Ipomoea* are sensitive to ethylene (Kende & Hanson, 1976). Even isolated corolla rib segments of *Ipomoea* behave in a manner identical to the intact flower. When floated on liquid, the inrolling of rib segments from flowers which have just opened is associated with increased production of ethylene, whilst exposure to ethylene hastens both inrolling and endogenous ethylene production. However, segments from flowers more than one day before opening are not sensitive to exogenous ethylene. The possible cause of such a change in sensitivity is the subject of another chapter (Whitehead, this volume).

The wilting of the *Ipomoea* corolla is accompanied by the breakdown of cellular integrity. In the mature bud and young flower, the mesophyll cells are the first to undergo autolysis and these cells are devoid of cytoplasmic contents the morning after flowering. The breakdown of the epidermal cells occurs slightly later, usually in the evening of the day of flowering and, unlike the mesophyll cells, the process continues during the next day. The vascular tissue structure remains apparently healthy, and presumably functional, for much longer, possibly to facilitate the translocation of the products of cellular autolysis (Matile & Winkenbach, 1971).

Although the flowers of several *Hemerocallis* species also wilt the process in these species does not seem to be influenced by ethylene, indeed treatment with aminooxyacetic acid (an ethylene biosynthesis inhibitor) or silver (an inhibitor of ethylene action) does not influence the time of petal wilting (Lukaszewski & Reid, 1989). Nevertheless, *Hemerocallis* flowers do produce ethylene and production does increase as the flowers wilt. However, the amounts produced are very small: the maximum rate of ethylene production from the flower was less than 1 nl g^{-1} fresh weight (Lay-Yee *et al.*, 1992), whilst in *Petunia* and *Dianthus* the rate exceeds 20 nl g^{-1} fresh weight (Whitehead & Halevy, 1989; Downs & Lovell, 1986). Cycloheximide (CHI) is very effective at preventing flower wilting in daylilies (Lukaszewski & Reid, 1989), hence protein synthesis appears to be essential for petal senescence in daylilies. The observation that CHI inhibits both petal wilting and protein degradation has led to the hypothesis that control of petal senescence in daylilies is dependant upon proteolysis. Electrophoretic separation of the petal proteins from daylily flowers of differing ages supports this contention (Lay-Yee *et al.*, 1992) as CHI effectively prevents the changes in the pattern of proteins observed in senescing flowers. Further studies have identified the presence of abundant ubiquitinated proteins and it has been suggested that the selective proteolytic breakdown of proteins may, in part at least, be via ubiquitin-specific proteases (Courtney, Rider & Stead, 1993). It is not known at present, however, how the process of protein ubiquitination is controlled within the plant tissue.

The appearance of the petal cells from *Hemerocallis* by both light and electron microscopy suggests that, as in *Ipomoea*, cellular senescence may commence even before the flower is fully open. In the young buds which are still six days from opening the petal mesophyll cells are spherical and close packed with little intercellular space between them (Fig. 1A). As the flower enlarges, the mesophyll cells enlarge and the intercellular spaces increase dramatically, even two days before flower opening, the mesophyll cell shape is distorted and it is only in the region of the vascular bundles that close packed cells can be seen (Fig. 1B). The cells of the upper epidermis expand and become irregularly shaped (Fig. 1B) whilst those of the lower epidermis remain more regular in shape even one day before opening (Fig. 1C). There is little change in the appearance of the tissue as the flower

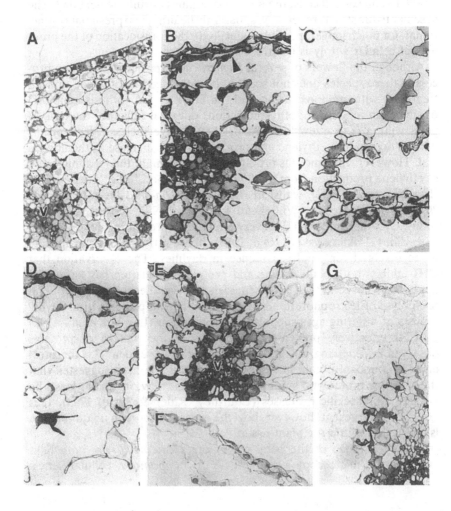

opens except that the intercellular spaces continue to enlarge (Fig. 1D). In the region of the vascular tissue where, possibly because of the greater continuity between cells, the epidermal cells appear to be pulled back from the surrounding tissue (Fig. 1E), leading to a regular ridged appearance of the petal surface. One day after opening the cellular integrity appears to be completely lost (Fig. 1F) and it is difficult to find any intact cells with the possible exception of the cells adjacent to the vascular tissue (Fig. 1G).

Electron microscopy studies of the petals six days before flower opening shows that the mesophyll cells are already very much larger (Fig. 2c) than the epidermal cells (Fig. 2a) and appear highly vacuolate with only a thin

Fig. 1. Light microscopy of *Hemerocallis* petals. Material for light and electron microscopy was fixed in 3% glutaraldehyde in 0.05 M phosphate buffer (pH 7.4); post-fixed in 1% aqueous osmium tetroxide; dehydrated through ethanol and embedded in Taab resin. For light microscopy, thick (c. 0.5 μm) sections were stained with toluidine blue, all micrographs X1000 magnification. A. Petal tissue from a flower approximately six days before flower opening. The epidermal cells are smaller than the mesophyll cells, except adjacent to the vascular tissue (v) but all the cells are more or less isodiametric in shape. B. Two days prior to opening the intracellular spaces between the mesophyll cells have expanded enormously. The mesophyll cells are no longer isodiametric and have few points of contact with adjacent cells, only the cells surrounding the vascular tissue remain small and closely packed. There is one, almost continual, layer of mesophyll cells below the epidermal cells. The upper epidermal cells have also expanded but only in width and breadth, not depth, and appear to present a rather irregular surface topography. C. In petals the day before opening the lower epidermal still remain regularly shaped and have one, more or less, complete layer of mesophyll cells underlying them. D. When fully open the mesophyll cells are large and irregularly shaped with few points of contact between cells. Unlike the mesophyll cells, the irregularly shaped upper epidermal cells still appear to have abundant cytoplasmic contents. E. In the region of the vascular tissue there remains good contact between the mesophyll cells and the epidermal cells, this appears to prevent the expansion of the tissue in this region thus producing ridges which run the length of the petal and correspond to the position of the vascular tissues. F. The epidermal cells from flowers one day after opening can still be distinguished, however, unlike the epidermal cells of younger flowers the contrast is very low due to the lack of cellular contents. G. One day after flower opening it is impossible to identify any internal detail within the mesophyll cells, many of which appear to have collapsed. In the region of the vascular tissue (v), however, the cells remain intact.

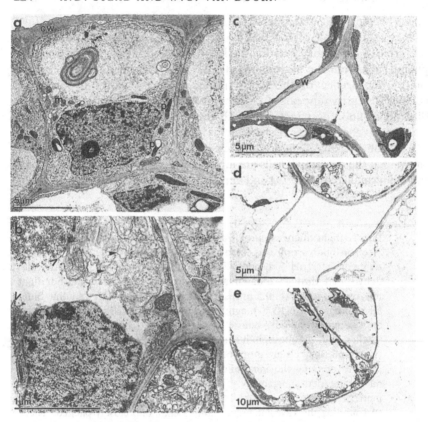

Fig. 2. Electron microscopy of *Hemerocallis* petals. a. The epidermal cells of flowers six days before opening contain abundant cytoplasm and a full range of organelles, including nucleus (n), plastids (p) and mitochondrion (m). b. The epidermal cells of fully open flowers also show signs of autolysis with discontinuities in the tonoplast (open arrow) and vesiculated cytoplasmic contents (closed arrow). c. Approximately six days before flower opening the mesophyll cells have only a thin peripheral layer of cytoplasm adjacent to the cell wall (cw). Although small, the plastids do contain starch (arrows). d. When fully open, the cytoplasmic contents of the mesophyll cells already show advanced signs of degeneration. e. One day after opening, the mesophyll cell contents have all but disappeared with only a few remnants of membranes remaining visible.

peripheral layer of cytoplasm. However, the normal complement of organ-elles is present and, although the plastids are very small, they contain accu-mulations of starch (Fig. 2c). By the time that the flower opens, this starch is almost totally absent and, in the mesophyll cells in particular, there is only a very thin peripheral layer of cytoplasm adjacent to the cell wall, due no doubt to the huge increase in cell volume, and this appears very degener-ated (Fig. 2d). Similarly, in the epidermal cells the tonoplast of many cells appears ruptured (Fig. 2b). One day later, when the petals have collapsed but not dried out, there are very few contents discernible in either the meso-phyll or epidermal cells (Fig. 2e), and autolysis of the cells is apparent. The senescence of the petal cells is therefore structurally very similar to that in *Ipomoea* (Matile & Winkenbach, 1971).

b) Flowers showing petal abscission

The flower buds of *Linum lewisii* open early in the morning and the petals abscise in mid-afternoon of the same day. Addicott (1977) con-cluded that, because of the speed and uniformity, petal abscission in this species must occur autonomously, without the influence of either pollination or fertilisation. His work further showed that the rate at which abscission occurred was accelerated after the application of either abscisic acid or indole acetic acid to the petals, but the role of ethylene was not investigated. Ultrastructural studies on the petal abscission zone in *Linum lewisii* have been reported by Wiatr (1978). The changes seen were typical of many leaf abscission cells in that there was an apparent increase in the abundance of the endoplasmic reticulum and the cell walls showed signs of degradation.

Flowers persisting for several days

In the case of the longer-lived flowers, that is to say those usually per-sisting for several days or, in the case of certain orchids, several weeks, longevity is often modified by pollination and/or fertilisation. A reduction in flower longevity, brought about by pollination, may have several bene-fits for a plant. Firstly, if sufficient pollen has been transferred to the stigma to bring about maximum seed set any further deposition of pollen is wasteful. Furthermore, the presence of excessive pollen tubes growing within the stylar tissues may result in competition for stylar food reserves. Thirdly, the maintenance of elaborate floral structures is a costly process, both in terms of respiratory substrate and the loss of water by transpir-ation. Lastly, the presence of large elaborate structures may make the plants more conspicuous and more liable to attack by herbivores (Dole, 1990). The senescence of floral tissues when they have completed their function must therefore be to the plants' advantage.

In most cases pollination accelerates the pattern of floral senescence seen in unpollinated flowers hence pollination may hasten petal wilting, abscission or changes in pigmentation. However, in some species (e.g. *Cyclamen*, *Potentilla*) there are distinct differences between pollination-induced and natural flower senescence (Gärtner, 1844; Zieslin & Gottsman, 1983; Halevy, Whitehead & Kofranek, 1984).

a) Flowers showing petal wilting

As with *Ipomoea* the wilting of carnation (*Dianthus caryophyllus*) petals commences with the inrolling of the petal margins. This process is accelerated by ethylene and pollination (Nichols, 1977). Pollination has been shown to cause a 10-fold increase in ethylene production from the stigma within 3 h (Nichols *et al.*, 1983) but increased production from the other floral organs only occurs much later. Since germinating pollen grains produce little or no ethylene when cultured *in vitro*, the increased ethylene production occurring as a result of pollination, must be the result of the interaction between the pollen tubes and the stigmatic tissues, since fertilisation could not occur so soon after pollination.

In *Petunia hybrida*, where pollination also accelerates the collapse of the corolla, early hypotheses suggested that the growth of the pollen tubes through the stigmatic tissues caused a wound-like response since corolla wilting also occurred if the stigma was damaged (Gilissen, 1977). Wounding and pollination also elicit a rapid increase in ethylene production.

Prior to the climacteric burst of ethylene production the epidermal and mesophyll petal cells of carnation were devoid of any vacuolar inclusions and, when viewed by cryo-scanning electron microscopy, the epidermal cells were closely packed and rounded in outline. Later, when ethylene production was maximal, the epidermal cells were completely collapsed (Smith, Saks & van Staden, 1992). During the senescence of the petals, extensive vacuolar and cytoplasmic vesiculation occurred and many of the cytoplasmic organelles became disrupted. Such ultrastructural changes are very similar to those seen in both short-lived, ethylene-sensitive flowers such as *Ipomoea* (Matile & Winkenbach, 1971) and short-lived ethylene-insensitive species such as *Hemerocallis* (see above).

b) Flowers showing petal abscission

Petal abscission occurs from the flowers of many species. In members of the Scrophulariaceae, pollination brings about rapid abscission of the corolla without any loss of either corolla fresh weight (FW) or dry weight (DW). Following pollination in *Digitalis*, for example, the corolla abscises within 24 h and at this time there is no detectable change in FW, protein content or membrane permeability between the corollas of pollinated and unpollinated flowers. In *Mimulus*, another member of the Scrophulariaceae, the corolla is abscised whilst still turgid and, unlike in *Digitalis*, corolla abscission may bring about self-pollination (Dole, 1990; Dudash & Ritland, 1991).

The corollas of unpollinated foxglove flowers remain attached for about six days following stigma opening and there is a gradual decline in corolla constituents and, especially in the last 24 h before corolla abscission, a small rise in membrane permeability and respiration (Stead & Moore, 1983). However, this increase in the leakage of solutes is very small as compared to those petals which wilt (van Doorn & Stead, this volume). Ethylene production rapidly increases after pollination and reduces the force required to detach the corolla, such that abscission is complete within 24 h of pollination, whereas the corollas of unpollinated flowers remain attached for several days after the stigmatic lobes separate and become receptive (Stead & Moore, 1983). Interestingly, the total amount of ethylene produced by pollinated and unpollinated foxglove flowers is approximately equal but in the case of unpollinated flowers it is produced over a period of about six days as opposed to one day in pollinated flowers (Stead, 1985). Observations such as this, and work with ethylene inhibitors, has led to the suggestion that ethylene co-ordinates the abscission but may not always induce it (Burdon & Sexton, 1990, 1993). Despite the question as to the precise role of ethylene, in all cases where petal abscission occurs, there is a dramatic increase in ethylene production concomitant with, or just prior to, abscission. In pollinated *Cyclamen* for example the increase is in excess of 60-fold (Halevy *et al.*, 1984) and in *Digitalis* the increase from pollinated flowers is more than 10-fold higher than that from similar-aged unpollinated flowers (Stead & Moore, 1983).

In at least one species, *Cyclamen*, the corolla may either wilt or abscise depending upon pollination. Ethylene production only increases after pollination and then corolla abscission occurs; in unpollinated flowers the corolla wilts without any climacteric-like increase in ethylene production (Halevy *et al.*, 1984). In *Potentilla argenta* and *P. nepalensis* the reverse is true: in pollinated flowers the petals wilt and later abscise but in unpollinated flowers they abscise whilst still fully turgid (Gärtner, 1844). It would be very interesting to measure ethylene production from these *Potentilla* species, as it would appear that ethylene is always involved in the induction of petal abscission whereas ethylene has been implicated in the control of petal wilting only in some, not all, species. A further distinction between flowers in which the petals wilt, as opposed to those in which the petals abscise, is that an increase in ethylene production from the wilting petals is observed, whereas, in *Digitalis* at least, the source of the increased ethylene production is not the petals since isolated petals produce very small amounts of ethylene.

The structure of the petal abscission zone has been studied in several species and seems to be essentially similar to that of leaf abscission zones. The putative abscission zone of young *Pelargonium* petals can be identified under the light microscope as the cells are considerably smaller than the adjacent cells (Evensen, Page & Stead, 1993). Examination by scanning electron microscopy of either the petal, or receptacle, surface after the petals are physically removed prior to abscission shows that the cells are ruptured but when examined immediately after abscission the cells, particularly on the side of the receptacle, are intact as the line of separation passes along the middle lamella. This can also be seen in *Rhododendron* (Fig. 3a–c).

During leaf abscission, it has been suggested that the weakening of the cell wall allows the turgor pressure to cause the cells to swell and therefore generate the force required to bring about abscission (Sexton, 1976). A similar mechanism probably operates in petals as after abscission the cells are clearly swollen (Evensen *et al.*, 1993). This is most apparent in the epidermal cells and it may be that, with the exception of the vascular tissue, the epidermal cells are the last to break away from one another.

The degradation of the cell wall appears to be similar in both leaf and petal abscission and increased amounts of endoplasmic reticulum have been reported in the abscission zone cells of the petals of *Linum ultissifolium* (Wiatr, 1978), *Geranium robertianum* (Sexton, Struthers & Lewis, 1983) and *Pelargonium* x hortorum (Evensen *et al.*, 1993). These changes

---

Fig. 3. Scanning electron microscopy of the abscission zone of *Rhododendron* petals. Material was fixed in 3% glutaraldehyde in 0.05M phosphate buffer (pH 7.4); dehdrated through ethanol and critical point dried. Specimens were sputter coated (Au:Pd) prior to examination. a. Immediately after abscission the petal base appears convex with no sign of cell rupture. b. At the same time the abscission zone on the receptacle is concave, but once again there is no indication of cell rupture as abscission has proceeded along the line of the middle lamellae. c. Enlargement of one cell from the abscission zone on the receptacle side immediately after abscission, the cell appears turgid and there is no sign that the cell integrity has been destroyed.

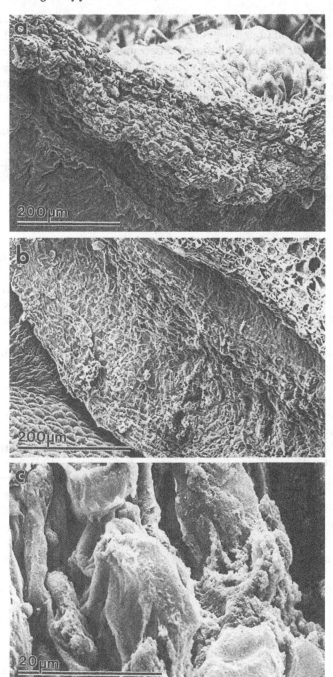

are thought to reflect increased secretory activity of the cells, in particular for the production of cell wall degrading enzymes. Similarly, in all of the species studied, the cell wall material becomes more diffuse and in some instances part of the protoplast appears to lack any cell wall material as it all has been degraded. Unlike during leaf abscission, however, increase in the activity of the dictyosome is less apparent in petals. This may be related to the differences in the speed of abscission which has been found between leaves and petals. In response to ethylene, complete weakening of the abscission zones of leaves and fruits takes 12–24 h, but petal abscission is usually faster. In *Digitalis* abscission weakening occurs within 8 h of ethylene treatment (Stead & Moore, 1983); in *Geranium robertianum* within 2.25 h (Sexton *et al.*, 1983) and in *Pelargonium* x hortorum within 1 h (Deneke, Evensen & Craig, 1990).

c) Flowers showing petal colour changes

Subtle colour changes of parts of the flower are very effective signals to potential pollination vectors. The phenomenon of perianth colour changes is very widespread; Weiss (1991) reports colour changes from at least 74 diverse Angiosperm families, and this has been recognised for over 200 years (Sprengel, 1793). To date, however, there have been no reported instances of colour changes in ephemeral flowers.

The significance of many of these colour changes is to direct pollination vectors to flowers which still have a reward and, although the colour changes are part of the normal development of the flowers, in some species at least the change is hastened by pollination. In at least one orchid, *Phalaenopsis lueddemanniana*, part of the perianth turns green and photosynthetic after pollination (Fitting, 1921; Curtis, 1943). In *Cymbidium* species the labellum turns pink or red after pollination or treatment with ethylene (Arditti, Horgan & Chadwick, 1973). In other orchid species, emasculation and pollination both result in increased ethylene production. Burg and Dijkman (1967) reported two peaks of ethylene production after pollination of *Vanda ptamboeran*; the first of these two peaks, 6 h after pollination, is now known to be associated with the labellum colour change and the second, some 40 h later, with the wilting of the petals and sepals (Woltering, Harren & Boerrigter, 1988).

In many *Lupinus* species the banner spot, an area at the base of the standard, changes from a pale colour to a darker one, owing to the accumulation of anthocyanins (Wainwright, 1978; Weiss, 1991). Again, this colour change is hastened by both ethylene and pollination (Stead & Reid, 1992). Therefore, as with those flowers in which the petals abscise, the role of ethylene appears critical in the induction of perianth colour changes. However, it is not always clear if the ethylene production rate from the tissues which change colour increases or whether these tissues respond to ethylene produced by other parts of the flower. For example, in *Lupinus albifrons* flowers, there was no significant increase in ethylene production from either the banner spot nor the adjacent tissues of the standard, at the time of the colour change. Instead, there was a 10-fold increase in ethylene production from the keel and pistil. Moreover, because of the very small size of the pistil, on a fresh weight basis the rate of production from the pistil was by far the highest (approximately 100 nl gF W^{-1} h^{-1}) (Stead & Reid, 1992). This situation is identical to other species in which the pistil, and in particular the style, has the highest rate of ethylene production per unit fresh weight of all the floral tissues (Hall & Forysth, 1967; Nichols, 1977; Suttle & Kende, 1978).

Pollination-induced signals

It is clear that, in many of these species, pollination hastens the normal pattern of petal senescence and in several studies it has been shown that the initiation of these changes occurs before fertilisation. In fact this distinction between pollination- and fertilisation-induced changes was recognised at least 150 years ago (Gärtner, 1844). Since the responses in the petal occur before fertilisation in many species, a signal must be translocated through the stylar tissue, ahead of the growing pollen tubes, to initiate these changes. Despite considerable research, however, the nature of this signal has not been conclusively identified in any species.

In carnations, it has been shown that following pollination a wave of increased potential for production of ethylene passes through the stylar tissue (Nichols *et al.*, 1983). Removal of various tissues following pollination also indicates that pollination-induced signals are generated within floral tissues and that these are transmitted through the style (Stead & Moore, 1983). In some cases the speed of the response is related to the amount of pollen applied and it seems likely, therefore, that the strength of any such signals is also probably related to the amount of pollen applied. Certainly the amount of ethylene produced by pollinated flowers of both *Digitalis* and *Nicotiana* is related to the amount of pollen applied (Stead, 1985; Hill, Stead & Nichols, 1987).

To date, there have been several suggestions as to the possible nature of the transmitted signal. However, each of the hypotheses has been subsequently shown to be flawed. The movement of pollen-held auxin was shown to be unlikely as radiolabelled indole acetic acid remained on the column when applied to orchid flowers (Burg & Dijkman, 1967). As an alternative they suggested that gaseous ethylene might move through the tissue, but it is difficult to envisage an effective means of translocating a gas through specific tissues.

With the discovery of 1-aminocyclopropane-1-carboxylic acid (ACC) as the immediate precursor of ethylene and its reported occurrence in the pollen of many species (Whitehead *et al.*, 1983; Stead, 1985) it was suggested that pollen-held ACC could be the transmissible signal in some species. However, the amount of ACC likely to be deposited on the stigma at pollination is small compared to the ethylene produced. In *Nicotiana*, for example, the amount of ACC derived from pollen would sustain floral ethylene production for no more than 5 h (Hill *et al.* 1987; Woodson, this volume). Further synthesis of ACC must therefore occur within either the stylar tissues or the growing pollen tubes. That *de novo* synthesis of ACC occurs in the style of *Petunia* after pollination has been established by Hoekstra and Weges (1986). Furthermore, Pech *et al.* (1987) have shown that pollen-held ACC is not eluted from pollen placed in a solution of equivalent osmotic potential to that of the stigmatic secretions.

More recently it has been reported that, in *Petunia* and carnation, the sensitivity to ethylene may be increased after pollination (Whitehead & Halevy, 1989) and this is discussed in the chapter by Whitehead in this volume. Lastly, in *Petunia*, Lovell, Lovell and Nichols (1987) suggested that the intact stigma may produce a compound which prevents senescence, and that it is the removal of the source of this compound, by either wounding or pollination, that brings about accelerated petal wilting.

Conclusions

The termination of the functional life-span of the flower can be brought about in several ways (petal wilting, abscission or corolla colour changes), each deters further visits by pollination vectors and reduces the wastage of pollen. In many, but not all, cases the strategy is correlated with taxonomic families. The selective advantages of the different strategies are as yet unknown. Both short- and long-lived flowers are found that show wilting and abscission type strategies but species in which the corolla changes colour are all relatively long lived (>1 day). It is not clear as to what the selective pressures are which influence flower longevity although pollinator type and density must be major factors.

The physiological control of petal wilting, petal abscission and petal colour changes is still unclear, but ethylene is certainly involved in the control of petal abscission in all species so far studied. Similarly, in all those species in which a corolla colour change occurs ethylene is involved, however, to date, relatively few species have been studied. In species showing petal wilting, ethylene is involved in some species but not in others. Structural studies show that cell autolysis occurs in wilting petals, the process often commencing before, or at least very soon after, flower opening. When petal abscission occurs, the structural changes in the abscission zone are similar to those reported for leaf abscission and include increases in the amount of endoplasmic reticulum and extensive cell wall degradation.

Many of these responses are the result of pollination, not fertilisation, and therefore some form of signal must pass through the style ahead of the growing pollen tubes but the identity of this signal remains to be identified.

Acknowledgements

Aspects of this work were supported by the AFRC (PMB programme) and the Nuffield Foundation. We are particularly grateful for the assistance given by Anton Page, Maura Fleming, Clare Steele, and the staff of the Electron Microscope Unit, Royal Holloway, for the ultrastructural investigations.

References

Addicott, F.T. (1977). Flower behaviour in *Linum lewisii*: some ecological and physiological factors in opening and abscission of petals. *American Midland Naturalist*, **97**, 321–32.

Arditti, J., Hogan, N.M. & Chadwick, A.V. (1973). Post-pollination phenomena in orchid flowers. IV. Effects of ethylene. *American Journal of Botany*, **60**, 883–8.

Burdon, J.N. & Sexton, R. (1990). The role of ethylene in the shedding of red raspberry fruit. *Annals of Botany*, **66**, 111–20.

Burdon, J.N. & Sexton, R. (1993). Ethylene co-ordinates petal abscission in red raspberry (*Rubus idaeus* L.) flowers. *Annals of Botany*, **72**, 289–94.

Burg, S.P. & Dijkman, M.J. (1967). Ethylene and auxin participation in pollen induced fading of *Vanda* orchid blossoms. *Plant Physiology*, **61**, 812–15.

Courtney, S.E., Rider, C.C. & Stead, A.D. (1993). Ubiquitination of proteins during floral development and senescence. In *Post-translational Modifications in Plants*, ed. N.H. Battey, H.G. Dickinson & A.M. Hetherington, pp. 285–303. Cambridge: Cambridge University Press.

Criley, R.A. & Broschat, T.M. (1992). *Heliconia*: botany and horticulture of a new floral crop. *Horticultural Reviews*, **14**, 1–55.

Curtis, J.T. (1943). An unusual pollen reaction in *Phalaenopsis*. *American Orchid Society Bulletin*, **11**, 258–60.

Darwin, C. (1891). *The Effects of Cross and Self Fertilisation in the Vegetable Kingdom*. London: Murray.

Deneke, C.F., Evensen, K.B. & Craig, R. (1990). Regulation of petal abscission in *Pelargonium* x hortorum. *HortScience*, **25**, 937–40.

Dobkin, D.S. (1984). Flowering patterns of long-lived *Heliconia* inflorescences: implications for visiting and resident nectarivores. *Oecologia*, **64**, 245–54.

Dobkin, D.S. (1987). Synchronous flower abscission in plants pollinated by hermit hummingbirds and the evolution of one-day flowers. *Biotropica*, **19**, 90–3.

Dole, J.A. (1990). Role of corolla abscission in delayed self-pollination of *Mimulus guttatus* (Scrophulariaceae). *American Journal of Botany*, **77**, 1505–7.

Downs, C.G. & Lovell, P.H. (1986). Carnations: relationship between timing of ethylene production and senescence of cut blooms. *New Zealand Journal of Experimental Agriculture*, **14**, 331–8.

Dudash, M.R. & Ritland, K. (1991). Multiple paternity and self-fertilization in relation to floral age in *Mimulus guttatus* (Scrophulariaceae). *American Journal of Botany*, **78**, 1746–53.

Evensen, K.B., Page, A.M. & Stead, A.D. (1993). Anatomy of ethylene-induced petal abscission in *Pelargonium* x hortorum. *Annals of Botany*, **71**, 559–66.

Fitting, H. (1921). Das Verblühen der Blüten. *Die Naturwissenschaften*, **9**, 1–9.

Gärtner, C.F. (1844). *Beiträge zur Kenntniss der Befruchtung der volkommeneren Gewächse. I. Theil. Versuche und Beobachtungen über die Befruchtungsorgane der vollkommeneren Gewächse und über die natürliche und künstliche Befruchtung durch den eigenen Pollen.* 644 pp. Stuttgart: E. Schweizebart Verlag.

Gilissen, L.J.W. (1977). Style-controlled wilting of the flower. *Planta*, **133**, 275–80.

Halevy, A.H., Whitehead, C.S. & Kofranek, A.M. (1984). Does pollination induce corolla abscission of cyclamen flowers by promoting ethylene production? *Plant Physiology*, **75**, 1090–3.

Hall, I.F. & Forsyth, F.R. (1967). Production of ethylene by flowers following pollination and treatments with water and auxin. *Canadian Journal of Botany*, **45**, 1163–6.

Hill, S.E., Stead, A.D. & Nichols, R. (1987). The production of 1-aminocyclopropane-1-carboxylic acid (ACC) by pollen of *Nicotiana tabacum* cv White Burley and the role of pollen-held ACC in pollination-induced ethylene production. *Journal of Plant Growth Regulation*, **6**, 1–13.

Hoekstra, F.A. & Weges, R. (1986). Lack of control by early pistillate ethylene of the accelerated wilting of *Petunia hybrida* flowers. *Plant Physiology*, **80**, 403–8.

Kende, H. & Hanson, A.D. (1976). Relationship between ethylene evolution and senescence in Morning Glory flower tissue. *Plant Physiology*, **57**, 523–7.

Kerner von Marilaun, A. (1891). *Pflanzenleben. Band 2.* Leipzig: *Verlag des Biblographisches Instituts*, 451 pp.

Kränzlin, F. (1910). Orchidaceae, Monandrae, Dendrobiinae. Pars 1. In: Das Pflanzenreich. ed. A. Engler. Heft 45 (IV.50. II.B.21), pp. 382, Leipzig: Wilhelm Engelmann Verlag.

Lay-Yee, M., Stead, A.D. & Reid, M.S. (1992). The effects of cycloheximide on protein changes in the petals of *Hemerocallis fluva* during development and senescence. *Physiologia Plantarum*, **86**, 308–14.

Lovell, P.H., Lovell, P.J. & Nichols, R. (1987). The importance of the stigma in flower senescence in petunia (*Petunia hybrida*). *Annals of Botany*, **60**, 41–7.

Lukaszewski, T.A. & Reid, M.S. (1989). Bulb-type flower senescence. *Acta Horticulturae*, **261**, 59–62.

Manning, K. (1985) The ethylene forming system in carnation flowers. In *Ethylene and Plant Development*, ed. J. Roberts & J. Tucker. pp 83–92. London: Butterworths.

Matile, Ph. & Winkenbach, F. (1971) Function of lysosomes and lysosomal enzymes in the senescing corolla of the Morning Glory (*Ipomoea purpurea*). *Journal of Experimental Botany*, **22**, 759–71.

Molisch, H. (1928). *Die Lebensdauer der Pflanzen.* Translated by E.H. Fulling (1939) as *The Longevity of Plants*, 226 pp. Lancaster, PA: Science Press.

Nichols, R. (1977). Sites of ethylene production in the pollinated and unpollinated senescing carnation (*Dianthus caryophyllus*) inflorescence. *Planta*, **135**, 155–9.

Nichols, R., Bufler, G., Mor, Y., Fujino, D.W. & Reid, M.S. (1983). Changes in ethylene and 1-aminocyclopropane-1-carboxylic acid content of pollinated carnation flowers. *Journal of Plant Growth Regulation*, **2**, 1–8.

Pech, J.-C., Latché, A., Larriigaudière, C. & Reid, M.S. (1987). Control of early ethylene synthesis in pollinated petunia flowers. *Plant Physiology and Biochemistry*, **25**, 431–7.

Primack, R.B. (1985). Longevity of individual flowers. *Annual Review of Ecology and Systematics*, **16**, 15–37.

Sexton, R. (1976). Some ultrastructural observations on the nature of foliar abscission in *Impatiens sultani*. *Planta*, **128**, 49–58.

Sexton, R., Struthers, W.A. & Lewis, L.N. (1983). Some observations on the very rapid abscission of the petals of *Geranium robertianum* L. *Protoplasma*, **116**, 179–86.

Smith, M.T., Saks, Y. & van Staden, J. (1992). Ultrastructural changes in the petals of senescing flowers of *Dianthus caryophyllus* L. *Annals of Botany*, **69**, 277–85.

Sprengel, C.K. (1793). *Das entdeckte Geheimniss der Natur im Bau und in der Berfruchtung der Blumen*, 447 pp. Berlin: Vieweg.

Stead, A.D. (1985). The relationship between pollination, ethylene production and flower senescence. In *Ethylene and Plant Development*, ed. J. Roberts & J. Tucker, pp. 71–81. London: Butterworths.

Stead, A.D. & Moore, K.G. (1979). Studies on flower longevity in *Digitalis*: I. Pollination induced corolla abscission in *Digitalis* flowers. *Planta*, **146**, 409–14

Stead, A.D. & Moore, K.G. (1983). Studies on flower longevity in *Digitalis*: II. The role of ethylene in corolla abscission. *Planta*, **157**, 15–21.

Stead, A.D. & Reid, M.S. (1992). The effect of pollination and ethylene on the colour change of the banner spot of *Lupinus albifrons* (Bentham) flowers. *Annals of Botany*, **66**, 655–63.

Stiles, F.G. (1975). Ecology, flowering phenology, and hummingbird pollination of some Costa Rican *Heliconia* species. *Ecology*, **56**, 285–301.

Stiles, F.G. (1978). Temporal organisation of flowering among the hummingbird foodplants of a tropical wet forest. *Biotropica*, **10**, 194–210.

Suttle, J.C. & Kende, H. (1978). Ethylene and senescence in petals of *Tradescantia*. *Plant Physiology*, **62**, 267–71.

Van Doorn, W.G. & Woltering, E.J. (1991). Developments in the use of growth regulators for the maintenance of post-harvest quality in cut flowers and potted plants. *Acta Horticulturae*, **298**, 195–208.

Veen, H. (1979). Effects of silver on ethylene synthesis and action in cut carnations. *Planta*, **145**, 467–70.

Veen, H. (1983). Silver thiosulphate: an experimental tool in plant science. *Scientia Horticulturae*, **20**, 211–224.

Wacker, H. (1911). Physiologische und morphologische Untersuchungen über das Verblühen. *Jahrbücher für wissenschaftliche Botanik* **49**, 522–78.

Wainwright, C.M. (1978). The floral biology and pollination ecology of two desert lupines. *Bulletin of the Torrey Botanical Club*, **105**, 24–38.

Weiss, M.A. (1991). Floral colour changes as cues for pollinators. *Nature*, **354**, 227–9.

Whitehead, C.S., Fujino, D.W. & Reid, M.S. (1983). Identification of the ethylene precursor, ACC, in pollen. *Scientia Horticulturae*, **21**, 291–7.

Whitehead, C.S. & Halevy, A.H. (1989). Ethylene sensitivity: the role of short-chain fatty acids in pollination-induced senescence of *Petunia hybrida* flowers. *Journal of Plant Growth Regulation*, **8**, 41–54.

Wiatr, S.M. (1978). *Physiological and structural investigation of petal abscission in* Linum lewisii. PhD thesis, Davis, University of California.

Woltering, E.J. & van Doorn, W.G. (1988). Role of ethylene in senescence of petals – morphological and taxonomic relationships. *Journal of Experimental Botany*, **39**, 1605–16.

Woltering, E.J., Harren, F. & Boerrigter, A.M. (1988). Use of a laser-driven photoacoustic detection system for the measurement of ethylene production in *Cymbidium* flowers. *Plant Physiology*, **88**, 506–10.

Zieslin, N. & Gottsman, V. (1983). Involvement of ethylene in the abscission of flowers and petals of *Leptospermum scoparium*. *Physiologia Plantarum*, **58**, 114–8.

W.G. VAN DOORN and A.D. STEAD

The physiology of petal senescence which is not initiated by ethylene

Es ist also kein Zweifel, dass schon mit dem Beginn des Aufblühens das Schicksal der Blütenblätter besiegelt ist und die Prozesse eingeschaltet werden, deren Ablauf nach kürzerer oder längerer Zeit das Leben der Blütenblätter begrenzt.

(There is, therefore, no doubt that the fate of the petals has already been sealed by the beginning of flower opening, and that the processes which limit the life of the petals, be it a shorter or longer period, have been initiated.)

W. Schumacher, 1953. Planta 42: 42–55

Introduction

Petals have an important function: the attraction of pollinators. Conspicuous petals are, therefore, only found in plants that are pollinated by animals. Attraction of pollinators is terminated by wilting of the petals, by abscission of turgid petals, or by petal colour changes. A change in colour, if it occurs, normally precedes wilting or abscission.

Schumacher (1953), quoted above, expressed what has become a widely accepted hypothesis: the functional life of the petals is genetically determined at an early phase of development, resulting in the expression of genes that are involved in cellular breakdown of the petals in the species that show wilting, in the changes in the abscission zone cells in the species that show abscission, and in the colour changes if these occur. The nature of the physiological processes leading to cellular degradation, and those leading to abscission is only partially known.

In flowers in which the calyx and the corolla are different in shape and colour, the corolla parts are called petals. In plants in which the calyx and the corolla are similar in shape (e.g. Amaryllidaceae, Liliaceae, Iridaceae, Orchidaceae and Euphorbiaceae), the perianth parts are called

Society for Experimental Biology Seminar Series 55: *Molecular and Cellular Aspects of Plant Reproduction*, ed. R.J. Scott & A.D. Stead.
© Cambridge University Press, 1994, pp. 239–254.

tepals. For the sake of simplicity, however, we will use the word petals throughout this chapter.

Results from studies in which exogenous ethylene is applied or the action of endogenous ethylene is blocked by the use of silver thiosulphate (STS), indicate that, in all plants studied, petal abscission is initiated by endogenous ethylene. The same type of experiments show that petal wilting is sensitive to exogenous ethylene in several species. In numerous other species, however, neither exogenous ethylene nor STS have an effect on petal wilting. Petal senescence in these species is, therefore, not initiated by endogenous ethylene (Woltering & van Doorn, 1988).

The wilting or the abscission of the petals, as the determinant of flower life, is often correlated to the family level, although there are exceptions to this rule (Stead and van Doorn, this volume). The plants in which wilting is sensitive to ethylene and those in which it is not are also generally found in different families (Woltering & van Doorn, 1988). Table 1 shows some plant families in which flower wilting is apparently not initiated by ethylene, as well as families in which it clearly is.

The physiology of senescence has predominantly been studied in species from the latter category, such as carnation (*Dianthus caryophyllus*) and *Gypsophila,* both from the Caryophyllaceae family, several orchid species (Orchidaceae), *Hibiscus* (Malvaceaea), as well as species from families not mentioned in Table 1, such as *Tradescantia*

Table 1. *Some plant families in which wilting of the petals is generally sensitive to exogenous ethylene, and families in which it is insensitive to ethylene*

	Sensitive to ethylene	Insensitive to ethylene
Monocotyledons	Orchidaceae	Amaryllidaceae Iridaceae Liliaceae
Dicotyledons	Campanulaceae Caryophyllaceae Cruciferae Dipsacaceae Malvaceae	Asteraceae Umbelliferae

In some Liliaceae and Orchidaceae species the petals become green and do not wilt nor abscise (Fitting, 1921; Pfeiffer, 1928).

(Commelinaceae), *Petunia* (Solanaceae) and *Ipomoea* (Convulvulaceae). Table 1, however, shows that flower wilting which is apparently not initiated by endogenous ethylene also occurs in several families, some of which contain large numbers of species, e.g. the Asteraceae (= Compositae). Petal wilting which is not initiated by ethylene is, therefore, widespread, yet rarely studied.

Flower senescence that is apparently not initiated by endogenous ethylene has been studied in detail in only a few species, such as *Gerbera* (Asteraceae), *Freesia* (Iridaceae) and *Hemerocallis* (Liliaceae). Petal wilting in *Gerbera* flowers was insensitive to exogenous ethylene (Woltering & van Doorn, 1988) and treatments with STS had no effect (Veken, unpublished observations). The senescence of *Freesia* florets was not affected by exogenous ethylene either (Spikman, 1986; Woltering & van Doorn, 1988), and both aminoethoxyvinylglycine, a potent inhibitor of ethylene synthesis, and STS did not delay wilting (Spikman, 1989). Cut *Hemerocallis fulva* flowers were also not affected by exogenous ethylene, nor by the inclusion of 1-aminocyclopropane-1-carboxylic acid (ACC), the precursor of ethylene, in the solution. Petal senescence was not delayed by aminooxyacetic acid, another inhibitor of ethylene synthesis, nor by STS (Lukaszewski & Reid, 1989). Observations have also been made on flowers from genera such as chrysanthemum, (*Dendranthema grandiflora*), *Dahlia* (Asteraceae), *Narcissus, Nerine* (Amaryllidaceae), *Iris* and *Lilium* (Liliaceae). Petal wilting in all species tested from these genera was insensitive to exogenous ethylene (Woltering & van Doorn, 1988).

The characteristics of the ethylene-insensitive senescence will be reviewed and compared to the wilting of the ethylene-sensitive species.

Time to petal wilting

The time between anthesis and visible wilting of the flowers may be long (2–3 weeks), as in chrysanthemum, short (about one week) such as in *Iris*, or very short such as in *Hemerocallis fulva*, where the time between flower opening and collapse is less than 24 h, hence its name daylily. The same is true for the species in which petal wilting is regulated by ethylene: time to wilting ranges from long as in several orchids, to short as in carnations (not treated with STS), and to very short as in *Ipomoea* and *Tradescantia*. Regulation by ethylene or not, therefore, apparently has little relationship with flower longevity.

The first stage of flower wilting is due to loss of turgor, followed by total collapse. These changes often start at the petal margins. Wilting may be preceded by a change in colour, such as in *Petunia* (Lovell,

Lovell & Nichols, 1987), or inrolling, for example in carnation (Nichols, 1966), although no reports refer to such colour changes occurring before, or concomitant with, the petal wilting in the species in which wilting is not initiated by ethylene. The apoplast may also become infiltrated with liquid, both in ethylene-sensitive flowers such as *Tradescantia* (Horie, 1961) and in ethylene-insensitive ones such as *Iris* (Bancher, 1938). Fitting (1921) noted that drops of liquid may even be hanging from the flowers of *Iris* and *Hemerocallis* species.

It would therefore appear that the time to petal wilting does not differ between species in which wilting is initiated by ethylene and in those in which it is not. Morphological changes are also similar, except for a colour change prior to wilting, which has apparently not been observed in the species in which petal wilting is not initiated by ethylene.

Time to wilting in cut and non-cut flowers

The early wilting of cut flowers, as compared to flowers left on the plant, may relate to the inability of the flower to take up water from the vase solution, which is often due to the presence of high numbers of bacteria and their extracellular products in the xylem vessels (van Doorn, Schurer & de Witte, 1989; van Doorn et al., 1991). This is especially critical in flowering stems such as *Gerbera* (van Meeteren, 1978) and some cultivars of chrysanthemum. Inclusion of antimicrobial compounds in the vase solution may prevent this premature petal wilting, albeit only when these compounds are not toxic to the flowers (Aarts, 1957a; van Doorn, de Witte & Perik, 1990). When taking adequate precautions, the time to wilting of cut flowers placed in aqueous solution can be compared with non-cut flowers held under similar temperatures. In *Hemerocallis fulva* the time to wilting was similar in cut and non-cut flowers (Reid, pers. comm. 1992), and similar observations were made in *Iris* x hollandica (van Doorn & Celikel, unpublished observations). In *Gerbera*, the petals of cut flowers wilted earlier than those of non-cut flowers (van Meeteren & van Gelder, 1980)

In ethylene-sensitive flowers, the time to wilting of the ephemeral species such as *Tradescantia* is apparently also similar in attached and in detached flowers (Suttle & Kende, 1978). In *Petunia*, on the other hand, petal wilting in attached flowers occurred after about 140 h and in cut flowers within 70 h (Hoekstra, pers. comm. 1993). In carnation, petal wilting of the cut flowers also occurs much earlier (by a factor of 2) than the wilting of uncut flowers, except in cultivars that are insensitive to ethylene or cultivars that are unable to produce an ethylene climacteric (Wu et al., 1989; Wu, van Doorn & Reid, 1991).

The time to wilting in cut, as compared to non-cut flowers, therefore is similar in species that are ephemeral, both in the species in which wilting is initiated by ethylene and in those in which it is not. In the longer-lived species the time to wilting is shorter in the cut flowers, at least in the ones in which senescence is initiated by ethylene.

Changes in fresh weight and dry weight

The visible wilting of *Hemerocallis* petals is accompanied by a sharp decrease in fresh weight (Lay-Yee, Stead & Reid, 1992), which was probably not due to a microbial occlusion of the xylem, as it happened within hours of anthesis. In *Gerbera* spikes in which vascular occlusion was delayed by including antimicrobial compounds in the vase water, a decrease in the petal fresh weight (FW) was noted after about 10–20 days of placing the scapes in water, depending on the cultivar. The decrease in FW started several days prior to visible wilting (van Meeteren, 1979).

Carnation flowers are harvested after petal elongation, when the outer petals are already horizontal, and the FW only increased for about 2 days after harvest, then decreased (Paulin, 1977). Similar changes in FW were found in other ethylene-sensitive flowers, placed in a solution with antimicrobial compounds (Aarts, 1957a,b).

In *Hemerocallis fulva*, flower dry weight (DW) increased during growth and opening, and decreased sharply (by a factor of 3 in 24 h) prior to wilting (Lay-Yee *et al.*, 1992). In *Gerbera* flowers, harvested at full bloom, and placed a solution preventing vascular occlusion, petal DW steadily decreased, and when the petals wilted it was about 20% of the DW at harvest (van Meeteren, 1979). Combes (1935a) also found a sharp decrease in DW, starting after full flower opening, in *Lilium croceum*.

In *Ipomoea*, petal DW decreased prior to visible wilting (Winkenbach, 1970a), similar to the decrease observed in *Hemerocallis fulva*, but in carnation flowers the petal DW remained constant after harvest, even when they showed inrolling and wilting (Paulin, 1977).

Petal FW, therefore, was found to decrease prior to the visible symptoms of wilting, both in ethylene-sensitive and in ethylene-insensitive species. Petal DW also often sharply decreases prior to wilting, in both types of species, carnation being a notable exception.

Respiration

Petal respiration has been studied by harvesting petals from the intact plant, at various stages of development, or taking them from cut flowers in the vase. Daylily petals harvested at the bud stage and held in water for 42–48 h showed a transient increase in the rate of respiration (expressed per gram fresh weight) prior to wilting (Bieleski & Reid, 1992). At its maximum the respiration rate was about 25% higher than the rate prior to the rise. Isolated flowers of *Dahlia variabilis*, placed in a solution that prevented microbial growth, wilted after five days and showed a decrease in respiration rate (expressed per flower) from harvest to day one, then a transient increase which peaked on day 4 (Aarts, 1957a). In chrysanthemum cut at various stages of development, the rate of respiration decreased to low levels and apparently showed a small transient rise prior to wilting (Nakamura, 1975), but flowers of *Iris germanica* (Carlier & Leblond, 1962) and *Narcissus pseudonarcissus* (Ballantyne, 1965), cut at various stages of development and immediately placed in a measuring chamber, did not show an increase in respiration rate.

The respiration rate of carnation flowers decreased after cutting, for about five days, then showed a transient increase prior to wilting. At its maximum, the rate was about 60% higher than prior to the increase (Maxie *et al.*, 1973). A similar peak was observed in *Dianthus plumarius*, in which the maximum was 50% higher than the mimimum before the rise (Aarts, 1957a). In *Matthiola incana* (Cruciferae), which was moderately sensitive to exogenous ethylene and in which STS delayed petal wilting (Woltering & van Doorn, 1988), a peak in respiration was also found prior to wilting (Aarts, 1957a).

Although the data are not always comparable, as they derive both from material held continuously in water for several days or from freshly cut material of different ages, the picture emerges that a transient increase in the rate of respiration always occurs in flowers in which petal senescence is controlled by ethylene and in some flowers in which it is not. When the transient occurs in the latter group, however, it is apparently smaller.

Ethylene production

Just before petal wilting a brief rise in ethylene production occurred in daylily flowers (Lay-Yee *et al.*, 1992). The base level ethylene production was $0.2 – 0.4$ nl flower^{-1} h^{-1} and the maximum rate 2.5 nl flower^{-1} h^{-1}, i.e. a factor of about 8. In chrysanthemum and *Narcissus* a surge of ethylene production prior to wilting was not observed (Nichols, 1966), but may have escaped the sensitivity of the technique then available. Cut freesia inflorescences did show a considerable peak in ethylene production (Spikman, 1986), but this was mainly due to the buds near the apex of the inflorescence which do not develop further when the stems are not fed with sugars. Ethylene production rates of the florets that were already fully open at cutting, however, also showed a small rise (Spikman, 1987).

The surge in ethylene production is much higher in ethylene-sensitive species. In carnation flowers it shows a maximum about 50–200 times the basal production rate (Nichols, 1966; Maxie *et al.*, 1973). Other ethylene-sensitive species in which such an ethylene-climacteric was observed include cotton (*Gossypium*; Morgan, Durham & Lipe, 1973), *Hibiscus* (Woodson, Hanchey & Chisholm, 1985), *Tradescantia* (Suttle & Kende, 1978) and *Ipomoea* (Kende & Baumgartner, 1974).

The results suggest that a rise in ethylene production accompanies the wilting of some species in which petal senescence is not initiated by ethylene as well as in those it is initiated by ethylene. The maximum is lower in the former group of species. It is still unclear what the function of the ethylene rise is in the species in which petal senescence is not initiated by ethylene.

Leakage of ions and sugars

In daylily petals, the rate of ion leakage, determined by measuring the electrical conductivity of deionised water in which petals of different developmental stages were placed, increased about four times just before the symptoms of wilting. The rate of sugar efflux was measured in petals from opening buds and from mature open flowers. In the latter, the efflux rate was much higher than in the former (Bieleski & Reid, 1992). In cut *Gerbera* spikes in which the development of a vascular occlusion was prevented by including antimicrobial compounds in the vase water, an increase in ion leakage occurred after 4–16 days, depending on the cultivar. In three cultivars tested the decrease in the water content of the petals started exactly at the time of the increase in ion leakage (van Meeteren, 1979).

Leakage of ions and sugars prior to petal wilting has also been observed in flowers in which the petal wilting is initiated by ethylene, such as *Tradescantia* (Suttle & Kende, 1980), *Ipomoea* (Hanson & Kende, 1975), and carnation (Eze *et al.*, 1986; Mayak, Vaadia & Dilley, 1977).

It is interesting to note that the increase in the leakage of ions does not occur, or is very small, prior to abscission of petals, e.g. in rose (Hoekstra, pers. comm. 1993) and *Digitalis* (Stead & Moore, 1983).

Nitrogen metabolism

The protein concentration of petals in intact daylily flowers, expressed per gram FW, also showed a rapid decrease as the buds grew and the flowers opened. During this period the FW increased by a factor of 4 and protein concentration decreased by a factor of 3, hence the protein content of the corolla was mainly diluted by growth. Therefore, only relatively little *de novo* synthesis occurred during this period. During the time between open bloom and wilting considerable net protein degradation occurred (Lay-Yee *et al.*, 1992). Combes (1935*b*) also noted that the decrease in protein content (expressed per flower) already started to decrease before full opening of the flowers of *Lilium croceum*, in the period when flower DW still increased. The protein content declined with the same rate in the period prior to, and during, wilting. The soluble nitrogen-compounds started to decrease after full flower opening (Combes, 1935*a*).

Schumacher (1932) also found a decrease in the protein and total N content in the petals of many flowers, including those in which the wilting is initiated by ethylene, such as *Ipomoea*. Winkenbach (1970*a,b*), in a detailed study on *Ipomoea,* found a rapid decrease in the level of RNA, amino acids and proteins, prior to, and concomitant with, a decrease in corolla FW. The level of DNA also decreased rapidly; starting even prior to the drop in FW. The activities of DNA and RNA degrading enzymes showed about a forty-fold increase, and this increase commenced before the flowers had fully opened, i.e. long before wilting. Protease activity did not clearly change during senescence. The work of Richner (1980) and Suchovsky (1980), however, show that the cellular compartmentation of the enzymes may render the interpretation of these enzyme activities ambiguous.

Phosphate metabolism

In the petals of daylily the incorporation of [^{32}P] into individual compounds was determined for 3 h in petals from pre-senescent fully opened flowers and petals sampled at the time coinciding with the respiration peak. No differences were found in incorporation into AMP, ADP or ATP, in the analogous UMP, UDP, and UTP, nor in a number of phospholipids such as phosphatidyl ethanolamine and phosphatidyl inositol. The incorporation of [^{32}P] into six different sugar phosphates, e.g glucose-6-phosphate and fructose-6-phosphate, however, was clearly lower at the second sampling time. The total incorporation into lipid precursors (phosphocholine, phosphoethanolamine and 1-phosphoglycerol) and one phospholipid (phosphatidylcholine) was also lower at incipient wilting.

As most phosphate esters in plants reach equilibrium labelling within 30 min, the amounts of each compound was considered to be approximately proportional to its radioactivity. Compared with other systems the total lipid precursor level was high already at the first measurement and even higher at the second, indicating that the synthesis of phospholipids was disrupted before visible wilting. This also followed from the phospholipid levels which were already low in the fully open flower (Bieleski & Reid, 1992).

Similar conclusions were reached when studying the petal wilting of carnation (Engelmann-Sylvestre *et al.*, 1989; Palyath & Thompson, 1990), *Tradescantia* (Suttle & Kende, 1980), and *Ipomoea* (Beutelmann & Kende, 1977), where reduced phospholipids levels were found because of enhanced degradation and reduced synthesis. The reduction in phospholipid levels was considered to be the main reason for decreased membrane fluidity, and could be a reason for the increase in leakiness (Thompson *et al.*, 1982).

Carbohydrate metabolism

In experiments in which microbial growth in the vase solution was inhibited by silver nitrate, Aarts (1957*b*) noted that inclusion of saccharose in the water delayed wilting in flowers of chrysanthemum, *Dahlia variabilis* and *Iris germanica*. Effects ranged from 20% (wilting after 15 days in controls and after 18 days in the treatment, in chrysanthemum) to 80% (in *Dahlia*). However, sugars did not delay wilting in the fully open or half open florets of the freesia cyme (Aarts, 1962). In *Nerine bowdenii*, which was insensitive to exogenous ethylene (Woltering & van Doorn, 1988), sugars seemingly also had little effect on wilting (Downs & Reihana, 1987). The endogenous sugar levels in the petals of the above species have apparently not been reported in the literature, but in *Lilium croceum* the levels of reducing, as well as non-reducing, sugars showed a rapid decrease starting directly after flower opening (Combes, 1936).

A marked delay of wilting by sugars has been observed in species in which wilting is initiated by ethylene. In cut carnation flowers, for example, which wilt after 7–9 days, sugars may increase the time to wilting to about 15 days (Nichols, 1973; Mayak & Dilley, 1976). Sugars reduced the sensitivity to exogenous ethylene and delayed the climacteric rise in ethylene production (Mayak & Dilley, 1976). Maintenance of adequate osmotic pressure by the sugars may, in part, explain the delay in petal wilting. Maintenance of the osmotic pressure by potassium salts also delayed wilting, but not as much as with added sugars. The reason for the specific effect of the sugars is as yet unknown (Halevy & Mayak,

1979; Mayak & Borochov, 1984). The sucrose concentration in the carnation petals was low at harvest and did not decrease until wilting; the level of reducing sugars was about three times higher and showed a rapid decline until only 20% of the initial level was left at incipient wilting (Nichols, 1973; Paulin, 1977).

Effects of metabolic inhibitors

The life of isolated petals of a number of flowers could be extended by placing them in an atmosphere of HCN. This was true for flowers in which wilting is apparently not initiated by ethylene such as *Iris pseudacoris*, and for flowers in which it is, such as *Ipomoea* and *Tradescantia* (Schumacher, 1953). The senescence of cut daylily flowers was delayed by about 6 days when cycloheximide was included in the vase water, whereas actinomycin D had no effect (Lukaszewski & Reid, 1989). As cycloheximide is considered to act mainly by inhibition of protein synthesis, the data imply that at least one protein, synthesised *de novo*, is involved in the processes leading to wilting.

Cycloheximide also delayed petal senescence in carnation flowers (Dilley & Carpenter, 1979). In isolated carnation petals (Wulster, Sacalis & Hanes, 1982) and *Ipomoea* flowers (Baumgartner, Hurter & Matile, 1975) cycloheximide also reduced the the sensitivity to exogenous ethylene, actinomycin D was ineffective in this respect (Wulster *et al.*, 1982).

Effects of hormones other than ethylene

Although wilting of *Gerbera* petals occurred earlier in cut flowers than in uncut ones, the levels of endogenous cytokinin activity (using the *Amaranthus* bioassay) were similar in cut and non-cut flowers, throughout their life. The time to wilting in petals of cut 'Citronella' *Gerbera* was similar to that of 'Wageningen Rood' flowers, but was much longer in 'Mini Wit' flowers. Levels of cytokinins at harvest were not correlated with the time to senescence in these cultivars (van Meeteren & van Gelder, 1980). However, immersing the cut 'Wageningen Rood' *Gerbera* flowers in an aqueous solution of the cytokinin 6-benzyladenin (BA) for 2 min, and subsequent placement of the scape in a solution which prevented vascular occlusion, delayed the time to petal wilting (van Meeteren, 1979).

The inclusion of BA in the vase solution had no effect on the wilting of daylily flowers (Lukaszewski & Reid, 1989) and a pulse treatment with this compound had no effect in *Narcissus* either (Ballantyne, 1965). Inclusion of isopentenyladenosine, another cytokinin, in the vase water did retard petal senescence of *Iris* flowers (Wang & Baker, 1979). Auxin (2,4-D) had no effect on *Narcissus,* but a combined treatment of BA and this auxin delayed wilting (Ballantyne, 1965). Whether the endogenous levels of these hormones play a role during the 'non-ethylene' wilting remains, however, unclear.

In carnation the time to petal wilting can be delayed by inclusion of cytokinins or gibberellins in the solution (Bossé & Van Staden, 1989; Eisinger, 1977; Garrod & Harris, 1978). Cytokinins delay the increase in ACC levels and block the conversion of ACC to ethylene (Mor, Spiegelstein & Halevy, 1983) apparently by delaying the *de novo* synthesis of ACC synthase and ACC oxidase. Although the levels of cytokinins in the petals are lower in cut as compared to non-cut flowers, and show a drop prior to wilting (Van Staden *et al.,* 1987), in carnation the role of endogenous levels of hormones other than ethylene has also not fully been established.

Conclusions

A major difference between species in which petal wilting is sensitive to ethylene and those in which it is not, is the smaller rise, or absence of the rise, in respiration rate and ethylene production in the latter group. Also, no changes in colour prior to wilting have been observed in the flowers in which wilting is not initiated by ethylene, while this is common in flowers in which it is ethylene regulated. Other characteristics of senescence, such as a decrease in nitrogen content, the leakage of ions and sugars, and the effects of inhibitors of transcription and translation seem essentially similar in both groups of plants. In both types of flowers, petal wilting is apparently dependent on the synthesis of proteins and occurs concomitant with increased leakiness for ions and sugars. The processes leading to this increase in leakiness have not been established. Among the possible causes are a reduction in the synthesis of phospholipids and/ or an increase in phospholipid breakdown.

Although the available data are as yet incomplete, they indicate that the underlying mechanisms leading to flower wilting are essentially similar in ethylene-sensitive and in ethylene-insensitive species.

References

Aarts, J.F.T. (1957*a*). Over de houdbaarheid van snijbloemen. *Mededelingen Landbouwhogeschool Wageningen,* **57** (9), 1–62.

Aarts, J.F.T. (1957*b*). De ontwikkeling en houdbaarheid van afgesneden bloemen. *Mededelingen van de Directeur van de Tuinbouw,* **20**, 690–701.

Aarts, J.F.T. (1962). The keepability of cut flowers. *XVI International Horticultural Congress, Brussels*, Vol. **5**, 46–53.

Ballantyne, D.J. (1965). Senescence of daffodil (*Narcissus pseudonarcissus* L.) cut flowers treated with benzyladenine and auxin. *Nature*, **205**, 809.

Bancher, E. (1938). Zellphysiologische Untersuchung über den Abblühvorgang bei *Iris* und *Gladiolus*. *Österreichische botanische Zeitschrift*, **87**, 221–38.

Baumgartner, B., Hurter, J. & Matile, P. (1975). On the fading of an ephemeral flower. *Biochemie und Physiologie der Pflanzen*, **168**, 299–306.

Beutelmann, P. & Kende, H. (1977). Membrane lipids in senescing flower tissue of *Ipomoea tricolor*. *Plant Physiology*, **59**, 888–93.

Bieleski R.L. & Reid, M.S. (1992). Physiological changes accompanying senescence in the ephemeral daylily flower. *Plant Physiology*, **98**, 1042–9.

Bossé, C.A. & van Staden, J. (1989). Cytokinins in cut carnation flowers. V. Effects of cytokinin type, concentration and mode of application on flower longevity. *Journal of Plant Physiology*, **135**, 155–9.

Carlier, G. & Leblond, C. (1962). Comportement respiratoire d'organes végétaux (feuilles, fleurs, fruits) à brève et longue échéange après leur séparation. *XVI International Horticultural Congress, Brussels*, Vol. **5**, 305–15.

Combes, R. (1935*a*). Étude biochimique de la fleur. La nutrition minérale de la corolle. *Comptes Rendus de l'Académie des Sciences, Paris. Série D*, **200**, 578–80.

Combes, R. (1935*b*). La nutrition azotée de la fleur. *Comptes Rendus de l'Académie des Sciences, Paris. Série D*, **200**, 1970–2.

Combes, R. (1936). La nutrition glucidique de la corolle. *Comptes Rendus de l'Académie des Sciences, Paris. Série D*, **203**, 1282–4.

Dilley, D.R. & Carpenter, W.J. (1979). The role of chemical adjuvants and ethylene synthesis in cut flower longevity. *Acta Horticulturae*, **41**, 117–32.

Downs, C. & Reihana, M. (1987). Extending vaselife and improving quality of nerine cut flowers with preservatives. *HortScience*, **22**, 670–1.

Eisinger, W. (1977). Role of cytokinins in cut carnation flower senescence. *Plant Physiology*, **59**, 707–9.

Engelmann-Sylvestre, I., Bureau, J.-M., Tremolieres A. & Paulin, A. (1989). Changes in membrane phospholipids and galactolipids during the senescence of cut carnations: connection with ethylene rise. *Plant Physiology and Biochemistry*, **27**, 931–7.

Eze, J.M.O., Mayak, S., Thompson, J.E. & Dumbroff, E.B. (1986). Senescence in cut carnation flowers: temporal and physiological relationships among water status, ethylene, abscisic acid and membrane permeability. *Physiologia Plantarum*, **68**, 323–8.

Fitting, H (1921). Das Verblühen der Blüten. *Die Naturwissenschaften*, **9**, 1–9.

Garrod, J.F. & Harris, G.P. (1978). Effect of gibberellic acid on senescence of isolated petals of carnation. *Annals of Applied Biology*, **88**, 309–11.

Goh, C.J., Halevy, A.H., Engel, R. & Kofranek, A.M. (1985). Ethylene evolution and sensitivity in cut orchid flowers. *Scientia Horticulturae*, **26**, 57–67.

Halevy, A.H. & Mayak, S. (1979). Senescence and postharvest physiology of cut flowers. Part I. *Horticultural Reviews*, **1**, 204–36.

Hanson, A.D. & Kende, H. (1975). Ethylene-enhanced ion and sucrose efflux in morning glory tissue. *Plant Physiology*, **55**, 663–9.

Horie, K. (1961). The behavior of the petals in the fading of the flowers of *Tradescantia reflexa*. *Protoplasma*, **53**, 377–86.

Kende, H. & Baumgartner, B. (1974). Regulation of ageing in flowers of *Ipomoea tricolor* by ethylene. *Planta*, **116**, 279–89.

Lay-Yee, M., Stead, A.D. & Reid, M.S. (1992). Flower senescence in daylily (*Hemerocallis*). *Physiologia Plantarum*, **86**, 308–14.

Lovell, P.H., Lovell, P.J. & Nichols, R. (1987). The importance of the stigma in flower senescence in petunia (*Petunia hybrida*). *Annals of Botany*, **60**, 41–7.

Lukaszewski, T. A. & Reid, M.S. (1989). Bulb-type flower senescence. *Acta Horticulturae*, **261**, 59–62.

Maxie, E.C., Farnham, D.S., Mitchell, F.G., Sommer, N.F., Parsons, R.A., Snyder, R.G., & Rae, H.L. (1973). Temperature and ethylene effects on cut flowers of carnation (*Dianthus caryophyllus* L.) *Journal of the American Society for Horticultural Science*, **98**, 568–72.

Mayak, S. & Borochov, A. (1984). Nonosmotic inhibition by sugars of the ethylene-forming activity associated with microsomal membranes from carnation petals. *Plant Physiology*, **76**, 191–5.

Mayak, S. & Dilley, D.R. (1976). Effects of sucrose in the response of cut carnations to kinetin, ethylene and abscisic acid. *Journal of the American Society for Horticultural Science*, **101**, 583–5.

Mayak, S., Vaadia, Y. & Dilley, D.R. (1977). Regulation of senescence in carnation (*Dianthus caryophyllus*) by ethylene. *Plant Physiology*, **59**, 591–3.

Mor, Y., Spiegelstein, H. & Halevy, A.H. (1983). Inhibition of ethylene biosynthesis in carnation petals by cytokinin. *Plant Physiology*, **71**, 541–6.

Morgan, P.W., Durham, J.I. & Lipe, J.A. (1973). Ethylene production by the cotton flower, role and regulation. In *Plant Growth Substances*, ed. Y. Sumiki, pp. 1062–8. Tokyo: Hirokawa Publ. Company.

Nakamura, R. (1975). Changes of the respiration rate in cut flowers. *Science Reports of the Faculty of Agriculture, Okayama University*, **46**, 29–37.

Nichols, R. (1966). Ethylene production during senescence of flowers. *Journal of Horticultural Science*, **41**, 279–90.

Nichols, R. (1973). Senescence of the cut carnation flower: respiration and sugar status. *Journal of Horticultural Science*, **48**, 111–21.

Palyath, G. & Thompson, J.E. (1990). Evidence for early changes in membrane structure during post-harvest development of cut carnation (*Dianthus caryophyllus* L.) flowers. *Physiologia Plantarum*, **114**, 555–62.

Paulin, A. (1975). Action exercée sur le métabolisme azoté de la fleur coupée d'*Iris germanica* L. par l'apport exogène de glucose ou de cycloheximide. *Physiologie Végétale*, **13**, 501–15.

Paulin, A. (1977). Métabolisme glucidique et protéique de la fleur d'oeillet alimentée ou non avec une solution de saccharose. *Acta Horticulturae*, **71**, 241–56.

Pfeiffer, H. (1928). *Die pflanzlichen Trennungsgewebe. Handbuch der Pflanzenanatomie, I Abteilung, 2 Teil: Histologie. Band V.*, ed. K. Linsbauer, pp. 1–236. Berlin: Gebrüder Borntraeger.

Richner, R.E. (1980). Aspekte der Alterung der Korolle von Ipomoea tricolor. PhD Thesis (number 6630), Eidgenössische Technische Hochschule, Zürich, Switzerland.

Schumacher, W. (1932). Über Eiweissumsetzungen in Blütenblatter. *Jahrbücher für wissenschaftliche Botanik*, **75**, 581–607.

Schumacher, W. (1953). Weitere Beobachtungen über das Welken ephemerer Blüten. *Planta*, **42**, 42–55.

Spikman, G. (1986). The effect of water stress on ethylene production and ethylene sensitivity of freesia inflorescences. *Acta Horticulturae*, **181**, 135–40.

Spikman, G. (1987). Ethylene production, ACC and MACC content of freesia buds and florets. *Scientia Horticulturae*, **33**, 291–7.

Spikman, G. (1989). Development and ethylene production of buds and florets of cut freesia inflorescences as influenced by silver thiosulphate, aminoethoxyvinylglycine, and sucrose. *Scientia Horticulturae*, **39**, 73–81.

Stead, A.D. & Moore, K.G. (1983). Studies of flower longevity in *Digitalis*: the role of ethylene in corolla abscission. *Planta*, **157**, 15–21.

Suchovsky, P. (1980). Zur Alterungsphysiologie der Zierwindeblüte. PhD Thesis (number 6568), Eidgenössische Technische Hochschule, Zürich, Switzerland.

Suttle, J.C. & Kende, H. (1978). Ethylene and senescence in petals of *Tradescantia*. *Plant Physiology*, **62**, 267–71.

Suttle, J.C. & Kende, H. (1980). Ethylene action and loss of membrane integrity during petal senescence in *Tradescantia*. *Plant Physiology*, **65**, 1067–72.

Thompson, J.E., Mayak, S., Shinitzky, M. & Halevy, A.H. (1982). Accerelation of membrane senescence in cut carnation flowers by treatment with ethylene. *Plant Physiology*, **69**, 859–63.

van Doorn W.G., de Stigter, H.C.M., de Witte, Y. & Boekestein, A. (1991). Micro-organisms at the cut surface and in xylem vessels of rose stems: a scanning electron microscope study. *Journal of Applied Bacteriology*, **70**, 34–9.

van Doorn, W.G., de Witte Y. & Perik, R.R.J. (1990). Effects of antimicrobial compounds on the number of bacteria in stems of cut rose flowers. *Journal of Applied Bacteriology*, **68**, 117–22.

van Doorn, W.G. & Reid, M.S. (1992). Role of ethylene in flower senescence of *Gypsophila paniculata* L. *Postharvest Biology and Technology*, **1**, 265–72.

van Doorn W.G., Schurer, K. & de Witte, Y. (1989). Role of endogenous bacteria in vascular blockage of cut rose flowers. *Journal of Plant Physiology*, **134**, 375–81.

van Meeteren, U. (1978). Water relations and keeping quality of cut gerbera flowers. I. The cause of stem break. *Scientia Horticulturae*, **8**, 65–74.

van Meeteren, U. (1979). Water relations and keeping quality of cut gerbera flowers. III. Water content, permeability and dry weight of the petals. *Scientia Horticulturae*, **10**, 261–9.

van Meeteren, U. & van Gelder, H. (1980). Water relations and keeping quality of cut gerbera flowers. V. Role of endogenous cytokinins. *Scientia Horticulturae*, **12**, 273–81.

van Staden, J., Featonbysmith, B.C., Mayak, S., Spiegelstein, H. & Halevy, A.H. (1987). Cytokinins in cut carnation flowers. 2. Relationship between endogenous ethylene and cytokinin levels in the petals. *Plant Growth Regulation* **5**, 75–86.

Wang, C.Y. & Baker, J.E. (1979). Vase life of cut flowers treated with rhizobitoxine analogs, sodium benzoate, and isopentenyl adenosine. *HortScience*, **14**, 59–60.

Whitehead, C.S., Halevy, A.H. & Reid, M.S. (1984). Roles of ethylene and 1-aminocyclopropane-1-carboxylic acid in pollination and wound-induced senescence of *Petunia hybrida* flowers. *Physiologia Plantarum*, **61**, 643–8.

Winkenbach, F. (1970a). Zum Stoffwechsel der aufblühenden und welkenden Korolle der Prunkwinde *Ipomoea purpurea*. I. Beziehungen zwischen Gestaltwandel, Stofftransport, Atmung und Invertaseaktivität. *Berichte der schweizerischen botanischen Gesellschaft*, **80**, 374–90.

Winkenbach, F. (1970b). Zum Stoffwechsel der aufblühenden und welkenden Korolle der Prunkwinde *Ipomoea purpurea*. II. Funktion und de novo Synthese lysosomaler Enzyme beim Welken. *Berichte der schweizerischen botanischen Gesellschaft*, **80**, 391–406.

Woltering, E.J. & van Doorn, W.G. (1988). Role of ethylene in senescence of petals – morphological and taxonomic relationships. *Journal of Experimental Botany*, **39**, 1605–16.

Woodson, W.R., Hanchey, S.H. & Chisholm, D.N. (1985). Role of ethylene in the senescence of isolated *Hibiscus* petals. *Plant Physiology*, **79**, 679–83.

Wu, M.J., van Doorn, W.G., Mayak, S. & Reid, M.S. (1989). Senescence of 'Sandra' carnation. *Acta Horticulturae*, **261**, 221–5.

Wu, M.J., van Doorn, W.G. & Reid, M.S. (1991). Variation in the senescence of carnation (*Dianthus caryophyllus* L.) cultivars. I. Comparison of flower life, respiration and ethylene biosynthesis. *Scientia Horticulturae*, **48**, 99–107.

Wulster, G., Sacalis, J. & Hanes, H. (1982). The effect of inhibitors of protein synthesis on ethylene-induced senescence in isolated carnation petals. *Journal of the American Society for Horticultural Science*, **107**, 112–15.

W.R. WOODSON

Molecular biology of flower senescence in carnation

Introduction

The programmed senescence of flower petals is a highly controlled developmental event and plays an important role in the overall reproductive strategy of many plants (see Stead & van Doorn, this volume). The flower is a terminally differentiated complex structure composed of many organs performing a variety of functions, the ultimate goal being successful sexual reproduction. These functions include the production of pollen, pollination, fusion of gametes, and the development and dispersal of viable seeds. In many species, the petals function in the attraction of insects for pollination. Consistent with the petals' short-lived role in reproduction, pollination often serves as a signal for the initiation of petal senescence (Stead, 1992). In the flowers of carnation, the phytohormone ethylene serves as a signal for the initiation of petal senescence following pollination (Nichols, 1977; Nichols et al., 1983). In the absence of pollination petal senescence in carnation still occurs and is mediated by the increased production of ethylene (Nichols, 1966; Wang & Woodson, 1989). This paper attempts to summarise recent data on the regulation of programmed organ death in carnation flower petals with a particular focus on the molecular events associated with the induction of senescence by ethylene. It is not intended as an exhaustive review of flower senescence, for which the reader is referred to papers by Borochov and Woodson (1989), Cook and van Staden (1988), and Reid and Wu (1992).

Ethylene and carnation flower senescence

Ethylene biosynthesis in senescing carnation flowers

The senescence of carnation flower petals is associated with a climacteric-like increase in the production of ethylene (Fig. 1). In pre-senescent

Society for Experimental Biology Seminar Series 55: *Molecular and Cellular Aspects of Plant Reproduction*, ed. R.J. Scott & A.D. Stead.
© Cambridge University Press, 1994, pp. 255–267.

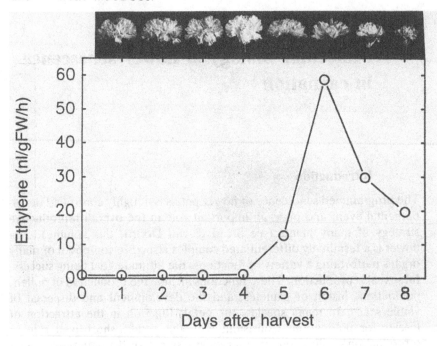

Fig. 1. Ethylene production by carnation flowers during senescence.

carnation petals, ethylene production is limited by low activities of both 1-aminocyclopropane-1-carboxylate (ACC) synthase and ACC oxidase, which convert AdoMet to ACC and ACC to ethylene, respectively (Woodson *et al.*, 1992). This is analogous to the situation in climacteric fruits such as tomato, which exhibit an increase in ethylene production during ripening (Theologis, 1992). In climacteric fruit and flower tissue, the basal level of ethylene production that precedes induction of senescence is referred to as system I ethylene, whereas the increased ethylene produced during senescence is system II ethylene (Kende, 1993; Yang & Hoffman, 1984). System II ethylene is autocatalytic in nature, i.e. ethylene stimulates its own synthesis. The induction of system II ethylene appears to be under the control of developmental signals. The transition to system II is characterised by an increase in the abundance of mRNAs encoding ACC synthase and ACC oxidase, with concomitant increases in the activity of both enzymes (Park, Drory & Woodson, 1992; Woodson *et al.*, 1992). Consistent with the autocatalytic nature of its regulation, the production of system II ethylene is completely dependent on the continued perception of ethylene. Treatment of flowers with 2,5-norbornadiene, a competitive inhibitor of ethylene action, results in a rapid inhibition of ethylene biosynthesis (Wang & Woodson, 1989). In addition, inhibition of ethylene action with 2,5-norbornadiene inhibits the activities of both ACC synthase and ACC oxidase (Wang & Woodson, 1989), and the expression of their corresponding mRNAs is prevented (Woodson *et al.*, 1992). Clearly, the expression of system II ethylene synthesis is under the regulation of ethylene. The question remains as to the developmental signals responsible for transition from system I to system II ethylene production during flower development. It is clear that system II ethylene in carnation is associated with the expression of a specific ACC synthase transcript (Woodson *et al.*, 1992). The cDNA clone representing this mRNA does not detect ACC synthase transcripts that accumulate in response to auxin, wounding or other signals that induce ethylene biosynthesis in carnation (Park *et al.*, 1992). The developmental signals responsible for the induction of system II ethylene likely involve a change in responsiveness to ethylene. Consistent with this carnation flower petals acquire the capacity to respond to ethylene by induction of system II ethylene as they mature (Lawton *et al.*, 1990). The nature of this change in responsiveness remains to be elucidated. The recent isolation of the senescence-related ACC synthase gene (Woodson *et al.*, 1993) will allow us to begin to examine the *cis*-elements and cellular factors involved in the regulation of autocatalytic ethylene during carnation flower senescence.

Role of ethylene in carnation flower senescence

The developmental expression of increased ethylene production in aging carnation petals plays a critical role in the initiation and coordination of the senescence processes. Chemicals that inhibit ethylene biosynthesis have been shown to delay or prevent the onset of petal senescence with the most effective inhibitors being those that inhibit ACC synthase (Reid & Wu, 1992). Recently, the expression of an antisense ACC oxidase transcript in transgenic carnations was found to inhibit ethylene production and delay senescence (Michael *et al.*, 1993). This result illustrates the possibility of extending the useful postharvest life of carnation flowers using a genetic engineering approach. In addition to the inhibition of ethylene biosynthesis, chemicals that inhibit the action of ethylene are effective in preventing petal senescence. These include the silver ion as a complex with sodium thiosulphate (Veen, 1979), 2,5-norbornadiene (Wang & Woodson, 1989), and diazocyclopentadiene (Sisler *et al.*, 1993). The conclusion drawn from studies using these inhibitors is that ethylene synthesis and action are strictly required for the initiation of petal senescence in carnation. In fact, it appears that ethylene not only serves to initiate the processes of programmed organ death, but also plays an active role in regulating these processes once initiated. For example, it has been shown that treatment of flowers with 2,5-norbornadiene following the onset of increased ethylene production and the appearance of visible symptoms of senescence results in a reversal of senescence symptoms and prolongs the useful life of the flowers (Wang & Woodson, 1989). This is in contrast to the view that the processes of senescence are irreversible and, once initiated, proceed in an uncontrolled manner.

Pollination-induced senescence

Pollination of carnation flowers leads to premature petal senescence (Nichols, 1977). Given the petals role of attracting insect pollinators, this trait likely evolved to prevent subsequent visits by pollinators and to conserve resources for the developing gynoecium. Pollination has been found to result in increases in ethylene production by styles, ovary, receptacles and petals (Nichols *et al.*, 1983). The increase in ethylene production by the various organs is sequential in nature (Fig. 2) with the styles producing ethylene most rapidly (within 30 min of pollination). At anthesis, ethylene production by carnation styles is limited by low activities of ACC synthase (Woodson *et al.*, 1992). In contrast to petals, styles exhibit high activities of ACC oxidase at anthesis. Given the rapid nature

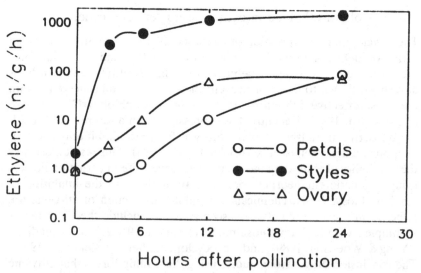

Fig. 2. Ethylene production by carnation floral organs following pollination.

of pollination-induced ethylene production by carnation styles it has been suggested that pollen-held ACC is responsible for the increase in ethylene following its diffusion to the stigma (Whitehead, Fujino & Reid, 1983). In carnation, the amount of pollen-held ACC (<10 nmol/g) is insufficient to account for the initial amounts of ethylene produced. For example, assuming a pollination with 0.1 mg of pollen and a pollen ACC content of 10 nmol/g, this would give rise to 0.022 nl of ethylene. Carnation styles produce as much as 1–2 nl of ethylene within the first 2 h of pollination. Pollination is associated with a rapid increase in the activity of ACC synthase (Woodson *et al.*, 1992), indicating a role for newly synthesised ACC in the increased ethylene. Finally, application of aminoethoxyvinyl glycine (AVG), an inhibitor of ACC synthase, completely blocks pollination induced ethylene by carnation (Woltering *et al.*, 1993). Taken together, these results clearly point to a role ACC synthase in the pollination-induced ethylene biosynthesis of carnation styles. The question remains as to how pollination leads to the induction of ACC synthase and ethylene production by the style.

The induction of premature petal senescence following pollination clearly involves interorgan communication. In this case, the sensing organ (style) must communicate with the petals that a successful pollination has occurred. This communication involves a sequential stimulation of ethylene biosynthesis, and suggests the involvement of a transmissible factor. Given the autocatalytic regulation of ethylene production in the petals, one possible explanation for the stimulation of ethylene production by the petals following pollination would be the exposure of petals to ethylene. Reid *et al.* (1984) reported that diffusion of ethylene through the atmosphere from pollinated stigmas was not responsible for the stimulation of ethylene biosynthesis by the petals. Another explanation to account for the stimulation of ethylene in the petals is that ethylene or its precursor ACC may diffuse or be transported to the petals from the stigma, where ethylene biosynthesis is stimulated by the pollen–pistil interaction. In support of this, Reid *et al.* (1984) found that radioactively labelled ACC applied to the stigmatic surface of carnation flowers resulted in the production of labelled ethylene by the petals. Ethylene, in some cases, has been shown to diffuse through tissues (Woltering, 1990) and this could account for the stimulation of autocatalytic ethylene production by petals following pollination. Alternatively, other transmissible factors that induce the production of ethylene or influence the petals' responsiveness to ethylene may be involved in interorgan communication following pollination. Recently, short-chain fatty acids have been implicated as playing a role in interorgan communication following pollination of petunia (Whitehead & Halevy, 1989). It was shown that these compounds accumulate in the eluate of pollinated *Petunia* styles and possess senescence-inducing properties when applied to corollas. However, recently a conflicting report indicated that these short-chain fatty acids did not induce senescence of *Petunia*, carnation or orchid flowers (Woltering *et al.*, 1993). The role of these or other compounds in the interorgan communication following pollination is discussed further in the chapter by Whitehead.

Expression of flower senescence-related genes

Molecular cloning of flower senescence-related mRNAs

The senescence of carnation petals in response to ethylene is associated with a series of biochemical changes including an increase in hydrolytic enzyme activity, degradation of macromolecules, increased respiratory activity, and ultimately a loss of cellular compartmentalisation leading to death (for review see Borochov & Woodson, 1989). Many of these processes are the result of active metabolism and involve the expression of senescence-related genes and new protein synthesis. Petal senescence is often prevented by treatment with chemical inhibitors of both transcription and translation (Wulster, Sacalis & Janes, 1982; Suttle & Kende, 1980). Changes in the pattern of protein synthesis has been correlated with the onset of petal senescence in carnation (Woodson, 1987). An analysis of *in vitro* translation products of mRNAs isolated from carnation petals at various stages of development revealed the onset of senescence is associated with the disappearance of several mRNAs and the appearance of new transcripts (Fig. 3). In an attempt to begin to elucidate the molecular mechanisms involved in the processes of petal senescence, we have isolated a number of cDNA clones representing senescence-related mRNAs from carnation petals (Lawton *et al.*, 1989; Wang & Woodson, 1991; Wang, Brandt & Woodson, 1993; Woodson *et al.*, 1993). Differential screening of a cDNA library prepared using poly(A)+ RNA isolated from senescing carnation petals has, to date, identified nine groups of cDNA clones that represent unique senescence-related (SR) mRNAs. These cDNA clones have been used to assess the relative abundance of their corresponding mRNAs in petals following harvest of carnation flowers. In all cases, the cDNA clones detected transcripts that increased in abundance at the onset of programmed organ death (Fig. 4). In most cases the increase in SR gene expression was not apparent until 5 days after harvest, which corresponds to the onset of increased ethylene biosynthesis. However, transcripts detected by hybridisation with pSR5 and pSR12 increased in abundance at least one day before

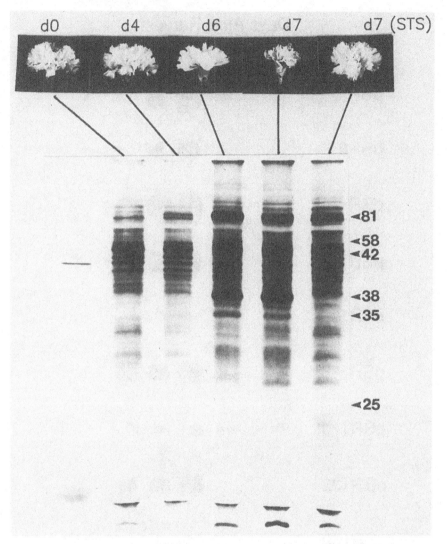

Fig. 3. Changes in mRNA populations during carnation flower petal senescence. Poly(A)$^+$ RNA isolated from flower petals at various times after harvest was translated *in vitro* using rabbit reticulocyte lysates in the presence of [^{35}S]-methionine. Major differences in translation products are indicated in kD to the right. Treatment of flowers with silver thiosulphate (STS) prevents petal senescence by inhibiting ethylene action and prevents the accumulation of senescence-related mRNAs. Lanes are d0, day of harvest; d4, 4 days after harvest; d6, 6 days after harvest; d7, 7 days after harvest; and d7(STS), 7 days after harvest and treatment with silver thiosulphate.

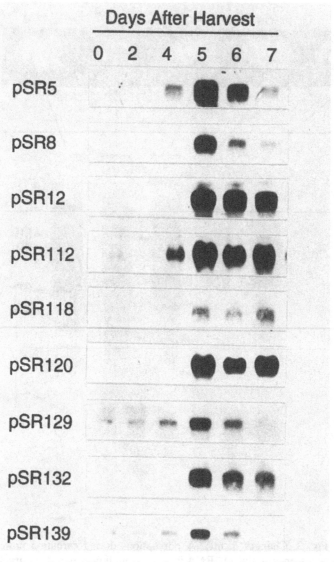

Fig. 4. Temporal expression of senescence-related mRNAs during carnation flower petal senescence. Total RNA was isolated from flower
petals at various times after harvest and subjected to RNA gel blot
analysis using SR cDNA clones as hybridisation probes.

the increase in ethylene biosynthesis. These transcripts are particularly interesting as possible candidates for playing a role in the transition from system I to system II ethylene. Examination of the spatial patterns of SR gene expression revealed maximum accumulation of these mRNAs in senescing petals (Lawton *et al.*, 1989; Woodson *et al.*, 1993). In most cases the expression of the SR transcripts is limited to floral organs during the ethylene climacteric (Lawton *et al.*, 1990). These expression patterns are consistent with a primary role for their protein products in cell death.

Function of SR proteins

The senescence of carnation flower petals is clearly a highly controlled process of regulated cell death. It is reasonable to suggest that some of these new senescence-related transcripts and the proteins they encode play functional roles in the processes of petal senescence. The nucleotide sequences of several SR cDNA clones described above have been determined and found to exhibit significant homologies with other genes encoding proteins with known functions. Recently, a search of the protein sequence databases with the predicted peptide sequences of pSR5 and pSR12 revealed homologies of greater than 50% with β-glucosidases and β-galactosidases, respectively. The expression of genes encoding these glycosidases is consistent with increased hydrolytic activity during petal senescence. Previously we reported that the protein encoded by pSR8 was a glutathione s-transferase (Meyer *et al.*, 1991). This enzyme catalyses the conjugation of the thiol group of glutathione to many different electrophilic compounds and aids in detoxification of many xenobiotics in animals. In plants, glutathione s-transferases (GSTs) have been shown to detoxify herbicides (Timmerman, 1983) and recently an auxin-binding protein was found to be a GST (Bilang *et al.*, 1993). The predicted protein of pSR120 shared greater than 70% identity with other known ACC oxidases from several plants (Wang & Woodson, 1991). The predicted protein of another SR cDNA clone, pSR132, was found to share significant homology with an enzyme from *Streptomyces*, carboxyphosphonoenolpyruvate mutase (Wang *et al.*, 1993). This enzyme is involved in the formation of phosphono (C–P) bonds in the biosynthetic pathway of bialaphos, an antibiotic synthesised in *Streptomyces* that is commonly used as a herbicide in plants. In addition the SR132 protein shared greater than 40% homology with a protein from the lower eukaryote *Tetrahymena*, phosphoenolpyruvate mutase that also catalyses the formation of a C–P bond in this organism. The formation of C–P bonds or their presence in higher plants has not been investigated.

Continued characterisation of the proteins encoded by these SR cDNAs may lead to a better understanding of the biochemistry of cell death.

Ethylene regulates the expression of senescence-related genes

Ethylene is a phytohormone with pleiotropic effects on plant growth and development. For example, ethylene is known to initiate many developmentally coordinated programs such as abscission, flower senescence, fruit ripening and seed germination (Mattoo & Suttle, 1991). In addition, ethylene is often regarded as a stress hormone in that increased ethylene production is associated with mechanical, environmental, and biotic stresses. There is growing evidence that many of these diverse responses to ethylene in plants are the result of changes in gene expression (Broglie & Broglie, 1991). The increase in abundance of the carnation flower SR transcripts concomitant with the increased production of ethylene suggests a possible role for ethylene in regulating their expression. Treatment of carnations with inhibitors of ethylene synthesis or action prevents the accumulation of many of the SR transcripts (Lawton *et al.*, 1989). In addition, treatment of pre-senescent carnation flower petals with ethylene results in increased abundance of SR transcripts (Lawton *et al.*, 1989, 1990). Nuclear run-on transcription studies revealed this increased abundance of SR mRNAs in response to ethylene was at least partially the result of increased transcription of these genes (Lawton *et al.*, 1990). In an attempt to begin to define the *cis*-elements and *trans*-acting factors that are responsible for ethylene regulation of SR genes we have focused on analysing the promoter of the *GST1* gene (Itzhaki & Woodson, 1993). This analysis has revealed that sequences responsible for ethylene-responsive expression are present in the 5′-flanking DNA of the *GST1* gene. The identification of DNA sequences responsible for the regulated expression of this gene represents an important first step in the biochemical and molecular dissection of signal transduction in response to ethylene during petal senescence.

Conclusions

The regulation of developmentally programmed senescence of carnation flower petals has received considerable attention and is of interest for both practical and scientific reasons. Extending the post-harvest longevity of cut carnation flowers is of significant value to producers, shippers, wholesalers and consumers of this attractive commodity. Knowledge of the role increased biosynthesis of ethylene plays in the regulation of carnation petal senescence has led to several chemical treatments that inhibit the synthesis and action of ethylene, which result in considerable delays in the senescence of flower petals. More recently, molecular cloning of genes encoding the enzymes in the ethylene biosynthetic pathway has led to improved flower longevity using a genetic engineering approach of expressing antisense RNAs for these genes. In addition to practical applications arising from an understanding of the regulation of carnation flower senescence, the study of senescence processes is of considerable interest to developmental biologists. Petal senescence represents an example of programmed organ death where developmental and hormonal signals lead to coordinated cell death involving new gene transcription and protein synthesis. Knowledge of the molecular mechanisms underlying the regulation of petal senescence will have implications for other developmental processes in plants involving programmed cell death such as differentiation of xylem elements.

Acknowledgements

Publication number 13,969 of the Purdue University Agricultural Experiment Station. Research in the author's laboratory is supported by the National Science Foundation and the United States Department of Agriculture National Research Initiative competitive grants program.

References

Bilang, J., Macdonald, H., King, P.J. & Sturm, A. (1993). A soluble auxin-binding protein from *Hyoscyamus muticus* is a glutathione s-transferase. *Plant Physiology*, **102**, 29–34.

Borochov, A. & Woodson, W.R. (1989). Physiology and biochemistry of flower petal senescence. *Horticultural Reviews*, **11**, 15–43.

Broglie, R. & Broglie, K. (1991). Ethylene and gene expression. In *The Plant Hormone Ethylene*, ed. A.K. Mattoo & J.C. Suttle, pp. 101–113. Boca Raton: CRC Press.

Cook, E.L. & van Staden, J. (1988). The carnation as a model for hormonal studies in flower senescence. *Plant Physiology and Biochemistry*, **26**, 793–807.

Itzhaki, H. & Woodson, W.R. (1993). Characterization of an ethylene-responsive glutathione s-transferase gene cluster in carnation. *Plant Molecular Biology*, **22**, 43–58.

Kende, H. (1993). Ethylene biosynthesis. *Annual Review of Plant Physiology and Plant Molecular Biology*, **44**, 283–307.

Lawton, K.A., Huang, B., Goldsbrough, P.B. & Woodson, W.R. (1989). Molecular cloning and characterization of senescence-related genes from carnation flower petals. *Plant Physiology*, **90**, 690–6.

Lawton, K.A., Raghothama, K.G., Goldsbrough, P.B. & Woodson, W.R. (1990). Regulation of senescence-related gene expression in carnation flower petals by ethylene. *Plant Physiology*, **93**, 1370–5.

Mattoo, A.K. & Suttle, J.C. (1991). *The Plant Hormone Ethylene*. Boca Raton: CRC Press.

Meyer, R.C., Goldsbrough, P.B. & Woodson, W.R. (1991). An ethylene-responsive flower senescence-related gene from carnation encodes a protein homologous to glutathione s-transferases. *Plant Molecular Biology*, **17**, 277–81.

Michael, M.Z., Savin, K.W., Baudinette, S.C., Graham, M.W., Chandler, S.F., Lu, C.-Y., Caesar, C., Gautrais, I., Young, R., Nugent, G.D., Stevenson, D.R., O'Conner, E.L.-J., Cobbett, C.S. & Cornish, E.C. (1993). Cloning of ethylene biosynthetic genes involved in petal senescence of carnation and petunia, and their antisense expression in transgenic plants. In *Cellular and Molecular Aspects of the Plant Hormone Ethylene*, ed. J.C. Pech, A. Latché & C. Balagué, pp. 298–303. Dordrecht: Kluwer Academic Publishers.

Nichols, R. (1966). Ethylene production during senescence of flowers. *Journal of Horticultural Science*, **41**, 279–90.

Nichols, R. (1977). Sites of ethylene production in the pollinated and unpollinated senescing carnation (*Dianthus caryophyllus*) inflorescence. *Planta*, **135**, 155–9.

Nichols, R., Bufler, G., Mor, Y., Fujino, D.W. & Reid, M.S. (1983). Changes in ethylene production and 1-aminocyclopropane-1-carboxylic acid content of pollinated carnation flowers. *Journal of Plant Growth Regulation*, **2**, 1–8.

Park, K.Y., Drory, A. & Woodson, W.R. (1992). Molecular cloning of an 1-aminocyclopropane-1-carboxylate synthase from senescing carnation flower petals. *Plant Molecular Biology*, **18**, 377–86.

Reid, M.S. & Wu, M.-J. (1992). Ethylene and flower senescence. *Plant Growth Regulation*, **11**, 37–43.

Reid, M.S., Fujino, D.W., Hoffman, N.E & Whitehead, C.S. (1984). 1-aminocyclopropane-1-carboxylic acid (ACC) – The transmitted stimulus in pollinated flowers? *Journal of Plant Growth Regulation*, **3**, 189–96.

Sisler, E.C., Blankenship, S.M., Fearn, J.C. & Haynes, R. (1993). Effect of diazocyclopentadiene (DACP) on cut carnations. In *Cellular and Molecular Aspects of the Plant Hormone Ethylene*, ed. J.C. Pech, A. Latché & C. Balagué, pp. 182–187. Dordrecht: Kluwer Academic Publishers.

Stead, A.D. (1992). Pollination-induced flower senescence: a review. *Plant Growth Regulation*, **11**, 13–20.

Suttle, J.C. & Kende, H. (1980). Ethylene action and loss of membrane integrity during petal senescence in *Tradescantia*. *Plant Physiology*, **65**, 1067–72.

Theologis, A. (1992). One rotten apple spoils the whole bushel: the role of ethylene in fruit ripening. *Cell*, **70**, 181–4.

Timmerman, K.P. (1983). Molecular characterization of corn gluta-thione s-transferase isozymes involved in herbicide detoxification. *Physiologia Plantarum*, **77**, 465–71.

Veen, H. (1979). Effects of silver on ethylene biosynthesis and action in cut carnations. *Planta*, **145**, 467–70.

Wang, H. & Woodson, W.R. (1989). Reversible inhibition of ethylene action and interruption of petal senescence in carnation flowers by norbornadiene. *Plant Physiology*, **89**, 434–8.

Wang, H. & Woodson, W.R. (1991). A flower senescence–related mRNA from carnation shares sequence similarity with fruit ripening-related mRNAs involved in ethylene biosynthesis. *Plant Physiology*, **96**, 1000–1.

Wang, H., Brandt, A.S. & Woodson, W.R. (1993). A flower senes-cence-related mRNA from carnation encodes a novel protein related to enzymes involved in phosphonate biosynthesis. *Plant Molecular Biology*, **22**, 719–24.

Whitehead, C.S. & Halevy, A.H. (1989). Ethylene sensitivity: the role of short-chain saturated fatty acids in pollination-induced senescence of *Petunia hybrida*. *Plant Growth Regulation*, **8**, 41–54.

Whitehead, C.S., Fujino, D.W. & Reid, M.S. (1983). Identification of the ethylene precursor, 1-aminocyclopropane-1-carboxylic acid (ACC), in pollen. *Scientia Horticulturae*, **21**, 291–7.

Woltering, E.J. (1990). Interorgan translocation of 1-aminocyclopro-pane-1-carboxylic acid and ethylene coordinates senescence in emas-culated *Cymbidium* flowers. *Plant Physiology*, **92**, 837–45.

Woltering, E.J., van Hout, M., Somhorst, D. & Harren, F. (1993). Roles of pollination and short-chain saturated fatty acids in flower senescence. *Plant Growth Regulation*, **12**, 1–10.

Woodson, W.R. (1987). Changes in protein and mRNA populations during the senescence of carnation petals. *Physiologia Plantarum*, **71**, 495–502.

Woodson, W.R., Park, K.Y., Drory, A., Larsen, P.B. & Wang, H. (1992). Expression of ethylene biosynthetic pathway transcripts in senescing carnation flowers. *Plant Physiology*, **99**, 526–32.

Woodson, W.R., Brandt, A.S., Itzhaki, H., Maxson, J.M., Wang, H., Park, K.Y. & Larsen, P.B. (1993). Ethylene regulation and function of flower senescence-related genes. In *Cellular and Molecular Aspects of the Plant Hormone Ethylene*, ed. J.C. Pech, A. Latché & C. Bal-agué, pp. 291–297. Dordrecht: Kluwer Academic Publishers.

Wulster, G., Sacalis, J. & Janes, H. (1982). Senescence in isolated carnation petals. Effects of indoleacetic acid and inhibitors of protein synthesis. *Plant Physiology*, **70**, 1039–43.

Yang, S.F. & Hoffman, N.E. (1984). Ethylene biosynthesis and its regu-lation in higher plants. *Annual Review of Plant Physiology*, **35**, 155–89.

C.S. WHITEHEAD

Ethylene sensitivity and flower senescence

Introduction

The hormonal theory of plant growth substances states that growth substances are limiting and, therefore, regulatory factors in plant growth and development. However, renewed interest in the role of the sensitivity of plant tissues to growth substances in growth and development was sparked of by a series of articles by Trewavas and his co-workers during the early 1980s (Trewavas, 1981, 1982; Trewavas & Jones, 1981). These articles pointed to the inadequacies of the hormonal theory, and emphasised the importance of sensitivity to growth substances as the primary limiting factor in the control of growth and development by these growth substances. The presence of both the growth substance and its receptor molecule is required for a biological response to occur. Since a biological response is caused by a receptor/growth substance complex, the response is dependent on the concentration of the growth substance as advocated by the hormonal theory. It is obvious that, if receptor availability changes during growth, and development, measurements of changes in growth substance concentration on its own has limited value.

The development of sensitivity to growth substances such as auxins, cytokinins and ethylene appears to precede the developmental process which it induces (Trewavas, 1982; Whitehead & Vasiljevic, 1993). In many flowers, ethylene plays an important role in the initiation and regulation of the processes that accompany corolla senescence. Senescence of flowers such as *Dianthus caryophyllus* (carnation) and *Petunia hybrida* is characterised by a climacteric rise in ethylene production during the final stages (Fig. 1) (Nichols *et al.*, 1983; Whitehead, Halevy & Reid, 1984) and an early increase in the sensitivity of the corolla to ethylene which precedes the onset of the ethylene climacteric (Whitehead & Halevy, 1989*a*; Whitehead *et al.*, 1984; Whitehead & Vasiljevic, 1993). This increase in sensitivity during the pre-climacteric phase is associated

Society for Experimental Biology Seminar Series 55: *Molecular and Cellular Aspects of Plant Reproduction*, ed. R.J. Scott & A.D. Stead.
© Cambridge University Press, 1994, pp. 269–284.

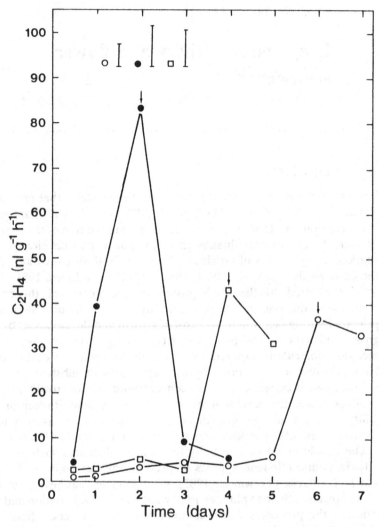

Fig. 1. Ethylene production by senescing unpollinated (○), pollinated (●) and octanoic acid treated (□) carnation flowers. The arrows show the time when petal wilting was first observed (Whitehead & Vasiljevic, 1993).

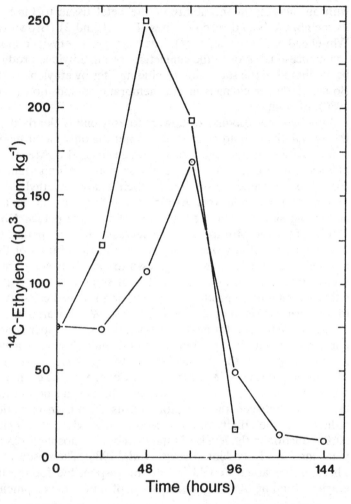

Fig. 2. [^{14}C]-Ethylene binding by senescing unpollinated (○) and octanoic acid treated (●) carnations. Non-specific binding was measured in the presence of excess unlabelled ethylene. Physiological binding was determined by calculating the difference between the means of three replicates measured in the absence and presence of unlabelled ethylene. (Whitehead & Vasiljevic, 1993).

with an increase in the ability of the petal tissue to bind ethylene to its membrane-bound receptor sites (Figs. 1 and 2) (Brown et al., 1986; Whitehead & Vasiljevic, 1993). The decrease in ethylene binding which commences just prior to the climacteric rise in ethylene production could be attributed to the saturation of binding sites by ethylene (Whitehead & Bossé, 1991), or changes in the membrane composition (Brown et al., 1986), or both.

Although the response of flowers to ethylene is clearly dependent on the availability of binding sites, this is not the only factor involved in the determination of the ability of the tissue to bind ethylene. The affinity of these binding sites for ethylene also plays an important role in the control of ethylene sensitivity. In their study on ethylene binding in senescing carnation flowers, Brown et al. (1986) have shown a single type of binding site in both pre- and post-climacteric petals. However, the affinity of the receptor for ethylene decreases in older petals. Differences in ethylene sensitivity due to differences in the affinity of the receptor molecules for ethylene are also known to exist between different carnation cultivars. The extended vase life of some cultivars is the result of either an inability to produce ethylene, or a reduced capacity to respond to the hormone (Brandt & Woodson, 1992; Wu, Zacarias & Reid, 1991).

The ethylene receptor molecule appears to be a copper-containing protein embedded in the lipid bilayers of the endoplasmic reticulum and, to a lesser extent, in the plasmalemma (Evans et al., 1982; Sisler, 1982; Thompson, Harlow & Whitney, 1983). Changes in the chemical composition or physical properties of these membranes are bound to affect the interaction between the receptor and its ligand. In carnation, a 50% reduction in the affinity of the receptor for ethylene is associated with a 50% reduction in the levels of total membrane phospholipids by the time petal inrolling is evident (Brown et al., 1986; Thompson et al., 1982). This reduced affinity could in part be responsible for the decrease in ethylene binding observed at this stage of flower development (Fig. 2).

The 'sensitivity factor'

Since tissue sensitivity to ethylene is an important factor in the control of flower senescence, an appreciation and understanding of the nature of the 'sensitivity factor' is central to our understanding of the mechanisms involved in the control of ethylene action. The involvement of ethylene in senescence varies greatly amongst the flowers of different species. Mature flowers of species such as carnation and petunia are sensitive to ethylene and exposure to ethylene will accelerate their senescence (Maxie et al., 1973; Whitehead et al., 1984; Whitehead & Vasiljevic, 1993). However, the flowers of some other species such as *Cyclamen persicum* produce very little ethylene and do not respond to the hormone (Halevy, Whitehead & Kofranek, 1984). Although these flowers differ in their response to ethylene during normal senescence, pollination induces accelerated senescence and a rapid increase in ethylene production and sensitivity (Halevy *et al.*, 1984; Nichols *et al.*, 1983; Whitehead *et al.*, 1984). In carnation and petunia, pollination results in a drastic increase in the synthesis of 1-aminocyclopropane-1-carboxylic acid (ACC, the immediate precursor to ethylene) by the stylar tissue. This ACC is transported rapidly to the corolla where it is immediately converted to ethylene which, in turn, stimulates autocatalytic ethylene production, leading to early corolla senescence (Hoekstra & Weges, 1986; Reid *et al.*, 1984; Whitehead *et al.*, 1984).

Although the production and transport of ACC is an important factor in the control of ethylene synthesis by the corolla tissue of pollinated flowers, ACC is not the only compound of interest transported to the corolla from the gynoecium following pollination. Treatment of pollinated flowers with an inhibitor of ACC synthesis such as aminoethoxyvinyl glycine (AVG) delays corolla senescence and abscission, indicating that ethylene synthesis is an important factor in the control of pollination-induced senescence (Lovell, Lovell & Nichols, 1987; Nichols & Frost, 1985). However, Gilissen and Hoekstra (1984) determined that pollination or wounding of the stigma of petunia leads to the production of a 'wilting factor' other than ACC in the styles. Even in ethylene-insensitive flowers such as cyclamen, pollination results in the production of an ethylene stimulating 'signal' and an ethylene 'sensitivity factor' in the gynoecium (Halevy *et al.*, 1984). Time-course studies have shown an increase in ethylene sensitivity to be the first detectable event occurring in the corollas of pollinated petunia flowers (7 hours after pollination), followed by an increase in ethylene synthesis (approximately 24 hours after pollination). This increase in ethylene sensitivity is independent from ethylene synthesis, since it occurs in pollinated flowers treated with AVG where little or no ethylene is produced (Whitehead & Halevy, 1989*b*). The observation made by Whitehead and Halevy (1989*b*) that application of ACC to the gynoecium of unpollinated flowers does not affect ethylene sensitivity provides further proof that the two processes occur independent of each other.

The existence of a transmissible 'sensitivity factor' was demonstrated in petunia when it was shown that application of an exudate collected from excised pollinated styles to the stigmas of unpollinated flowers resulted in accelerated senescence and a drastic increase in the sensitivity of the flowers to ethylene when compared with flowers treated with an exudate from unpollinated styles (Whitehead & Halevy, 1989a). Abscisic acid has been shown to increase the sensitivity of detached rice leaves to ethylene (Kao & Yang, 1983). However, in petunia application of abscisic acid to the gynoecium does not affect the sensitivity of the corolla to ethylene (Whitehead & Halevy, 1989b). Analysis of the trimethylsilyl ethers prepared from the exudate obtained from excised pollinated carnation and petunia styles revealed the presence of large quantities of short-chain saturated fatty acids ranging in chain length from C_7 to C_{10} (Whitehead & Halevy, 1989a; Whitehead & Vasiljevic, 1993). Application of these acids to the stigmas of unpollinated flowers results in the stimulation of ethylene sensitivity in the corollas similar to that observed in the corollas of pollinated flowers. By using [14C]-labelled octanoic and decanoic acids, Whitehead and Halevy (1989) were able to demonstrate that these acids move rapidly from the stigma to the corolla without accumulating in any of the other floral parts along the way. The similarity between the short-chain fatty acid composition of exudates from pollinated styles and that of petals from pollinated flowers also indicate that these acids are transported from the style to the corolla in pollinated flowers (Whitehead & Vasiljevic, 1993). In carnation nonanoic and decanoic acids are the most abundant acids present in the stylar excudates and petals of pollinated flowers. However, in unpollinated flowers octanoic acid appears to be the most abundant fatty acid present in the petal tissue during senescence (Tables 1,2 and 3). The accumulation of these acids during the early stages of senescence in the corollas of both pollinated and unpollinated flowers is not the result of the degradation

Table 1. *Changes in the short-chain saturated fatty acid composition of outer whorl petals from senescing unpollinated carnation flowers*

Time after harvest (days)	Fatty acid ($\mu g \ g^{-1}$ FW)			
	C_7	C_8	C_9	C_{10}
0	0.50	1.25	0.90	0.80
1	0.64	1.06	0.36	0.28
3	1.16	2.08	0.54	0.36
5	1.42	3.16	0.64	0.48
7	0.82	2.34	0.70	0.48
LSD	0.18	0.29	0.19	0.19

Means of three replicates. LSD ($p = 0.05$) for each acid is indicated in the Table (Whitehead & Vasiljevic, 1993).

Table 2. *Changes in the short-chain saturated fatty acid composition of outer whorl petals from pollinated carnation flowers*

Time after pollination (hours)	Fatty acid (μg g^{-1} FW)			
	C_7	C_8	C_9	C_{10}
0	0.50	1.25	0.90	0.80
6	0.55	1.38	0.90	0.90
12	1.08	1.25	3.38	3.54
24	0.74	2.44	1.34	0.66
48	0.76	0.96	0.84	0.46
LSD	0.16	0.28	0.31	0.39

Means of three replicates. LSD ($p = 0.05$) for each acid is indicated in the Table (Whitehead & Vasiljevic, 1993).

Table 3. *Effect of pollination on the production of short-chain saturated fatty acids by excised carnation styles*

Treatment	fatty acid production (ng per flower)			
	C_7	C_8	C_9	C_{10}
Unpollinated	10.7 ± 1.76	7.7 ± 1.41	30.2 ± 1.20	11.8 ± 1.39
Pollinated	7.9 ± 0.61	8.7 ± 0.89	90.4 ± 21.79	13.1 ± 1.15

Styles from individual flowers were excised, pollinated and placed in separate glass tubes containing 200 μl distilled water. Exudates were collected after 48 h and analysed for fatty acids. Means of three replicates ± SD (Whitehead & Vasiljevic, 1993).

of existing membrane lipids, but their synthesis along the usual fatty acid pathway (Whitehead & Halevy, 1989a).

In cut carnation, changes in ethylene sensitivity is associated with changes in the levels of C_7 to C_{10} short-chain saturated fatty acids (Fig. 3). Expansion of the flower to full size during the first day after harvest is accompanied by a decline in the levels of these acids, while no increase in ethylene sensitivity and binding occurs (Figs. 2 and 3). However, the subsequent increase in ethylene sensitivity during the next few days prior to the onset of the climacteric rise in ethylene production is associated with a marked increase in the endogenous levels of the acids and the

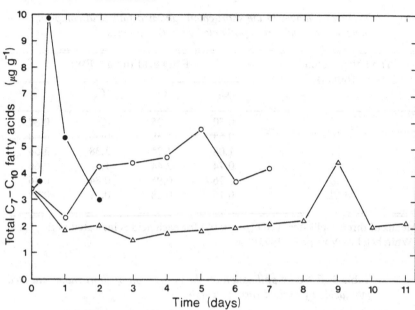

Fig. 3. Changes in the total amount of C_7 to C_{10} short-chain saturated fatty acids in the petals of senescing unpollinated (○), pollinated (●) and STS-treated (△) carnation flowers (Whitehead & Vasiljevic, 1993).

ability of the tissue to bind ethylene. Direct evidence for the involvement of these acids in increasing ethylene sensitivity by increasing ethylene binding is found in the fact that application of octanoic acid to the styles of unpollinated flowers results in an increase in ethylene sensitivity and binding by the petals (Table 4; Fig. 2). The relationship between short-chain saturated fatty acids and ethylene sensitivity is confirmed by the rapid increase in short-chain fatty acid levels which accompanies a sudden increase in ethylene sensitivity during the first 12 hours following pollination (Fig. 3). This increase in saturated fatty acid levels appears to be the result of their increased production by the stylar tissue from where they are rapidly transported to the corolla as discussed above (Whitehead & Halevy, 1989a; Whitehead & Vasiljevic, 1993).

The reaction of carnation flowers to ethylene is suppressed by treatment with silver ions (Veen, 1979; Veen & Van de Geijn, 1978). It is known that silver acts by preventing the binding of ethylene to its membrane associated receptor sites (Beyer, 1976; Goren, Matto & Anderson., 1984). In cut carnation, the suppression of ethylene sensitivity caused by treatment with silver thiosulphate is accompanied by a drastic

Table 4. *Effect of 2,5-norbornadiene, octanoic acid, pollination and STS on ethylene sensitivity*

Treatment	Flower longevity (% of the control)
Unpollinated	100
Ethylene	63
Pollinated	50
Pollinated + ethylene	30
Octanoic acid	75
Octanoic acid + ethylene	37
NBD	150
NBD + ethylene	140
NBD + octanoic acid + ethylene	60
Octanoic acid + NBD	90
Octanoic acid + NBD + ethylene	90
STS	170
STS + ethylene	163
STS + pollinated + ethylene	63
STS + octanoic acid + ethylene	72

Flowers were treated and exposed to ethylene for 6 h. Longevity was measured from the end of ethylene treatment until petals showed signs of desiccation. Mean longevity of 10 flowers from each treatment was expressed as a percentage of the mean longevity of 10 control flowers not exposed to ethylene (Whitehead & Vasiljevic, 1993).

suppression of C_7 to C_{10} fatty acid synthesis (Fig. 3; Table 4). However, even in silver-treated flowers pollination or treatment with octanoic acid results in a marked increase in ethylene sensitivity, suggesting that fatty acids produced during the early stages of senescence, or transported to the corolla after pollination or application to the styles, results in an increase in the ability of the tissue to bind ethylene. The cyclic olefin 2,5-norbornadiene (NBD) is known to inhibit ethylene action by competitively binding to the ethylene binding sites, thus counteracting ethylene induced responses (Hyodo, Terada & Noda et al., 1990; Peiser, 1989; Raskin & Beyer, 1989; Sisler, Blankenship & Guest, 1990; Wang & Woodson, 1989). Treatment of cut carnation with NBD results in a marked suppression of ethylene sensitivity as a result of its ability to block the ethylene binding sites (Sisler et al., 1990). However, application of octanoic acid to NBD-treated flowers results in a drastic increase in ethylene sensitivity, indicating that the fatty acid causes an increase in the ability of the tissue to bind ethylene (Table 4). Furthermore, Whitehead and Vasiljevic (1993) have shown that the increase in ethylene sensitivity caused by treatment with octanoic acid is completely suppressed by exposure to NBD prior to ethylene treatment, suggesting that NBD prevents ethylene binding to additional binding sites made available by treatment with the acid.

Cellular membranes appear to be the site of action for short-chain saturated fatty acids. It is known that treatment of plant tissues with these acids could result in changes in the physical properties of cellular membranes which could lead to an increase in membrane permeability (Babiano et al., 1984; Jackson & Taylor, 1970; Stewart & Berrie, 1979; Takenaka et al., 1981). In carnation, petal senescence is characterised by an increase in fatty acid levels concomitant with a gradual increase in membrane permeability during the pre-climacteric stage, while stylar application of octanoic acid results in a marked increase in petal membrane permeability similar to pollination (Whitehead & Vasiljevic, 1993).

Ethylene sensitivity and other growth substances

It is widely accepted that the different naturally occurring growth substances interact to control plant growth and development. Growth substances such as gibberellins, cytokinins and polyamines are known to affect ethylene synthesis and action, and to delay senescence of various plant tissues. In Ipomoea nil changes in the responsiveness of expanding corolla tissue to ethylene have been attributed in part to the effect of gibberellins on ethylene sensitivity (Raab & Koning, 1987). Treatment with gibberellic acid results in a marked reduction in ethylene sensitivity in ripening persimmon fruit (Ben-Arie et al., 1989).

Cytokinins such as benzyladenine (BA) are also known to delay flower senescence (Cook, Rasche & Eisinger, 1985; Eisinger, 1977; Van Staden & Joughin, 1988). The delay of flower senescence by BA is associated with an inhibition of ACC synthesis and the ability of the tissue to convert ACC to ethylene (Cook et al., 1985; Eisinger, 1977, 1982). In Ipomoea the inhibition of ethylene synthesis by BA is accompanied by a marked decrease in the rate of phospholipid breakdown (Beutelmann & Kende, 1977), indicating that treatment with BA results in the stabilisation of cellular membranes. Aside from the suppression of ethylene synthesis, treatment of carnation flowers with BA also results in a marked decrease in ethylene sensitivity which could be ascribed to the suppression of ethylene binding by the tissue (Nel, 1991). However, treatment with BA does not result in a complete inhibition of ethylene sensitivity, possibly because ethylene binding is suppressed to a limited extent.

Although BA reduces the increase in ethylene sensitivity caused by octanoic acid, it is not as effective as NBD, indicating that it does not interact directly with the ethylene binding sites. Its effect on ethylene binding appears to be the result of its effect on the stabilisation of cellular membranes which will affect ethylene binding indirectly. Analysis of changes in endogenous cytokinin activity in senescing carnation petals indicate that the endogenous cytokinin levels decrease drastically during the pre-climacteric stage concomitant with the increase in ethylene sensitivity and prior to the climacteric rise in ethylene synthesis (Fig. 4).

There is evidence that the diamine putrescine and the polyamines spermidine and spermine can delay senescence in a variety of plant tissues by inhibiting ACC synthesis and/or the conversion of ACC to ethylene (Apelbaum *et al.*, 1981; Fuhrer *et al.*, 1982; Roberts *et al.*, 1984; Suttle, 1981), possibly as a result of their stabilising effect on cellular membranes (Ben-Arie, Lurie & Mattoo, 1982; Roberts, Dumbroff & Thompson, 1986). However, in *Petunia* it was shown that polyamines could also act by decreasing the sensitivity of the corolla to ethylene

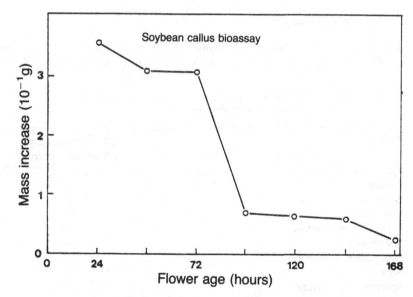

Fig. 4. Changes in cytokinin activity in petals from senescing unpollinated carnation flowers. Cytokinins were extracted from an equal mass of tissue obtained from flowers of different ages. Extracts were tested for cytokinin activity using the soybean callus bioassay. The Figure shows a drastic decline in cytokinin activity during the pre-climacteric stage.

(Table 5). As is the case with cytokinins, the levels of polyamines decrease rapidly during the early stages of senescence (Botha & Whitehead, 1992). The pollination-induced acceleration of senescence and stimulation of ethylene sensitivity in petunia is accompanied by a very rapid decline in polyamine levels during the first 12 hours which coincides exactly with an increase in short-chain saturated fatty acid levels (Whitehead & Halevy, 1989a).

It is clear from the evidence presented above that growth substances which are known to promote growth and delay senescence could be involved to some extent in the control of ethylene sensitivity in senescing flower tissue. The levels of these compounds decrease concomitant with the senescence associated increase in ethylene sensitivity. However, it appears that these substances are not the primary factors involved in the control of ethylene sensitivity, since their application does not result in a complete suppression of ethylene sensitivity and binding. Furthermore, pollination results in the translocation of a 'sensitivity factor' from the gynoecium to the corolla which is not the case with any of the growth substances discussed above.

Conclusions

The discussions in this paper emphasise the importance of an increase in ethylene sensitivity in the control of ethylene action during flower

Table 5. *Effect of putrescine, spermidine and spermine on ethylene sensitivity in senescing petunia flowers*

Treatment	Flower longevity (% of the control)
Control	100
Ethylene	33
Putrescine	100
Putrescine + ethylene	39
Spermidine	100
Spermidine + ethylene	39
Spermine	100
Spermine + ethylene	39

Flowers were placed in 10 μM solutions and exposed to ethylene continuously for the duration of the experiment. Longevity was measured from the start of ethylene treatment until the corolla had turned completely blue. Mean longevity of 10 flowers from each treatment was expressed as a percentage of the mean longevity of 10 control flowers not exposed to ethylene.

senescence. Changes in the levels of 'anti-ethylene' growth substances are not the primary mechanism by which ethylene sensitivity is controlled. It appears that short-chain saturated fatty acids ranging in chain length from C_7 to C_{10} are the 'sensitivity factor' responsible for the increase in ethylene sensitivity during flower senescence. The increase in ethylene sensitivity caused by these acids is related to their ability to increase ethylene binding by altering certain membrane properties which could lead to an increase in the availability of ethylene binding sites.

References

Apelbaum, A., Burgoon, A.C., Anderson, J.D. & Lieberman, M. (1981). Polyamines inhibit biosynthesis of ethylene in higher plant tissue and fruit protoplasts. *Plant Physiology*, **68**, 453–6.

Babiano, M.J., Aldasoro, J.J., Hernàndez-Nistal, J., Rodriquez, D., Matilla, A. & Nicholàs, G. (1984). Effect of nonanoic acid and other short-chain fatty acids on exchange properties in embryonic axes of *Cicer arietinum* during germination. *Physiologia Plantarum*, **61**, 391–5.

Ben-Arie, R., Lurie, S. & Mattoo, A.K. (1982). Temperature-dependent inhibitory effects of calcium and spermine on ethylene biosynthesis in apple discs correlate with changes in microsomal membrane microviscosity. *Plant Science Letters*, **24**, 239–47.

Ben-Arie, R., Roisman, Y., Zuthi, Y. & Blumenfeld, A. (1989). Gibberellic acid reduces sensitivity of persimmon fruits to ethylene. In *Biochemical and Physiological Aspects of Ethylene Production in Lower and Higher Plants*, ed. H. Clijsters, M. de Proft, R. Marcelle & M. van Poucke, pp. 165–171. Dordrecht: Kluwer Academic Publishers.

Beutelmann, P. & Kende, H. (1977). Membrane lipids in senescing flower tissue of *Ipomoea tricolor*. *Plant Physiology*, **59**, 888–93.

Beyer, E.M. (1976). A potent inhibitor of ethylene action in plants. *Plant Physiology*, **58**, 268–71.

Botha, M.L. & Whitehead, C.S. (1992). The effect of polyamines on ethylene synthesis during normal and pollination-induced senescence of *Petunia hybrida* L. flowers. *Planta*, **188**, 478–83.

Brandt, A.S. & Woodson, W.R. (1992). Variation in flower senescence and ethylene biosynthesis among carnations. *HortScience*, **27**, 1100–2.

Brown, J.H., Legge, R.L., Sisler, E.C., Baker, J.E. & Thompson, J.E. (1986). Ethylene binding in senescing carnation petals. *Journal of Experimental Botany*, **37**, 526–34.

Cook, D., Rásche, M. & Eisinger, W. (1985). Regulation of ethylene biosynthesis and action in cut carnation flower senescence by cytokinins. *Journal of the American Society for Horticultural Science*, **110**, 24–7.

282   C.S. WHITEHEAD

Eisinger, W. (1977). Role of cytokinins in carnation flower senescence. *Plant Physiology*, **59**, 707–9.

Eisinger, W. (1982). Regulation of carnation flower senescence by ethylene and cytokinins. *Plant Physiology*, **69**, 136.

Evans, D.E., Bengochea, T., Cairns, A.J., Dodds, J.H. & Hall, M.A. (1982). Studies on ethylene binding by cell-free preparations from cotyledons of *Phaseolus vulgaris* L.: subcellular localization. *Plant Cell and Environment*, **5**, 101–7.

Fuhrer, J., Kaur-Sawhney, R., Shih, L.M. & Galston, A.W. (1982). Effects of exogenous 1,3-diaminopropane and spemidine on senescence of oat leaves. 2. Inhibition of ethylene biosynthesis and possible mode of action. *Plant Physiology*, **70**, 1597–600.

Gilissen, L.J.W. & Hoekstra, F.A. (1984). Pollination-induced corolla wilting in *Petunia hybrida*: rapid transfer through the style of a wilting substance. *Plant Physiology*, **75**, 496–8.

Goren, R., Mattoo, A.K. & Anderson, J.D. (1984). Ethylene binding during leaf development and senescence and its inhibition by silver nitrate. *Journal of Plant Physiology*, **177**, 243–8.

Halevy, A.H., Whitehead, C.S. & Kofranek, A.M. (1984). Does polliantion induce corolla abscission of cyclamen flowers by promoting ethylene production? *Plant Physiology*, **75**, 1090–3.

Hoekstra, F.A. & Weges, R. (1986). Lack of control by early pistillate ethylene of the accelerated wilting of *Petunia hybrida* flowers. *Plant Physiology*, **80**, 403–8.

Hyodo, H., Terada, Y. & Noda, S. (1990). Effects of 2,5-norbornadiene and ethylene on the induction of activity of 1-aminocyclopropane-1-carboxylate (ACC) synthase, and on increases in the ACC content and the rate of ethylene production in petals of cut carnation flowers during senescence. *Journal of the Japanese Society for Horticultural Science*, **59**, 151–6.

Jackson, P.C. & Taylor, J.M. (1970). Effects of organic acids on ion uptake and retention in barley roots. *Plant Physiology*, **46**, 538–42.

Kao, C.H. & Yang, S.F. (1983). Role of ethylene in the senescence of detached rice leaves. *Plant Physiology*, **73**, 881–5.

Lovell, P.J., Lovell, P.H. & Nichols, R. (1987). The control of flower senescence in petunia (*Petunia hybrida*). *Annals of Botany*, **60**, 49–59.

Maxie, E.C., Farnham, D.S., Mitchell, F.G., Sommer, N.F., Parsons, R.A., Snyder, R.G. & Rae, H.L. (1973). Temperature and ethylene effects on cut flowers of carnation (*Dianthus caryophyllus* L.). *Journal of the American Society for Horticultural Science*, **98**, 568–72.

Nel, G.P. (1991). Etileensensitiwiteit in verouderende angelierblomme: Die interaksie tussen bensieladenien en kortketting versadigde vetsure. *MSc Thesis*, Johannesburg: Randse Afrikaanse Universiteit.

Nichols, R., Bufler, G., Mor, Y., Fujino, D.W. & Reid, M.S. (1983). Changes in ethylene production and 1-aminocyclopropane-1-carboxylic acid content in pollinated carnation flowers. *Journal of Plant Growth Regulation*, **2**, 1–8.

Nichols, R. & Frost, C.E. (1985). Wound-induced production of 1-aminocyclopropane-1-carboxylic acid and accelerated senescence of *Petunia* corollas. *Scientia Horticulturae*, **26**, 47–55.

Peiser, G. (1989). Effect of 2,5-norbornadiene upon ethylene biosynthesis in mid-climacteric carnation flowers. *Plant Physiology*, **90**, 21–4.

Raab, M.M. & Koning, R.E. (1987). Changes in responsiveness to ethylene and gibberellin during corolla expansion of *Ipomoea nil*. *Journal of Plant Growth Regulation*, **6**, 121–31.

Raskin, I. & Beyer, E.M. (1989). Role of ethylene metabolism in *Amaranthus retroflexus*. *Plant Physiology*, **90**, 1–5.

Reid, M.S., Fujino, D.W., Hoffman, N.E. & Whitehead, C.S. (1984). 1-Aminocyclopropane-1-carboxylic acid (ACC) – the transmitted stimulus in pollinated flowers? *Journal of Plant Growth Regulation*, **3**, 189–96.

Roberts, D.R., Dumbroff, E.B. & Thompson, J.E. (1986). Exogenous polyamines alter membrane fluidity in bean leaves – a basis for potential misinterpretation of their physiological role. *Planta*, **167**, 395–401.

Roberts, D.R., Walker, M.A., Thompson, J.E. & Dumbroff, E.B. (1984). The effects of inhibitors of polyamine and ethylene biosynthesis on senescence, ethylene production and polyamine levels in cut carnation flowers. *Plant and Cell Physiology*, **25**, 315–22.

Sisler, E.C. (1982). Properties of a Triton X-100 extract from mung bean sprouts. *Journal of Plant Growth Regulation*, **1**, 211–18.

Sisler, E.C., Blankenship, S.M. & Guest, M. (1990). Competition of cyclooctenes and cyclooctadienes for ethylene binding and activity in plants. *Plant Growth Regulation*, **9**, 157–64.

Stewart, R.R.C. & Berrie, A.M.M. (1979). Effects of temperature on the short-chain fatty acid-induced inhibition of lettuce seed germination. *Plant Physiology*, **63**, 61–2.

Suttle, J.C. (1981). Effect of polyamines on ethylene production. *Phytochemistry*, **20**, 1477–80.

Takenaka, T., Horie, H., Hori, H., Yoshioka, T. & Iwanami, Y. (1981). Inhibitory effects of myrmicacin on the sodium channel of the squid giant axon. *Proceedings of the Japanese Academy*, **57**, 314–17.

Thompson, J.S., Harlow, R.L. & Whitney, J.F. (1983). Copper (I)-olefin complexes. Support for the proposed role of copper in ethylene effects in plants. *Journal of the American Chemical Society*, **105**, 3522–7.

Thompson, J.E., Mayak, S., Shinitzky, M. & Halevy, A.H. (1982). Acceleration of membrane senescence in cut carnation flowers by treatment with ethylene. *Plant Physiology*, **69**, 859–63.

Trewavas, A.J. (1981). How do plant growth substances work? *Plant Cell and Environment*, **4**, 203–8.

Trewavas, A.J. (1982). Growth substances sensitivity: the limiting factor in plant development. *Physiologia Plantarum*, **55**, 60–72.

Trewavas, A.J. & Jones, A.M. (1981). Consequences of hormone-binding studies for plant growth substance research. *What's New in Plant Physiology*, **12**, 5–8.

Van Staden, J. & Joughin, J.I. (1988). Cytokinins in cut carnation flowers. IV. Effects of benzyladenine on flower longevity and the role of different longevity treatments on its transport following application to the petals. *Plant Growth Regulation*, **7**, 117–28.

Veen, H. (1979). Effects of silver on ethylene synthesis and action in cut carnations. *Planta*, **145**, 467–70.

Veen, H. & Van de Geijn, S.C. (1978). Mobility and ionic form of silver as related to longevity of cut carnations. *Planta*, **140**, 93–6.

Wang, H. & Woodson, W.R. (1989). Reversable inhibition of ethylene action and interruption of petal senescence in carnation flowers by norbornadiene. *Plant Physiology*, **89**, 434–8.

Whitehead, C.S. & Bossé, C.A. (1991). The effect of short-chain saturated fatty acids on ethylene sensitivity and binding in ripening bananas. *Journal of Plant Physiology*, **137**, 358–62.

Whitehead, C.S. & Halevy, A.H. (1989a). Ethylene sensitivity: the role of short-chain saturated fatty acids in pollination-induced senescence of *Petunia hybrida* flowers. *Plant Growth Regulation*, **8**, 41–54.

Whitehead, C.S. & Halevy, A.H. (1989b). The role of octanoic acid and decanoic acid in ethylene sensitivity during pollination-induced senescence of *Petunia hybrida* flowers. *Acta Horticulturae*, **261**, 151–6.

Whitehead, C.S., Halevy, A.H. & Reid, M.S. (1984). Roles of ethylene and 1-aminocyclopropane-1-carboxylic acid in pollination and wound-induced senescence of *Petunia hybrida* flowers. *Physiologia Plantarum*, **61**, 643–8.

Whitehead, C.S. & Vasiljevic, D. (1993). Role of short-chain saturated fatty-acids in the control of ethylene sensitivity in senescing carnation flowers. *Physiologia Plantarum*, **88**, 243–50.

Wu, M.J., Zacarias, L. & Reid, M.S. (1991). Variation in the senescence of carnation (*Dianthus caryophyllus* L.) cultivars. II. Comparison of sensitivity to exogenous ethylene and of ethylene binding. *Scientia Horticulturae*, **48**, 109–16.

E.J. WOLTERING, A. TEN HAVE,
P.B. LARSEN and W.R. WOODSON

Ethylene biosynthetic genes and inter-organ signalling during flower senescence

Systems for inter-organ communication

In order to successfully survive, all living organisms require a system to respond to the continuously changing environment, and to regulate the various developmental processes in cells and organs at different locations in the body. The following discussion highlights the communication systems in plants with special emphasis on reproductive aspects.

In animals, two major systems are involved in inter-organ communication, i.e. the nervous system and the hormone system. In addition, the immune system uses different classes of chemical messengers (e.g. interleukins, interferons, nitric oxide) to coordinate the defence response (Norman & Litwack, 1987; Roitt, Brostaff & Male, 1993). Although the metabolism of individual plant and animal cells may show many similarities, there is no *a priori* reason why the mechanisms, involved in the responses to environmental cues and in communication between different organs should in any way be the same. However, systems showing some similarities to the animal nervous system have been found in plants. For instance, mechanisms have evolved to respond within seconds to mechanical perturbation in *Drosera* (sundew) and *Mimosa pudica*. The movements of the leaves of *Mimosa pudica* are regulated by motor organs (pulvini) and are the result of turgor variations in the cortical (motor) cells of these organs. It has been shown that ionic migration, in particular of K^+ and Cl^-, appears in the pulvinus, concomitant with the turgor changes (Satter & Galston, 1981; Fleurat-Lessard *et al.*, 1988). Recently it has been shown that mechanically wounding of the cotyledon of a young tomato plant results in the slow transmission (1–4 mm/s) of an action potential out of the cotyledon and into the first leaf. This signal is thought to be responsible for the systemic induction of proteinase inhibitors following mechanical damage due to insect attack (Wildon *et al.*, 1992).

Society for Experimental Biology Seminar Series 55: *Molecular and Cellular Aspects of Plant Reproduction*, ed. R.J. Scott & A.D. Stead.
© Cambridge University Press, 1994, pp. 285–307.

Upon infection with microbes, most plants accumulate anti-microbial compounds of low molecular weight at the site of infection. At present, more than 100 of such plant-protecting substances, called phytoalexins, have been isolated from a large number of plant species. These compounds (e.g. sesquiterpenoids, isoflavonoids) are rapidly produced by the host cells and their accumulation is the result of *de novo* synthesis of mRNAs and enzymes which lead to their production (Darvill & Albersheim, 1984; Érsek & Király, 1986).

The molecules that signal plants to start the process of phytoalexin synthesis are called elicitors. Elicitors may be constituents of microbes or enzymes secreted by the infecting microbes or may be released by the cell wall itself. Furthermore, abiotic factors such as heavy metals or UV-light may also elicit phytoalexin production. It is currently not clear as to how elicitors turn on the expression of selected genes and how they may inhibit the expression of other genes (Davis & Hahlbrock, 1987; Baldwin & Pressey, 1988).

In interactions between host plants and herbivores, microbes or viruses, induced systemic resistance in non-affected plant parts may be found following the initial infection. This has been associated with the synthesis of 'pathogenesis-related' proteins in the non-affected areas. Both volatile (ethylene and jasmonic acid) and soluble chemicals (salicylic acid and the 18-amino-acid-peptide 'systemin') have been implicated as translocatable chemical messengers responsible for systemic resistance (Ryan, 1992; Enyedi *et al.*, 1992). Besides communication between different parts of the same plant, interplant communication by means of volatile chemicals has also been established. For instance, *Artemisia tridentata* (sagebrush) plants produce methyl jasmonate in quantities sufficient to induce a defence response in other species (Farmer & Ryan, 1990) and *Acacia* trees may produce sufficient quantities of ethylene to elicit a defence response in trees growing as much as 50 metres down wind (Van Hoven, 1991).

At present, it is generally accepted that plants contain a system that has similarities to the animal hormone system. It involves chemicals of small molecular weight that have a pronounced effect on developmental processes such as growth, flowering, ripening and senescence. In contrast to animal hormones, which are predominantly produced in specific organs (Norman & Litwack, 1987), plant hormones are produced by many individual cells. Therefore, translocation may not be a necessary feature. However, xylcm and phloem sap may contain different classes of plant hormones, and it is likely that most plant hormones, are indeed translocated as messengers within the plant (King, 1976; Bradford & Yang, 1980).

The mode of action of plant hormones at the sub-cellular level is still a matter of debate. Although at present no single receptor protein has been identified with certainty, it is generally believed that hormone perception in plants shows many similarities to the mechanism in animals. The hormone is thought to bind to a receptor protein, after which the hormone–receptor complex or a second messenger modifies the transcription process, leading to changes in the production of specific mRNAs (Moore, 1989). Unlike animal hormones, it is most likely that a physiological response to a plant hormone is achieved not predominantly from a change in hormone concentration but also from a change in sensitivity of the responding tissue (Trewavas, 1981; Starling, Jones & Trewavas, 1984). The sensitivity may depend on the amount or accessibility of receptor proteins.

The above examples show that, in plants, both chemical messengers and transmissible action potentials provide for inter-organ communication. The basic ideas therefore are similar as in animals. However, as plants and animals are structurally different, the routes by which the signals travel are quite distinct. In animals, the vascular and nervous systems provide excellent ways for rapid inter-organ communication; action potentials in neurons may travel up to 100 m/s. In plants, which lack such specialised systems for inter-organ communication, most signals travel at a much slower rate presumably mainly from cell to cell.

Ethylene biosynthesis and action

Ethylene (C_2H_4) is a gaseous plant hormone that plays an important role in many developmental processes. In growing organs, it generally inhibits growth, while, in mature organs, ripening, senescence and abscission are promoted. In addition, ethylene may stimulate seed germination and the formation of female flowers in some plant families (Abeles, Morgan & Saltveit, 1992). Its pronounced effect on flower senescence has been extensively studied, and evidence has been presented that ethylene and its soluble precursor 1-aminocyclopropane-1-carboxylic acid (ACC), may play an important role in inter-organ communication in flowers.

Ethylene biosynthesis

Ethylene is synthesised from methionine through the intermediates S-adenosyl methionine (SAM) and the cyclic amino acid 1-aminocyclopropane-1-carboxylic acid (ACC) (Fig. 1). The conversion of SAM to ACC and methylthioadenosine (the latter being recycled into methionine) is generally believed to be the rate-limiting step in ethylene biosynthesis. The enzyme involved, ACC synthase, is located in the cyto-

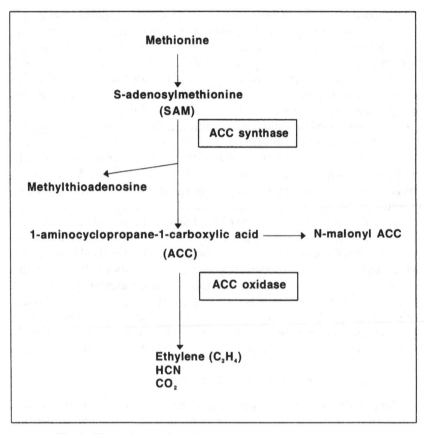

Fig. 1. Biosynthesis of ethylene.

plasm, requires pyridoxal phosphate and is readily induced by various stress conditions including exposure to ethylene (Yang & Hoffman, 1984; Kende, 1993).

The conversion of ACC to ethylene, HCN and CO_2 is catalysed by ACC oxidase. It has recently been found that ACC oxidase, like ACC synthase, is most probably located in the cytoplasm and that it requires ascorbate and Fe^{2+}. Furthermore, the enzyme requires carbon dioxide for activation (Ververidis & John, 1991; Smith & John, 1993). ACC oxidase may be induced by ethylene (Kende, 1993).

In many situations, ACC is converted into N-malonyl ACC (MACC). The enzyme responsible is not specific but converts all D-amino acids into their malonyl derivatives. The lack of stereochemistry of ACC is the reason that it is also 'detoxified'. The malonyl transferase was shown to be regulated by ethylene and by light. MACC is believed to be an inactive end product predominantly transported into the vacuole. It is thought not to be converted into ACC under natural conditions (Amrhein *et al.*, 1982; Van Loon & Fontaine, 1984; Pech *et al.*, 1989).

An interesting feature of the ethylene biosynthetic pathway is that the enzymes involved may be stimulated by ethylene, leading to a sudden upsurge in ethylene production known as autocatalysis (Yang & Hoffman, 1984; Kende, 1993). The enhanced level of endogenous ethylene may surpass threshold values required for the induction of such phenomena as climacteric respiration and enzyme production in, e.g. ripening fruits. However, in some cases, autoinhibition through suppression of ACC synthase activity or stimulation of malonyltransferase activity may also occur (Riov & Yang, 1982; Philosoph-Hadas, Meir & Aharoni, 1985).

ACC synthase genes

ACC synthase cDNA clones have been isolated from a large number of plant species (*Arabidopsis*, tomato, mung bean, winter squash, apple, tobacco, avocado, melon, orchid flower, carnation flower and peach). It has become clear that ACC synthases are encoded by a highly divergent gene family and that expression of the different isoforms is, in many cases, organ and stimulus specific (e.g. Rottmann *et al.*, 1991; Olson *et al.*, 1991; Mori *et al.*, 1993).

For example, in light grown *Arabidopsis* plants, five different ACC synthase genes were identified. Expression of these genes was not uniform over the different organs. One of the genes (ACC 4) was expressed in roots, leaves and flowers but not in stems and siliques. Another gene (ACC 5) was expressed only in flowers and siliques. In etiolated seedlings wounding stimulated expression of ACC 2 and ACC 4 genes but suppressed expression of ACC 5. Auxin treatment induced only the ACC 4 gene (Liang *et al.*, 1992). In another report, expression of an *Arabidopsis* ACC synthase gene was studied in plants transformed with an ACC synthase promoter fused to the β-glucuronidase reporter gene (Rodrigues-Pousada *et al.*, 1993). These authors showed that expression of the gene was associated with developing tissues like young floral buds, shoot or inflorescence meristems, young developing leaves and early stages of lateral root formation. Wounding did not induce expression but exposure to ethylene did.

The amino acid sequence identities of the different ACC synthases vary from 48 to 97%. In the vast majority of cases studied so far, ACC synthase activity is regulated at the transcriptional level and the different

ACC synthase genes are differentially regulated in response to a variety of developmental, environmental and chemical factors (Kende, 1993). In many cases, detailed expression data are not yet available or inadequate to group the different genes according to their roles in plant development.

ACC oxidase genes

Using the antisense technique it was discovered that the tomato ripening related gene pTOM13 encoded a component of the ethylene biosynthesis route (Hamilton, Lycett & Grierson, 1990). Functional expression of the gene in yeast revealed that the gene encoded ACC oxidase (Hamilton, Bouzayen & Grierson, 1991). Comparison of the nucleotide sequence with other genes showed that the sequence shared significant homology with a flavanone 3-hydroxylase gene from petunia. This led to the development of methods to isolate the protein and study its characteristics *in vitro*. Subsequently, several pTOM 13 homologues were isolated from a variety of plant species (apple, peach, carnation flowers, avocado, melon, petunia, pea, kiwi, orchid flower).

The ACC oxidase genes from different species were found to be much more homologous than the ACC synthase genes. Generally more than 80% homology at the protein level is found among ACC oxidases (Kende, 1993).

In tomato, three different ACC-oxidase genes were isolated and, using gene-specific probes, it was shown that the genes were differentially expressed in ripening fruits and wounded leaves (Holdsworth, Schuch & Grierson, 1988). In *Petunia* flowers, four different ACC oxidase genes were isolated. Three of these genes were found to be expressed while the fourth had characteristics of a pseudogene (Tang *et al.*, 1994).

Ethylene perception

Once produced, ethylene is thought to bind, presumably in a reversible way, to a receptor protein. The ethylene–receptor complex leads to the transcription of specific genes and production of enzymes that are responsible for the physiological effects (Sisler & Goren, 1981; Woodson & Lawton, 1988; Lawton *et al.*, 1990). Recently, elements of the primary perception and signal transduction system have been identified using *Arabidopsis* mutants (Kieber & Ecker, 1993; Theologis, 1993). Firstly, a candidate for the ethylene receptor (*ETR1* gene) has been identified (Bleecker *et al.*, 1988). The (*ETR1* gene product exhibits a striking similarity to the so-called two-component signal sensing elements exhibiting histidyl kinase activity in bacteria (Chang *et al.*, 1993).

Secondly, the isolation and characterization of the *CTR1* gene revealed that it encodes a putative serine/threonine protein kinase closely related to the Raf protein kinase family in other organisms (Kieber *et al.*, 1993; Heidecker *et al.*, 1992). The *CTR1* gene acts as a negative regulator of the ethylene response in a later stage of signal transduction. Thus, the ethylene signal transduction pathway may include a cascade of protein kinases with similarities to signal transduction pathways involved in the modulation of cellular responses to extracellular signals in other organisms, including animals and insects.

Inter-organ signalling in flowers

A flower is a complex structure composed of several different organs, each of which has a variety of functions essential for successful reproduction. Pollination and, in orchids, emasculation often leads to rapid changes in flower parts distinct from the site of action thereby providing an excellent model to study the possible involvement of translocatable factors in the coordination of the post-pollination and post-emasculation events.

Pollination and emasculation

In many species, pollination induces rapid changes in the perianth. This may include changes in pigmentation, senescence and abscission (Stead, 1992). In orchid flowers, emasculation (removal of the pollinia and the anther cap), which is a natural part of the pollination event, often has a comparable effect. Pollination- or emasculation-induced changes in pigmentation have been reported in flowers of *Cymbidium* (Woltering & Somhorst, 1990) and *Lupinus albifrons* (Stead & Reid, 1990) and examples of over 50 angiosperm families exhibiting subtle pigmentation changes have been described (M. Weiss, pers. comm.). Given the perianth's role in attraction of pollinators, pollination-induced pigmentation changes have likely evolved to deter further visits by pollinators.

Premature senescence of petals is another common response to pollination (Borochov & Woodson, 1989) and has been studied in detail in the flowers of carnation (Nichols, 1977; Nichols *et al.*, 1983), *Petunia* (Gilissen, 1976; Gilissen & Hoekstra, 1984), and orchids (Burg & Dijkman, 1967).

Pollination-induced corolla abscission has been reported for flowers of *Digitalis* (Stead & Moore, 1979), *Cyclamen* (Halevy, Whitehead & Kofranek, 1984), and *Pelargonium* (Wallner *et al.*, 1979). Both premature senescence and abscission of flower petals have evolved as mechanisms to deter further visits by pollinators and also to deter feeding by herbivores, thereby protecting the developing fruit. Besides changes in the perianth, pollination often induces ovary growth (Nichols, 1971) and may initiate ovule development (Zhang & O'Neill, 1993).

An increase in the production of ethylene is often a very early post-pollination event (Hoekstra & Weges, 1986; Pech *et al.*, 1987). Generally, a very rapid response as in *Petunia* and carnation flowers is observed (Fig. 2). In *Cymbidium* flowers, a delay of several hours was observed between pollination and the start of the ethylene production. The reason for this delayed reaction may be the existence of a considerable amount of a viscous fluid on the stigmatic surface that may initially prevent physical contact between the applied pollen and the stigma (Woltering *et al.*, 1993*b*). During this increase in ethylene production, the styles were identified as major producers. In carnation styles, pollination-induced ethylene is defined by three distinct peaks (Fig. 3). The first peak occurs within the first 2 hours post-pollination, and before the germination and penetration of the pollen tubes. The second peak of ethylene production occurs approximately 12 hours after pollination, and the third peak

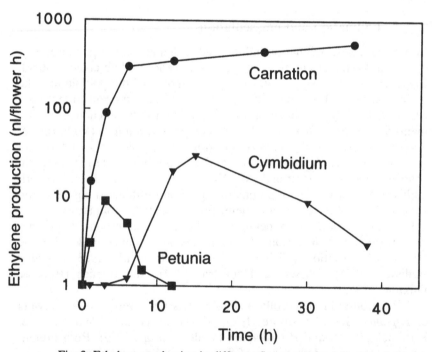

Fig. 2. Ethylene production in different flower species following pollination at *t*=0.

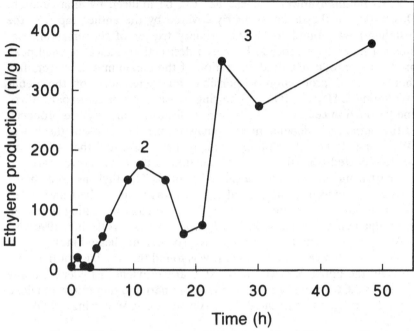

Fig. 3. Ethylene production of carnation styles following pollination at $t=0$. Ethylene production was measured by enclosing styles, excised at different times after pollination, in a 25 ml glass vial for 0.5 h and analysing the accumulated ethylene by gas chromatography.

occurs between 24 and 48 hours post-pollination and is associated with the deterioration of the style (Larsen, Woltering & Woodson, 1993). Also emasculation in orchid flowers rapidly leads to changes in ethylene production and there is considerable evidence that most of the pollination- and emasculation-induced events are triggered by style-produced ethylene (Woltering & Harren, 1989).

ACC and ethylene as translocatable factors

Circumstantial evidence has accumulated over the years that, in pollinated flowers, the direct precursor of ethylene, ACC, may play a role in the coordination of the senescence process (Nichols *et al.*, 1983; Hoekstra & Weges, 1986). Reid *et al.* (1984) provided direct proof for the translocation of ACC between flower parts. These authors treated the carnation stigma with ACC and were able to recover a burst of ethylene from the petals. In addition, petals produced radiolabelled ethylene from stigma-applied radiolabelled ACC.

In *Cymbidium* flowers, senescence can be induced by emasculation. Desiccation of the tissue, formerly covered by the anther cap (i.e. the rostellum) was found to be the primary trigger of the emasculation-induced senescence process. Localised desiccation induces a small peak in ethylene production, red coloration of the labellum and triggers further ethylene production responsible for senescence of the petals (Woltering & Harren, 1989). Following excision of a perianth part, ethylene production ceased within 10 min, indicating that ethylene produced in the perianth is dependent on signals from other parts of the flower (Woltering, 1990b). Several experiments have indicated that ACC may be translocated from the site of production to the other flower parts.

Apart from the possible translocation of ACC, ethylene itself, being primarily produced in the central column, was found to be translocated in significant amounts to the other flower organs indicating a role in inter-organ communication in *Cymbidium* flowers (Woltering, 1990a,b).

ACC and ethylene translocation has also been implicated in the senescence of isolated carnation petals. It was found that the base of the petal provides the upper part with both ACC and ethylene. Ethylene initially promotes ACC oxidase activity in the upper part and thereafter, ethylene is produced from translocated ACC (Overbeek & Woltering, 1990).

Other translocatable signals

It has been suggested that, besides ACC and ethylene, other wilting factors (i.e. sensitivity factors) may be translocated in pollinated flowers. In *Cyclamen* flowers, pollination-induced corolla abscission was ascribed to the action of a pollination-induced ethylene-sensitivity factor as abscission in pollinated flowers could be prevented by the inhibitor of ethylene action, silver thiosulphate, whereas it could not be induced by ethylene or ACC in unpollinated flowers (Halevy *et al.*, 1984).

Eluates from pollinated *Petunia* styles were shown to possess wilt-inducing properties when re-applied to the stigma of fresh flowers. These eluates did not contain any detectable ACC (Gilissen & Hoekstra, 1984). GC–MS analysis of the composition of the *Petunia* style eluate showed that a number of short-chain saturated fatty acids were present (Whitehead & Halevy, 1989). These fatty acids exhibited wilt-inducing properties and it was shown that they were produced in the pollinated style and subsequently transported to the corolla. In the corolla, they are thought to render the tissue more sensitive to ethylene thereby causing an acceleration of the wilting process. Also, in pollinated carnation flowers an increase in the levels of these fatty acids in the petals was found. It was shown that treatment with these chemicals increased the binding of ethylene to its putative receptor which may explain their effect on ethylene sensitivity (Whitehead & Vasiljevic, 1993 and this volume). Other authors, in contrast, could not demonstrate any effect of these acids on senescence of *Petunia, Cymbidium* or carnation flowers (Woltering *et al.*, 1993*b*). The role of these compounds in inter-organ communication during flower senescence therefore remains uncertain.

In orchid flowers, in contrast to carnation and *Petunia* flowers, application of auxin to the stigma leads to an increase in ethylene production, senescence of the perianth and development of the ovary in a way comparable to pollination (Arditti & Knauft, 1969). Since orchid pollinia are known to contain auxin (Stead, 1992) and ovary development is dependent on auxin (Zhang & O'Neill, 1993) it may be hypothesised that, in orchids, auxin may serve as a translocatable factor following pollination. However, in *Vanda* orchids stigma-applied auxin was found to be translocated very poorly (Burg & Dijkman, 1967), leaving a possible role for auxin in inter-organ communication still to be proven.

Ethylene biosynthetic genes and inter-organ communication in flowers

There is now a vast amount of evidence that inter-organ communication occurs in flowers. This phenomenon has only been studied in ethylene sensitive flowers. The exact signal(s), although probably related to ethylene, remain obscure. Study of the expression patterns of the genes coding for the key enzymes in ethylene biosynthesis will be helpful to identify the sites where ethylene and ACC are initially synthesised and to identify tissues that are able to synthesise ACC and ethylene as opposed to tissues that are dependent on translocated substrate.

Pollination-induced senescence

O'Neill *et al.* (1993) isolated ACC synthase and ACC oxidase genes from *Phalaenopsis* flowers and studied their expression in different flower parts following pollination. Although all the flower parts (gynoecium, petals, sepals, labellum) produced significant amounts of ethylene following pollination, ACC synthase transcripts were only detectable in the gynoecium (stigma and ovary) and in the labellum. In the petals and sepals, only ACC oxidase transcripts were detected. In line with the absence of ACC synthase transcripts, no ACC synthase activity was detected in the petals.

This indicates that ethylene produced by the petals and sepals is synthesised from translocated substrate. The observation that ACC oxidase gene expression in these organs could be induced by treatment with

ethylene indicates that ethylene may be the primary factor being translocated and that ACC may be secondary.

Further investigations showed that auxin, which is known to be present
in orchid pollinia, is able to induce ACC synthase gene expression in the
stigma. Treatment with aminoethoxyvinylglycine (AVG), an inhibitor of
ACC synthase activity, completely blocked the effect of auxin. This
indicates that the observed expression of ACC synthase genes, in
response to auxin, was mediated by ethylene. Pollination or auxin treatment presumably induces another, not yet identified, ACC synthase gene
responsible for a first small increase in ethylene production (O'Neill *et
al.*, 1993).

In carnation flowers there is evidence that pollination-induced ethylene
serves to coordinate the subsequent wilting process. Following pollination, three distinct peaks in ethylene production are apparent in the styles
(Fig. 3). The second peak was found to be associated with the expression
of a style-specific ACC synthase gene (CARAS1; Henskens, Somhorst &
Woltering, 1993) and massive accumulation of ACC oxidase (pSR120;
Woodson *et al.*, 1992) transcripts. During the third peak, no expression
of the two known carnation ACC synthase genes was observed, whereas
the ethylene production was very high (unpublished observations). This
has been interpreted as evidence that, at this stage, the styles are provided with ACC from the other parts. The amount of pollen-held ACC
was found to be insufficient to cause the initial small peak in ethylene
production in the style and pre-treatment with AVG completely blocked
pollination-induced ethylene production indicating that another ACC
synthase gene that is specifically induced by pollination, is involved in
the initial response (Woltering *et al.*, 1993*b*; Larsen *et al.*, 1993). Some 6
to 12 hours after pollination, transcripts of a petal specific ACC synthase
(CARACC3; Park, Drory & Woodson, 1992) and of ACC oxidase
(pSR120) are detected in the petals concomitant with an increased ethylene production in these organs. Several experiments have indicated that
the third peak in stylar ethylene production is regulated through the
ovary, and the same is true for the increase in petal ethylene production
probably through translocation of ACC and ethylene (Larsen *et al.*,
1993).

Senescence of unpollinated flowers

During senescence of unpollinated carnation flowers, an early increase
in ethylene production is observed in the ovary. A little later the other
flower parts (styles, petals) start to produce ethylene (Fig. 4). Expression
studies of ACC synthase and ACC oxidase genes revealed that, initially,

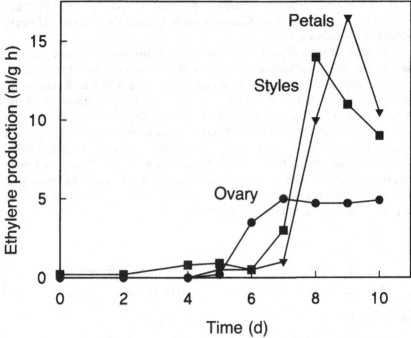

Fig. 4. Ethylene production of different flower parts of cut carnation flowers during the vase life. Ethylene production was measured by enclosing styles, ovary and petals from the outer worl, excised at different times, in a 25 ml glass vial for 0.5–1 h and analysing the accumulated ethylene by gas chromatography.

ACC synthase transcripts hybridising with the cDNA probe CARAS1 are detected in the gynoecium (ovary and styles) and, later on, transcripts hybridizing with the ACC synthase cDNA probe CARACC3 occur in the ovary, the styles and the petals. ACC oxidase transcripts were first detected in the ovary, and later on in the styles and petals (unpublished observations). These data indicate that, during senescence of unpollinated flowers, the ovary is the site from which the senescence process is mediated.

Isolated carnation petals, when placed with their cut base in water, go through the senescence processes in a similar way to petals in intact flowers; however, the life-span of isolated petals is much longer. During senescence of isolated petals, ACC oxidase transcripts are detected in upper and basal parts of the petal. ACC synthase transcripts, in contrast, are only detectable in the basal part of the petals. In the upper part of the petal ACC oxidase is initially induced by translocated ethylene. Thereafter, translocated ACC may serve to sustain the response (Drory, Mayak & Woodson, 1993).

Although in unpollinated flowers there is no direct reason to start developmental processes associated with pollination and fertilisation, ovary growth is often observed during senescence and was found to be markedly stimulated by ethylene and inhibited by pretreatment of the flowers with STS (Fig. 5). Besides ethylene, other plant hormones, e.g auxins, cytokinins and giberellins, are known to stimulate ovary growth in carnation as well as in many other species (e.g. Nichols, 1971; Vercher & Carbonell, 1991; Woodson & Brandt, 1991). It may therefore be argued that the senescence process is regulated through the ovary and that subtle changes in hormone levels may be the initiating factors both in pollinated and in unpollinated flowers.

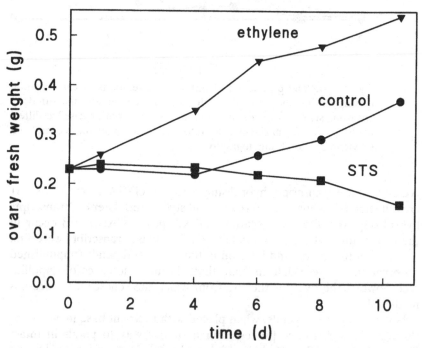

Fig. 5. Increase in fresh weight of ovaries during the vase life of carnation flowers. Flowers were pre-treated overnight with water (control), 0.2 mM silver thiosulphate (STS) or 10 µl/l ethylene (ethylene) and thereafter placed in water.

A model for inter-organ communication during senescence

Based on our own experiments, and a vast amount of data derived from the literature, a model for inter-organ relations during senescence of carnation flowers has been established (Fig. 6). This model places the ovary in a central position from which most events are regulated. During senescence of unpollinated flowers, a developmentally regulated change in, e.g. cytokinin or auxin levels in the ovary or ovules, stimulates expression of a specific ACC synthase gene resulting in the production of ACC and ethylene. The produced ethylene further stimulates ethylene-responsive ACC synthase and ACC oxidase genes leading to autocatalysis.

Some of the produced ACC and ethylene is translocated to the petal base. Here, ethylene induces expression of ethylene responsive ACC synthase and ACC oxidase genes and translocated ACC is converted into ethylene. In the petal base, autocatalysis occurs leading to the accumulation of ACC and ethylene and subsequent translocation to the upper part of the petal. In the upper part, ethylene induces ACC oxidase gene expression by which the translocated ACC is converted into ethylene. The vast amount of ethylene produced by the petal leads, presumably through gene expression (Lawton *et al.*, 1990) to a cascade of biochemical events eventually leading to the observed senescence symptoms.

During senescence of unpollinated flowers, the increase in ethylene production and senescence of the styles may be triggered by ACC and ethylene translocation from the ovary. Following pollination, it is most likely that pollen-derived factors (e.g. elicitors) induce a specific ACC synthase gene in the style leading to increased ethylene production and subsequent ethylene-induced expression of ACC synthase and ACC oxidase genes in the styles and translocation of ethylene to the ovary. Given the extremely high ACC oxidase activity in pollinated styles (Nichols *et al.*, 1983), ACC is not likely to accumulate and may not be able to escape to other flower parts. In the ovary, ethylene presumably induces ethylene-responsive ACC synthase and ACC oxidase genes leading to accumulation and translocation of ACC and ethylene to the styles and petals.

Conclusions

Similar to animal systems, inter-organ communication by means of chemical messengers and transmissible action potentials plays an important role in the coordination of a variety of developmental processes and environmental response mechanisms in plants. During flower senescence, translocation of the gaseous plant hormone ethylene and its soluble pre-

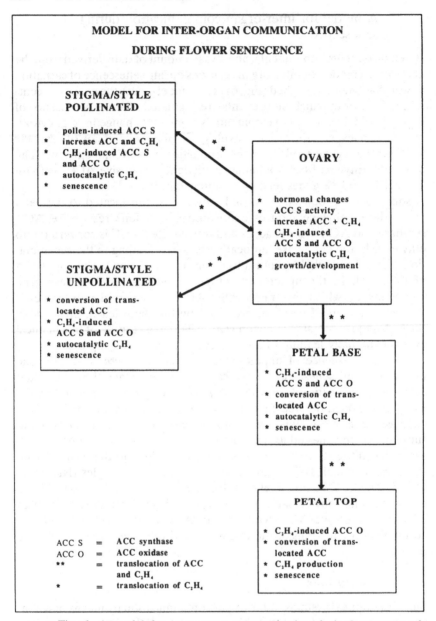

Fig. 6. A model for inter-organ communication during senesence of pollinated and unpollinated carnation flowers.

cursor, ACC, apparently play major roles as signal molecules responsible for petal senescence and ovary development. Studies of the expression pattern of ethylene biosynthetic genes in carnation revealed that the different ACC synthase genes are expressed in a stimulus- and organ-specific way, and that the developing ovary apparently plays a major role in the coordination of the senescence processes that occur in the other flower parts. In line with the proposed regulatory role of the ovary in flower senescence, it was found that carnation cultivars with less developed ovaries (e.g. cvs. Chinera and Epomeo) generally have a longer vase life than 'normal' cultivars (Woltering, Somhorst & De Beer, 1993*a*).

A lot is yet to be learnt about the mechanisms by which different environmental and endogenous factors initiate ethylene biosynthesis and the routes or mechanisms by which the chemical messengers are translocated. The isolation, characterisation and organ-specific expression patterns of the genes responsible for ethylene biosynthesis will no doubt shed more light on the complex relations between different flower parts and will open new ways to manipulate flower senescence by means of genetic engineering.

Acknowledgements

This work was supported by the US Department of Agriculture National Research Initiative Comparative Grants Program and the National Science Foundation (W.R.W.) and by the Commodity Board for Ornamental Horticulture in The Netherlands (E.J.W.). The authors are grateful to Truus de Vrije for critically reading the manuscript.

References

Abeles, F.B., Morgan, P.W. & Saltveit, M.E. Jr. (1992). *Ethylene in Plant Biology*. 2nd edn., 414 pp. San Diego: Academic Press.

Amrhein, N., Breuing, F., Eberle, J., Skorupka, H. & Tophof, S. (1982). The metabolism of 1-aminocyclopropane-1-carboxylic acid. In *Plant Growth Substances*, ed. P.F Warering, pp. 249–58. London/ New York: Academic Press.

Arditti, J. & Knauft, R. (1969). The effects of auxin, actinomycin D, ethionine, and puromycin on post-pollination behaviour by *Cymbidium* (Orchidaceae) flowers. *American Journal of Botany*, **56**, 620–8.

Baldwin, E.A. & Pressey, R. (1988). Tomato polygalacturonase elicits ethylene production in tomato fruits. *Journal of the American Society of Horticultural Science,* **113**, 92–5.

Bleecker, A.B., Estelle, M.A., Somerville, C. & Kende, H. (1988). Insensitivity to ethylene conferred by a dominant mutation in *Arabidopsis thaliana*. *Science*, **241**, 1086–9.

Borochov, A. & Woodson, W.R. (1989). Physiology and biochemistry of flower petal senescence. *Horticultural Reviews*, **11**, 15–43.

Bradford, K.J. & Yang, S.F. (1980). Xylem transport of 1-aminocyclopropane-1-carboxylic acid, an ethylene precursor, in waterlogged tomato plants. *Plant Physiology*, **65**, 322–6.

Burg, S.P. & Dijkman, M.J. (1967). Ethylene and auxin participation in pollen induced fading of *Vanda* orchids. *Plant Physiology*, **42**, 1648–50.

Chang, C., Kwok, S.F., Bleecker, A.B. & Meyerowitz, E.M. (1993). *Arabidopsis* ethylene-response gene *ETR1*: similarity of product to two-component regulators. *Science*, **262**, 539–99.

Darvill, A.G. & Albersheim, P. (1984). Phytoalexins and their elicitors – A defence against microbial infection in plants. *Annual Review of Plant Physiology*, **35**, 243–75.

Davis, K.R. & Hahlbrock, K. (1987). Induction of defense responses in cultured parsley cells by plant cell wall fragments. *Plant Physiology*, **85**, 1286–90.

Drory, A., Mayak, S. & Woodson, W.R. (1993). Expression of ethylene biosynthetic pathway mRNAs is spatially regulated within carnation flower petals. *Journal of Plant Physiology*, **141**, 663–7.

Enyedi, A.J., Yalpani, N., Silverman, P. & Raskin, I. (1992). Signal molecules in systemic plant resistance to pathogens and pests. *Cell*, **70**, 879–86.

Érsek, T. & Király, Z. (1986). Phytoalexins: warding-off compounds in plants? *Physiologia Plantarum*, **68**, 343–6.

Farmer, E.E. & Ryan, C.A. (1990). Interplant communication: airborne methyl jasmonate induces synthesis of proteinase inhibitors in plant leaves. *Proceedings of the National Academy of Sciences, USA*, **87**, 7713–16.

Fleurat-Lessard, P., Roblin, G., Bonmort, J. & Besse, C. (1988). Effects of colchicine, vinblastine, cytochalasin B and phalloidin on the seismonastic movement of *Mimosa pudica* leaf and on motor cell ultrastructure. *Journal of Experimental Botany*, **39**, 209–21.

Gilissen, L.J.W. (1976). The role of the style as a sense-organ in relation to wilting of the flower. *Planta*, **131**, 201–2.

Gilissen, L.J.W. & Hoekstra, F.A. (1984). Pollination-induced corolla wilting in *Petunia hybrida*. Rapid transfer through the style of a wilting-inducing substance. *Plant Physiology*, **75**, 496–8.

Halevy, A.H., Whitehead, C.S. & Kofranek, A.M. (1984). Does pollination induce corolla abscission of cyclamen flowers by promoting ethylene production? *Plant Physiology*, **75**, 1090–3.

Hamilton, A.J., Bouzayen, M. & Grierson, D. (1991). Identification of a tomato gene for the ethylene-forming enzyme by expression in yeast. *Proceedings of the National Academy of Sciences, USA*, **88**, 7434–7.

Hamilton, A.J., Lycett, G.W. & Grierson, D. (1990). Antisense gene that inhibits synthesis of ethylene in transgenic plants. *Nature*, **346**, 284–7.

Heidecker, G., Kolch, W., Morrison, D.K. & Rapp, U.R. (1992). The role of Raf-l phosphorylation in signal transduction. *Advances in Cancer Research*, **58**, 53–73.

Henskens, J.A.M., Somhorst, D. & Woltering, E.J. (1993). Expression of two ACC synthase mRNAs in carnation flower parts during aging and following treatment with ethylene. In *Cellular and Molecular Aspects of the Plant Hormone Ethylene*, ed. J.C. Pech *et al.*, pp. 323–4. Kluwer Academic Publishers, Boston.

Hoekstra, F.A. & Weges, R. (1986). Lack of control by early pistillate ethylene of the accelerated wilting in *Petunia hybrida* flowers. *Plant Physiology*, **80**, 403–8.

Holdsworth, M.J., Schuch, W. & Grierson, D. (1988). Organisation and expression of a wound/ripening-related small multigene family from tomato. *Plant Molecular Biology*, **11**, 81–8.

Kende, H. (1993). Ethylene biosynthesis. *Annual Review of Plant Physiology and Plant Molecular Biology*, **44**, 283–307.

Kieber, J.J. & Ecker, J.R. (1993). Ethylene gas: It's not just for ripening any more! *Trends in Genetics*, **9**, 356–62.

Kieber, J.J., Rothenberg, M., Roman, G., Feldmann, K.A. & Ecker, J.R. (1993). CTR1, a negative regulator of the ethylene response pathway in *Arabidopsis*, encodes a member of the Raf family of protein kinases. *Cell*, **72**, 427–41.

King, R.W. (1976). Implications for plant growth of the transport of regulatory compounds in phloem and xylem. In *Transport and Transfer Processes in Plants*, ed. I.F. Wardlaw & J.B. Passioura, pp. 415–31. New York: Academic Press.

Larsen, P.B., Woltering, E.J. & Woodson, W.R. (1993). Ethylene and interorgan signaling in flowers following pollination. In *Plant Signals in Interactions with Other Organisms*, ed. I. Raskin & J. Schultz, pp. 112–122. *American Society of Plant Physiologists.*

Lawton, K.A., Raghothama, K.G., Goldsbrough, P.G. & Woodson, W.R. (1990). Regulation of senescence-related gene expression in carnation flower petals by ethylene. *Plant Physiology*, **93**, 1370–5.

Liang, X., Abel, S., Keller, J.A., Shen, N.F. & Theologis, A. (1992). The 1-aminocyclopropane-1-carboxylate synthase gene family of *Arabidopsis thaliana*. *Proceedings of the National Academy of Sciences, USA*, **89**, 11046–50.

Moore, T.C. (1989). *Biochemistry and Physiology of Plant Hormones* 2nd edn. 330 pp., Springer Verlag, New York.

Mori, H., Nakagawa, N., Ono, T., Yamagishi, N. & Imaseki, H. (1993). Structural characteristics of ACC synthase isozymes and differential expression of their genes. In *Cellular and Molecular Aspects of the Plant Hormone Ethylene*, ed. J.C. Pech *et al.*, pp. 1–6. Kluwer Academic Publishers, Boston.

Nichols, R. (1971). Induction of flower senescence and gynaecium development in the carnation (*Dianthus caryophyllus*) by ethylene and 2-chloroethylphosphonic acid. *Journal of Horticultural Science*, **46**, 323–32.

Nichols, R. (1977). Sites of ethylene production in the pollinated and unpollinated senescing carnation (*Dianthus caryophyllus*) inflorescence. *Planta*, **135**, 155–9.

Nichols, R., Bufler, G., Mor, Y., Fujino, D.W. & Reid, M.S. (1983). Changes in ethylene production and 1-aminocyclopropane-1-carboxylic acid content of pollinated carnation flowers. *Journal of Plant Growth Regulation*, **2**, 1–8.

Norman, A.W. & Litwack, G. (1987). *Hormones*. 806 pp. San Diego: Academic Press.

O'Neill, S.D., Nadeau, J.A., Zhang, X.S., Bui, A.Q. & Halevy, A.H. (1993). Interorgan regulation of ethylene biosynthetic genes by pollination. *Plant Cell*, **5**, 419–32.

Olson, D.C., White, J.A., Edelman, L., Harkins, R.N. & Kende, H. (1991). Differential expression of two genes for 1-aminocyclopropane-1-carboxylate synthase in tomato fruits. *Proceedings of the National Academy of Sciences, USA*, **88**, 5340–4.

Overbeek, J.H.M. & Woltering, E.J. (1990). Synergistic effect of 1-aminocyclopropane-1-carboxylic acid and ethylene during senescence of isolated carnation petals. *Physiologia Plantarum*, **79**, 368–76.

Park, K.Y., Drory, A. & Woodson, W.R. (1992). Molecular cloning of an 1-aminocyclopropane-1-carboxylate synthase from senescing carnation flower petals. *Plant Molecular Biology*, **18**, 377–86.

Pech, J.C., Bouzayen, M., Alibert, G. & Latché, A. (1989). Subcellular localization of 1-aminocyclopropane-1-carboxylic acid metabolism in plant cells. In *Biochemical and Physiological Aspects of Ethylene Production in Lower and Higher Plants*, ed. Clysters *et al.*, pp. 33–40. Boston: Kluwer Academic Publishers.

Pech, J.C., Latché, A., Larrigaudière, C., & Reid, M.S. (1987). Control of early ethylene synthesis in pollinated petunia flowers. *Plant Physiology and Biochemistry*, **25**, 413–37.

Philosoph-Hadas, S., Meir, S. & Aharoni, N. (1985). Auto inhibition of ethylene production in tobacco leaf discs: enhancement of 1-aminocyclopropane-1-carboxylic acid conjugation. *Physiologia Plantarum*, **63**, 431–7.

Reid, M.S., Fujino, D.W., Hoffman, N.E. & Whitehead, C.S. (1984). 1-aminocyclopropane-1-carboxylic acid (ACC). The transmitted stimulus in pollinated flowers. *Journal of Plant Growth Regulation*, **3**, 189–96.

Riov, J. & Yang, S.F. (1982). Effects of exogenous ethylene on ethylene production in tobacco leaf discs: Enhancement of 1-amino-cyclopropane-1-carboxylic acid conjugation. *Plant Physiology*, **70**, 136–41.

Rodrigues-Pousada, R.A., van der Straeten, D., Dedonder, A. & van Montagu, M. (1993). Cloning and expression analysis of an *Arabidopsis thaliana* 1-aminocyclopropane-1-carboxylate synthase gene: pattern of temporal and spatial expression. In *Cellular and Molecular Aspects of the Plant Hormone Ethylene* ed. J.C. Pech *et al.*, pp 24–30. Boston: Kluwer Academic Publishers.

Roitt, I.M., Brostaff, J. & Male, D.K. (1993). *Immunology* 3rd ed. *Mosby, St. Louis*, 399 pp.

Rottmann, W.H., Peter, G.F., Oeller, P.W., Keller, J.A., Shen, N.F., Nagy, B.P., Taylor, L.P., Campbell, A.D. & Theologis, A. (1991). 1-Aminocyclopropane-1-carboxylate synthase in tomato is encoded by a multigene family whose transcription is induced during fruit and floral senescence. *Journal of Molecular Biology*, **222**, 937–61.

Ryan, C.A. (1992). The search for the proteinase inhibitor-inducing factor, PIIF. *Plant Molecular Biology*, **19**, 123–33.

Satter, R.L. & Galston, A.W. (1981). Mechanisms of control of leaf movements. *Annual Review of Plant Physiology*, **32**, 83–110.

Sisler, E.C. & Goren, R. (1981). Ethylene binding – The basis for hormone action in plants? *What's New in Plant Physiology*, **32**, 83–110.

Smith, J.J. & John, P. (1993). Activation of 1-aminocyclopropane-1-carboxylate oxidase by bicarbonate/carbon dioxide. *Phytochemistry*, **32**, 1381–6.

Starling, R.J., Jones, A.M. & Trewavas, A.J. (1984). Binding sites for plant hormones and their possible roles in determining tissue sensitivity. *What's New in Plant Physiology*, **15**, 37–40.

Stead, A.D. (1992). Pollination-induced flower senescence: a review. *Plant Growth Regulation*, **11**, 13–20.

Stead, A.D. & Moore, K.G. (1979). Studies on flower longevity in *Digitalis*. Pollination induced corolla abscisson in *Digitalis* flowers. *Planta*, **146**, 409–14.

Stead, A.D. & Reid, M.S. (1990). The effect of pollination and ethylene on the colour change of the banner spot of *Lupinus albifrons* (Bentham) flowers. *Annals of Botany*, **66**, 655–63.

Tang, X., Wang, H., Brandt, A.S. & Woodson, W.R. Organization and structure of the 1-aminocyclopropane-1-carboxylate oxidase gene family from *Petunia hybrida*. *Plant Molecular Biology* (in press).

Theologis, A. (1993). What a gas! *Current Biology*, **3**, 369–71.

Trewavas, A. (1981). How do plant growth substances work? *Plant Cell and Environment*, **4**, 203–28.

Van Hoven, W. (1991). Mortalities in kudu populations related to chemical defence in trees. *Journal of African Zoology*, **105**, 141–5.

Van Loon, L.C. & Fontaine, J.H.H. (1984). Accumulation of 1-(malonyl-amino)cyclopropane-1-carboxylic acid in ethylene synthesizing tobacco leaves. *Plant Growth Regulation,* **2**, 227–34.

Vercher, Y. & Carbonell, J. (1991). Changes in the structure of ovary tissues and in the ultra structure of mesocarp cells during ovary senescence or fruit development induced by plant growth substances in *Pisum sativum. Physiologia Plantarum,* **81**, 518–26.

Ververidis, P. & John, P. (1991). Complete recovery *in vitro* of ethylene-forming enzyme activity. *Phytochemistry,* **30**, 725–7.

Wallner, S., Kassalen, R., Burgood, J. & Craig, R. (1979). Pollination, ethylene production and shattering in geraniums. *HortScience,* **14**, 446.

Whitehead, C.S. & Halevy, A.H. (1989). Ethylene sensitivity: The role of short-chain saturated fatty acids in pollination-induced senescence of *Petunia hybrida. Plant Growth Regulation,* **8**, 41–54.

Whitehead, C.S. & Vasiljevic, D. (1993). Role of short-chain saturated fatty acids in the control of ethylene sensitivity in senescing carnation flowers. *Physiologia Plantarum,* **88**, 243–50.

Wildon, D.C., Thain, J.F., Minchin, P.E.H., Gubb, I.R., Reilly, A.J., Skipper, Y.D., Doherty, H.M., O'Donnell, P.J. & Bowles, D.J. (1992). Electrical signalling and systemic proteinase inhibitor induction in the wounded plant. *Nature,* **360**, 62–5.

Woltering, E.J. (1990a). Interrelationship between the different flower parts during emasculation-induced senescence in *Cymbidium* flowers. *Journal of Experimental Botany,* **41**, 1021–4.

Woltering, E.J. (1990b). Interorgan translocation of 1-aminocyclopropane-1-carboxylic acid and ethylene coordinates senescence in emasculated *Cymbidium* flowers. *Plant Physiology,* **91**, 837–45.

Woltering, E.J. & Harren, F. (1989). Role of rostellum desiccation in emasculation-induced phenomena in orchid flowers. *Journal of Experimental Botany,* **40**, 907–12.

Woltering, E.J. & Somhorst, D. (1990). Regulation of anthocyanin synthesis in *Cymbidium* flowers: Effects of emasculation and ethylene. *Journal of Plant Physiology,* **136**, 295–9.

Woltering, E.J., Somhorst, D. & De Beer, C.A. (1993a). Roles of ethylene production and sensitivity in senescence of carnation flower (*Dianthus caryophyllus*) cultivars White Sim, Chinera and Epomeo. *Journal of Plant Physiology,* **141**, 329–35.

Woltering, E.J., Van Hout, M., Somhorst, D. & Harren, F. (1993b). Roles of pollination and short-chain saturated fatty acids in flower senescence. *Plant Growth Regulation,* **12**, 1–10.

Woodson, W.R. & Brandt, A.S. (1991). Role of the gynoecium in cytokinin-induced carnation petal senescence. *Journal of the American Society of Horticultural Science,* **116**, 676–9.

Woodson, W.R. & Lawton, K.A. (1988). Ethylene-induced gene expression in carnation petals. Relationship to autocatalytic ethylene production and senescence. *Plant Physiology,* **87**, 498–503.

Woodson, W.R., Park, K.Y., Drory, A., Larsen, P.B. & Wang, H. (1992). Expression of ethylene biosynthetic pathway transcripts in senescing carnation flowers. *Plant Physiology,* **99**, 526–32.

Yang, S.F. & Hoffman, N.E. (1984). Ethylene biosynthesis and its regulation in higher plants. *Annual Review of Plant Physiology,* **35**, 155–89.

Zhang, X.S. & O'Neill, S.D. (1993). Ovary and gametophyte development are coordinately regulated by auxin and ethylene following pollination. *Plant Cell,* **5**, 403–18.

Index

nb – page numbers in *italics* refer to figures and tables